Polyethylene Based Blends and Composites

Polyethylene Based Blends and Composites

Dr. Vikas Mittal
Editor and Lead Author

Central West Publishing

This edition has been published by Central West Publishing, Australia

For more information about the books published by Central West Publishing, please visit https://centralwestpublishing.com

Disclaimer

NATIONAL LIBRARY OF AUSTRALIA

A catalogue record for this book is available from the National Library of Australia

ISBN (print): 978-1-922617-11-8

Contents

Preface

Polymer blends and composites represent functional materials generated for combining useful properties of their components. Effective mixing of the components in these materials usually results in synergistic enhancements in mechanical, thermal, flow, barrier and flammability performances. This tuning and control on the material properties leads to the applications of these functional blends and composites in many applications. A great degree of advancements in these materials have been achieved in the recent years on all aspects related to material combinations, synthesis routes, organic-inorganic phase compatibilization methodologies, etc. This text aims to present a number of such blend and composite systems based on polyethylene (PE). For this reason, a majority of the chapters presented in this text focus on experimental details as well as accurate material characterizations to present deep insights into controlling structure-property correlations.

In Chapter 1, the effect of particle compatibilization of polyethylene and polystyrene blends on the composite properties was elucidated by using a variety of 0-D and 2-D nanomaterials, viz. silica, silicate and thermally reduced graphene, in the absence of any compatibilizer. In Chapter 2, the effect of the hybrid nanofillers (based on MOF and graphene) on the thermal, mechanical and rheological properties of the high density polyethylene (HDPE) matrix has been analyzed. In Chapter 3, morphological characterization of the PE + chlorinated PE (CPE) blends and PE + graphene oxide (GO) + CPE nanocomposites, generated using CPE of two different chlorination contents, was carried out. In Chapter 4, the thermo-mechanical properties of HDPE/PA1010 blend nanocomposites with carbon nanotubes (CNTs) have been studied, at varying amounts of CNTs (1, 2, 3 and 5 wt%), so as to further develop this important class of blend composites for various applications. On the other hand, Chapter 5 reports the mechanical, thermal and structural characteristics of graphene nanoplatelets (GNPs) filled PA1010/HDPE nanocomposites. In Chapter 6, high molecular weight polyethylene materials of different molecular weights and molecular weight distributions were analyzed for their structural properties using a multitude of characterization methods such as tensile testing, thermogravimetry, scanning calorimetry, diffraction, scanning electron microscopy, scattering, nuclear magmatic resonance, etc. In Chapter 7, the effect of a variety of polymeric modifications physically adsorbed on the sepiolite surface was studied on the thermal, mechanical and rheological properties of the HDPE nanocomposites. Chapter 8 reviews the recent studies reporting the development of polyethylene/graphene nanocomposites with a focus on the composite properties, especially morphological, mechanical and thermal properties. In Chapter 9, efforts have been made to gain insights into the various aspects related to the composites of polyethylene with a variety of 0-D, 1-D and 2-D fillers as well as their hybrids. Chapter 10 explores the effectiveness of the masterbatch approach over direct melt mixing method in improving the filler dispersion and resulting properties of the polyethylene composites. In Chapter 11, the effect of amine functionalized reduced graphene oxide (r-AGO) on the crystallization, thermal stability and mechanical properties of HDPE has been studied, using CPE as compatibilizer. In order to further explore the development of polyethylene based functional materials, especially for low temperature flexibility, blends of polyethylene with a variety of copolymers have been generated in Chapter 12.

The support of all the chapter contributors is greatly acknowledged towards the successful accomplishment of the book. The book is dedicated to my family for their constant support.

Vikas MITTAL

1

Filler Distribution and Properties of Polyethylene/ Polystyrene Blend Composites with 0-D and 2-D Nanomaterials

1.1 Introduction

Blending or alloying of polymers is an increasingly common practice for the development of new compounds having improved and diverse properties, however, phases of the mostly immiscible systems need to be efficiently mixed [1-4]. Polyolefins, especially polyethylene (PE), have outstanding properties including low cost, recyclability, good processability, non-toxicity and biocompatibility, but suffer from poor barrier properties, shrinkage and relatively low stiffness [5-10]. To overcome these limitations, blends of polyolefins with stiff polymers like polystyrene (PS) represent effective alternatives, however, their phase domains due to unfavorable enthalpy of mixing need efficient mixing [11,12]. Significant efforts have been devoted in order to modify the macrophase distribution of polymers by lowering the interfacial tension and improving the adhesion using compatibilizers like block copolymers [12-17]. The drawbacks associated with the usage of polymeric compatibilizers are their low molecular weight, hindrance to polymer crystallization and weakening of mechanical properties of the blends [18]. Some studies have also dealt with compatibilization of the polyolefin and PS blends by employing the reactive blending approach [11,19], however, the processing requires efficient control on the reaction progress as no side reactions can be tolerated, which otherwise lead to a significant deterioration in the properties.

Polymers reinforced with nano-fillers, such as layered silicates, carbon nanotubes (CNTs) and graphene nano-sheets have attracted much research attention due to significant enhancements in the polymer performance [20-31]. Among these fillers, layered silicates have also been used to improve the blend miscibility by enhancing the

Vikas Mittal[a], Ali U. Chaudhry[a,], Fakhruddin Patwary[a] and Nadejda Matsko[b]*
[a]Khalifa University of Science and Technology, Abu Dhabi, UAE; [b]Graz Centre for Electron Microscopy, Graz, Austria
**Current address: Texas A&M University, Qatar*

weak interfacial adhesion between the component phases in PE/PS [32], polyamide/polypropylene (PP) [33], elastomers/PP [34], PP/ethylene propylene rubber [35] blends, among others [36]. Along with the morphological variations leading to effective phase distribution, layered silicate particles also exhibited reinforcing effect on the blend properties. Also, graphene sheets are one-atom-thick two-dimensional layers of sp^2-bonded carbon and exhibit a variety of remarkable properties such as electrical and thermal conductivity, gas impermeability and mechanical properties, etc. [37,38]. Recently, Cao *et al.* used the amphiphilic nature of graphene oxide (GO) and reported its compatibilizing behavior towards immiscible polymer blend of polyamide/polyphenylene oxide. The authors observed remarkable reduction in the size of microphase and enhanced mechanical properties. The behavior was due to coupling effect of GO owing to the presence of carboxylic acid at the edges as well as phenol, hydroxyl and epoxide groups at basal planes [39]. In addition, there is a need to further evaluate the role of particles like graphene on the compatibilization and properties of the immiscible blends, in comparison with other 0-D and 2-D particles like silica and alumino-silicates.

Various mechanisms have been described in the literature for morphologically stabilized blends of immiscible polymers using particles with different shapes, sizes and surface characteristics. For instance, one of the mechanisms is the accumulation of filler at the interface of dispersed phases which suppresses the coalescence and phase separation [40]. It was reported that the particles at the interface act as a bridge which was observed by gel behavior at low frequency during rheological measurements. The gel behavior was the consequence of the formation of particle network between the viscous fluids having different phases [41]. Thus, it is important to tune the localization of the filler in the polymer phases. Pawar *et al.* [42] also reported co-continuous morphology of poly(3-hydroxybutyrate-co-3-hydroxyvalerate) (PHBV) and poly(butylene adipate-co-terephthalate) (PBAT) in the blend nanocomposites, though the filler (graphene and vermiculite platelets) was localized in the tougher PBAT phase, as shown in Figure 1.1 [42]. Reduction of microphase size of the polymer phases on particle incorporation is another common mechanism of stabilization [39]. Tanaka *et al.* reported for the stabilization behavior of binary immiscible polymer mixtures containing monodisperse glass spheres that coarsening of phases was affected by the mobile glass particles due to the dynamic interactions between

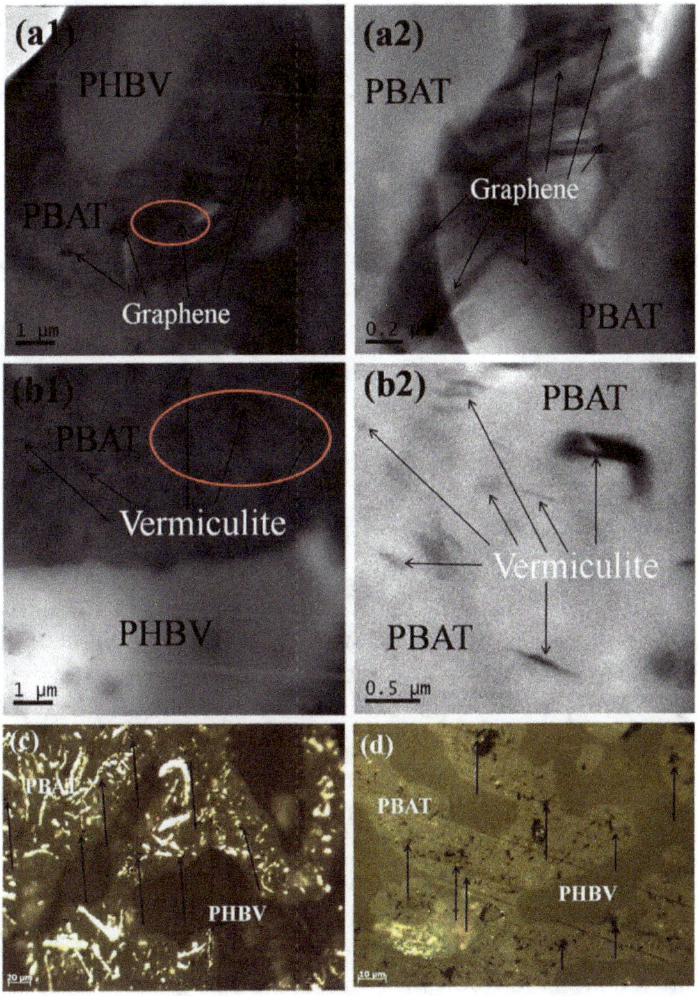

Figure 1.1 TEM micrographs of PHBV/PBAT composites (a1 and a2) with 5 wt% graphene and (b1 and b2) with 5 wt% vermiculite. Optical images of PHBV/PBAT composites (c) with 5 wt% graphene and (d) with 5 wt% vermiculite. Reproduced from Reference 42 with permission from Springer.

phase separation and wetting [43]. Similar results were also confirmed by using simulation of the motion of spherical particles in binary systems [44]. Thermodynamic parameter such as interaction parameter has also been reported to play a significant role in lowering the total Gibbs free energy (ΔG_{mix}) of blends containing the inorganic solid filler phase depending on the interaction and preferential adsorption of the components [45,46]. Further, surface features such as attached functional groups with polymer or filler have also been reported to help in compatibilization and reducing surface tension between the immiscible phases [47].

In this study, the effect of particle compatibilization of PE and PS blends on the composite properties was elucidated by using a variety of 0-D and 2-D nanomaterials, viz. silica, silicate and thermally reduced graphene, in the absence of any compatibilizer. The effect of the nano-filler morphology on its distribution and subsequent composite properties was quantified as a function of blend composition.

1.2 Experimental

1.2.1 Materials

High density polyethylene BB2581 was received from Abu Dhabi Polymers Company Limited (Borouge), UAE. It was in the form of white pellets with specific gravity 0.958. Polystyrene (PS) with average molecular weight (M_w) of 400,000 was procured from Sigma. The polymers were used as obtained. Synthetic silicon dioxide powder (ZEOFREE® 5161 S) and synthetic alumino-silicate (ZEOLEX® 23) were supplied by J. M. Huber Private Limited, India. Thermally reduced graphene was produced by thermal exfoliation of precursor graphite oxide using modified Hummer's method, as reported earlier [31,48].

1.2.2 Generation of Blends and Blend Composites

Melt mixing using mini-twin conical screw extruder (MiniLab HAAKE Rheomex CTW5, Germany) was carried out to generate PS/PE blends and nanocomposites. The polymers and fillers were dried at 60° C overnight under vacuum so as to remove any surface adsorbed moisture which can lead to unwanted degradation during the melt compounding process. Blends of three compositions were generated: 25 wt% PE + 75 wt% PS, 50 wt% PE + 50 wt% PS and 75 wt% PE + 25 wt% PS, which were correspondingly designated as $PE_{25}PS_{75}$, $PE_{50}PS_{50}$ and $PE_{75}PS_{25}$, respectively. The filler content was fixed to 7 wt% in the composites. Blends and blend composites were prepared by pre-mixing of the polymers with fillers and melt mixing these together at a temperature of 205 °C and rotational speed of 80 rpm for 7 min. To study the effect of processing conditions on the filler distribution and resulting properties, the composites were also prepared by (i) melting PE and PS, followed by filler addition, (ii) melting PE, followed by addition of graphene and PS, (iii) melting PS, followed by addition of graphene and PE, (iv) melt mixing the components together using a rotational speed of 120 rpm or using a mixing time of 15 min. Tetrahedron MTP-10 hot press was used to compression mold the blend and composite sheets. A temperature of 205 °C and a pressure of 800 bar were used (holding pressure 290 bar). The samples were cooled to 50 °C at a rate of 5 °C/min. Sheets of 12*12 cm with a thickness of 1.5 mm were produced and disc, dumbbell and bar-shaped test specimens for rheological and mechanical characterization were stamped out from these compression molded sheets.

1.2.3 Materials Characterization

Differential scanning calorimetric (DSC) measurements on the polymer blends and nanocomposites were performed using a Netzsch DSC under nitrogen atmosphere. The scans were obtained from 30-250-30 °C using heating and cooling rates of 5 °C/min. Thermal properties of the samples were analyzed using Netzsch thermogravimetric analyzer (TGA) using nitrogen as carrier gas. The scans were obtained from 50 to 700 °C at a heating rate of 10 °C/min.

Universal testing machine Testometric M-350 10CT was used to analyze the mechanical performance of the samples. The dumbbell-shaped samples of 53 mm length, 4 mm width and 2 mm thickness (gauge length was 26 mm) conforming to standard ASTM-638-V were used. A loading rate of 1 mm/min was used and the tests were carried out at room temperature. Win Test Analysis software was used for the calculation of tensile modulus and other properties. An average of five values was calculated.

DMA (dynamic mechanical analyzer) RSA3 of TA Instruments was used to study the change in modulus with temperature (25-120 °C). Dynamic temperature ramp test was used at a frequency of 5 rad/s on bar-shaped samples, using a ramp rate of 3 °C/min and 1% strain for all samples. AR 2000 rheometer from TA Instruments was used to characterize rheological properties such as storage modulus (G'), loss modulus (G''), viscosity (η') and elasticity (η''), along with complex modulus (G^*) and complex viscosity (η^*). Disc shaped samples of 25 mm diameter and 2 mm thickness were characterized at 205 °C using a gap opening of 1 mm. The samples were subjected to a strain sweep at 205 °C and angular frequency of 1 rad/s. Subsequently, the frequency sweep analysis (dynamic testing) of all blends and blend nanocomposite samples was carried out at 1% strain from ω = 0.01 to 100 rad/s.

For the microscopy analysis, the samples were mounted in special holders that at the same time fit in the microtome and were also suitable for the examination of the block face by atomic force microscopy (AFM). Ultrathin sections (30-90 nm) as well as block faces of the composite samples were obtained using a Leica Ultracut E microtome (Leica, Austria) equipped with a diamond knife (Diatome, Switzerland) at -120 °C. The so-obtained sections for transmission electron microscopy (TEM) analysis were collected on 400 mesh formvar electron microscopy grids (coated with holey carbon film), and were examined in a Philips CM 20 (Philips/FEI, Germany) electron microscope at 200 kV at room temperature without additional staining. TEM image processing was performed using DigitalMicrograph software (Gatan, USA). All energy-dispersive X-ray spectroscopy (EDXS) experiments on blends and composites were performed in STEM mode with a probe current of 20 nA and a beam diameter of 100 nm. EDXS spectra were collected using a HPGe detector (solid angle 0.13 sr). For the calculation of elemental distribution images, jump ratio method was used and EDXS spectra have been visualized using DigitalMicrograph software (Gatan, USA). The block faces of specimens after cryo-ultramicrotomy were investigated at ambient conditions using Dimension 3100 AFM/SPM (Veeco, USA) atomic force microscope. AFM images were collected in tapping mode using silicon nitride cantilevers with natural frequencies in the 300 kHz range (force constant = 20 N m^{-1}, tip radius = 10 nm (NT-MDT, Russia)). AFM image processing was performed using Nanoscope v720 software (Veeco, USA). The block faces of specimens after cryo-ultramicrotomy were also investigated at ambient conditions using ZEISS Axioplan light microscope (Zeiss, Germany) equipped with Zeiss Axio Cam ICc 1 CCD camera in reflected polarized light.

1.3 Results and Discussion

In the current study, different fillers with 0-D and 2-D nanostructure were added to PE/PS blends and their effect on the morphology and physical properties of the composites was studied as a function of blend composition. Initially, all the blend composites were generated using one-step melt mixing process where all the components were added to the compounder together. In the later part of the study, variations in the mixing protocols were made to study the corresponding effect on the filler distribution and composite properties. The effect of mixing time and rotational speed in the compounder was also analyzed. However, the compounding temperature was not exceeded beyond 205 °C so as to avoid any extensive thermal degradation of the polymers. The filler extent was also kept constant at 7 wt% so as to relate the variation in the performance of the composites to the type of the filler and blend composition. Also, the filler fraction of 7 wt% was chosen, as this was reported to an optimum content in an earlier study [49].

Figure 1.2 demonstrates the TGA thermograms of polymer blends and blend composites with silica, silicate and graphene, as a function of blend composition. As observed in Figure 1.2a, the blends exhibited two step degradation behavior, where the low temperature step

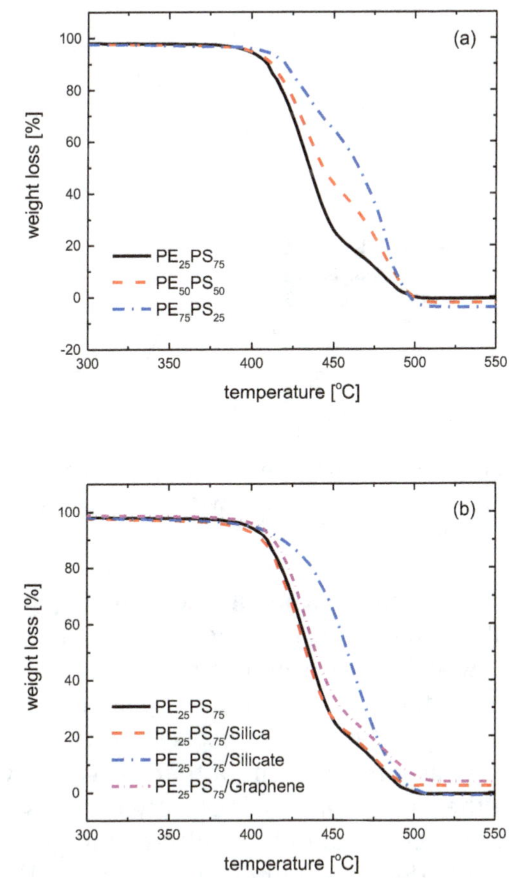

Figure 1.2 (continued on next page)

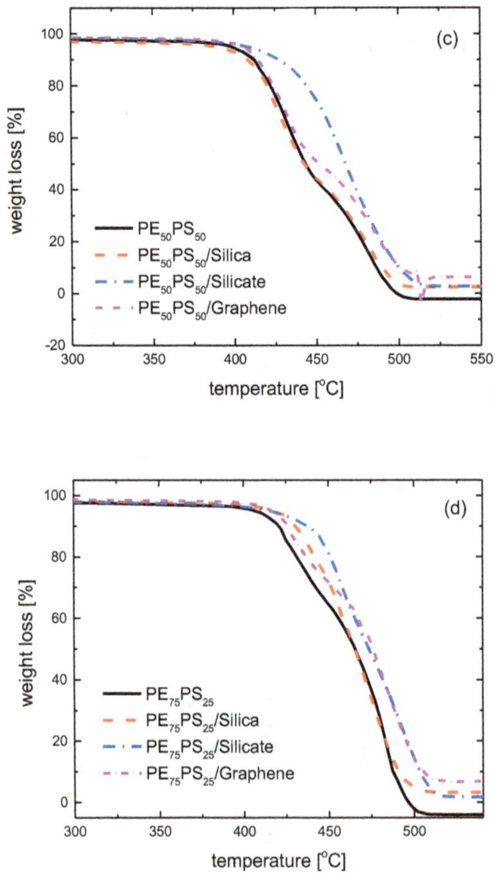

Figure 1.2 TGA thermograms of polymer blends (a) and silica, silicate and graphene composites with polymer blends with compositions of (b) $PE_{25}PS_{75}$, (c) $PE_{50}PS_{50}$ and (d) $PE_{75}PS_{25}$.

step corresponded to PS phase, whereas the high temperature degradation step was attributed to PE. Presence of the individual degradation patterns of PE and PS also indicated the immiscibility of the phases. On increasing the amount of PE in the blend, both onset and peak degradation temperature were observed to shift to higher values, thus, indicating the enhancement of thermal stability on the addition of PE. The thermal stability of the polymer was observed to largely improve in the composites with silicate and graphene fillers. The silicate filler provided maximum enhancement in the onset and peak degradation temperatures at all blend compositions. Addition of silica did not impact the polymer degradation temperatures. In addition, the blend composites with silica and graphene also exhibited two step degradation behavior as the pure blends owing to different PE and PS phases. In contrast, the degradation patterns of composites with silicate filler did not exhibit two step degradation pattern, and rather a smooth single step degradation behavior was observed. Probably, silicate provided effective phase compatibilization between PS and PE, thus, leading to a reduction in their domain sizes in the composites.

Figure 1.3 and Table 1.1 also present the calorimetric properties of the blends and nanocomposites. Second heating curves, demonstrated in Figure 1.3, were used for the analysis of peak melting temperature as well as melt enthalpy. In these blends, no melt transition

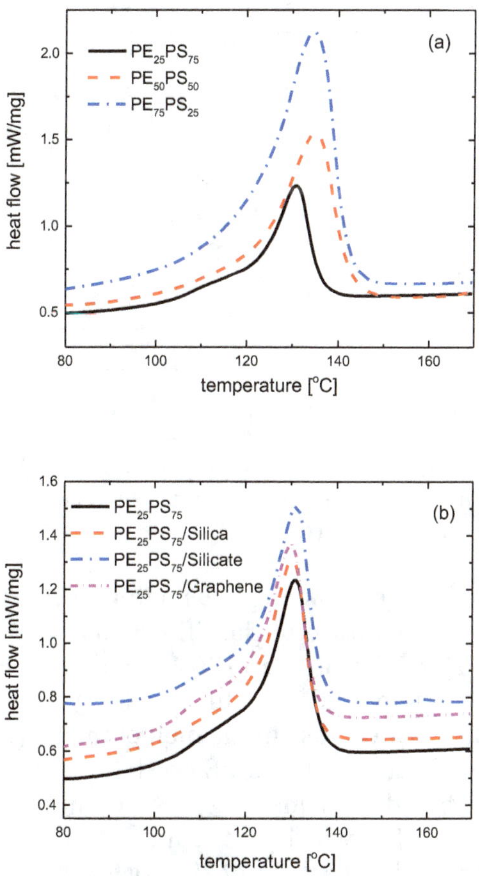

Figure 1.3 (continued on next page)

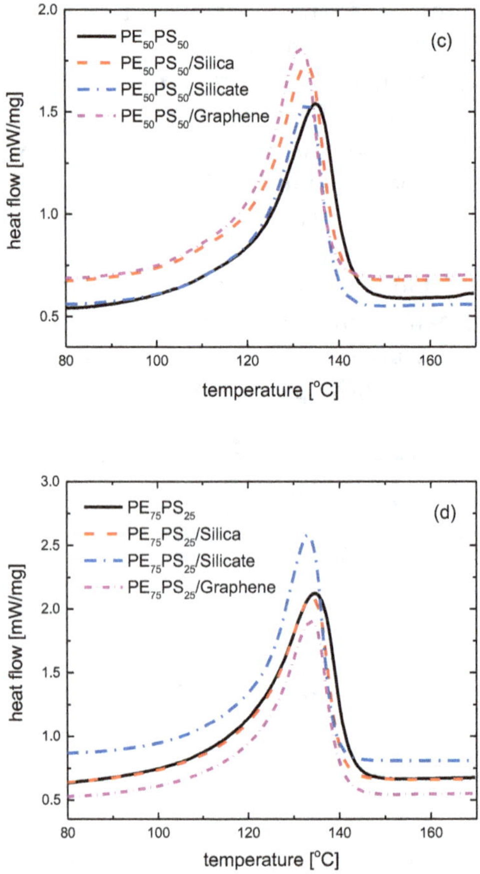

Figure 1.3 DSC melting thermograms of polymer blends (a) and silica, silicate and graphene composites with polymer blends with compositions of (b) $PE_{25}PS_{75}$, (c) $PE_{50}PS_{50}$ and (d) $PE_{75}PS_{25}$.

was observed for PS due its amorphous nature. On the other hand, the peak melting temperature of the PE was observed to decrease from 134.6 °C for 75% PE content to 130.6 °C when the PE extent was 25%, indicating possible hindrance to the PE crystallization by the majority PS phase. The decrease in the melt enthalpy (normalized to PE content) also indicated reduced PE crystallinity with decreasing PE content in the blends. For instance, melt enthalpy for the blend with 75% PE was 156 J/g, which was reduced to 132 J/g in the 25% PE blend. Comparing with the value of 293 J/g for the theoretical melt enthalpy of 100% crystalline PE sample [50], degree of crystallinity

Table 1.1 Calorimetric properties of the PE/PS blends and composites

	Peak melting temperature T_m, °C	Melt enthalpy ΔH_m, J/g	Peak crystallization temperature T_c, °C	Crystallization enthalpy ΔH_c, J/g
$PE_{25}PS_{75}$	130.9	132	115.7	128
$PE_{50}PS_{50}$	133.1	134	115.4	136
$PE_{75}PS_{25}$	134.6	156	115.7	151
$PE_{25}PS_{75}$/Silica	130.2	163	116.5	168
$PE_{25}PS_{75}$/Silicate	131.3	163	115.4	168
$PE_{25}PS_{75}$/Graphene	130.9	159	115.2	151
$PE_{50}PS_{50}$/Silica	133.1	140	115.2	146
$PE_{50}PS_{50}$/Silicate	133.2	137	115.5	142
$PE_{50}PS_{50}$/Graphene	133.1	144	116.0	148
$PE_{75}PS_{25}$/Silica	134.1	148	116.5	145
$PE_{75}PS_{25}$/Silicate	133.2	141	115.7	145
$PE_{75}PS_{25}$/Graphene	134.0	124	115.2	123

of 53% and 45% respectively was observed. Similar trend was also exhibited by the crystallization enthalpy of the PE phase in the blends (Table 1.1). The melt enthalpy was significantly affected by the filler addition, which was further a function of blend composition. In the blend with 25% PE, addition of filler resulted in significant enhancement of melt enthalpy, though peak melting temperature was not affected. The peak crystallization temperature was also observed to be similar to the blend indicating the absence of any nucleation effect by the filler. In addition, all fillers exhibited similar enhancement in the melt enthalpy of PE. The blend with 50% PE also exhibited increment in the melt enthalpy with the addition of filler, though the extent of enhancement was reduced in comparison with the composite of the blend with 25% PE content. The blend with 75% PE exhibited a reduction in the melt enthalpy on incorporation of fillers, with the maximum reduction observed in the case of graphene. Thus, depending on the blend compositions, the fillers had a variable effect on the PE crystallinity, though there was no indication of the nucleation effect by the fillers. The observed effect would have resulted due to the specific filler dispersion in the blend composites generated as a function of blend composition. As the PE crystallinity is significantly affected on addition of the fillers, it is also envisaged that the filler particles may have been predominantly localized in the PE phase. The blend and composites with 75% PS also exhibited an endothermic signal near 110 °C (Figure 1.3), probably indicating the glass transition temperature of polystyrene. The signal was not observed in the other materials owing to constraining of the PS phase inside PE, thus, hindering the movement of PS chains.

Figure 1.4 shows the TEM micrographs of the $PE_{50}PS_{50}$ composites with silica, silicate and graphene fillers. In the images, the darker phase represented PE, whereas lighter phase contained PS. The filler was observed to be specifically localized in the PE phase, with PS phase completely devoid of it. Similar morphology was observed for all filler types and supported the findings from calorimetric analysis where the observed effect on PE crystallinity was suspected to occur due to specific localization of filler in PE. In certain instances, the filler particles were also present at the interface of the PE and PS phases. As PE chains are not expected to have any specific interaction with the fillers, thus, the observed composite morphology would not result due to thermodynamic mixing of the phases. In contrast, the kinetic factors seem to have resulted in the observed morphology, due to the processing method used to generate the nanocomposites. As both PS and PE were added together in the compounder along with the filler, thus, the early melting of PE phase would have resulted in its mixing with the filler before the PS melt could be formed. Thus, the PS melt would have phase mixed with PE-filler mixture, rather than PE and filler individually. Thus, in the scenario of formation of the melt from both PE and PS phases before the addition of filler, the distribution of filler in both phases can be expected. Also, it should be noted that no significant interface stabilization occurred in these composites due to the filler largely embedded in the PE phase. However, the domains of PE and PS were reduced in the composites as compared to pure blends, which indicated that the filler particles still exhibited stabilization effect, even though most of the filler was present in the PE phase. Also, the phases did not coalesce with time, further confirming the phase stability in the composites. Similar effect was also observed earlier by Pawar *et al.* [42], where specific localization of filler in PBAT phase in the PHBV/PBAT blend composite led to the synergistic enhancement of the composite properties. In fact, the observed morphology of filler localization in the flexible phase (PE) is also beneficial from the perspective of physical properties of the composites, as distribution of the filler in the stiffer phase (PS) would have resulted in even further decrease in the toughness of the composites. In addition, though TGA analysis had signaled the possibility of better phase

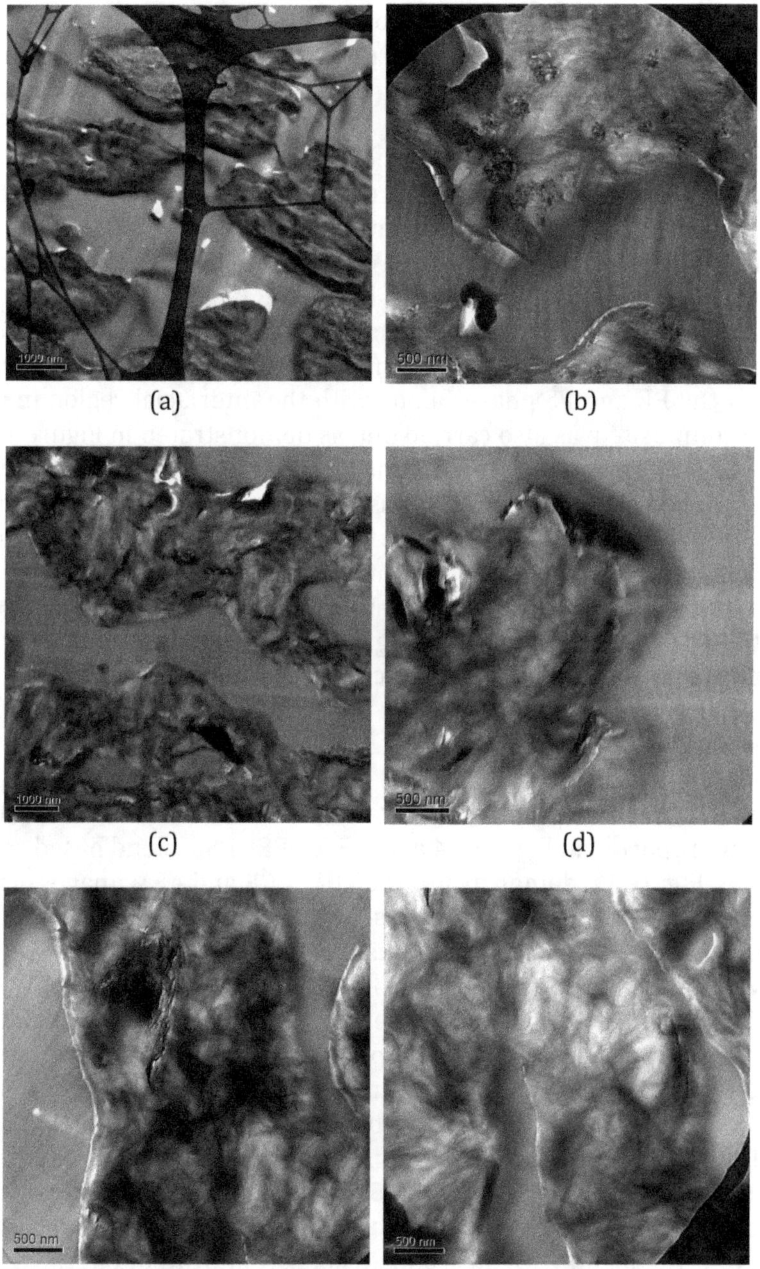

Figure 1.4 TEM micrographs of $PE_{50}PS_{50}$ composites at two magnifications depicting the phase distribution; (a),(b) using silica as filler, (c),(d) using silicate as filler and (e),(f) using graphene as filler.

mixing in the case of silicate composites due to the presence of a single degradation step, however, morphological analysis did not exhibit any difference in the morphology of the composites with different fillers.

Figure 1.5 depicts further morphological analysis of the $PE_{50}PS_{50}$ composites with silica and graphene. Figures 1.5a and b depict the TEM and PCR (plasmon to carbon ratio) images of the $PE_{50}PS_{50}$ composite with silica. As observed in Figure 1.4, the TEM image confirmed the smaller domain sizes of PS and PE phases. In addition, the distribution of silica solely in PE phase was also confirmed in the corresponding PCR image (seen as white spherical domains). EDX analysis of the PE and PS phases along with the interfacial region in the silica composite was also carried out, as demonstrated in Figure 1.5c and d. The PE phase exhibited a sharp signal for Si, thus, confirming the presence of silica in the PE phase. On the other hand, no Si signal was observed in the PS phase. The interfacial region also exhibited a weak signal of Si, thus, also indicating the presence of a fraction of filler at the interface between PE and PS. Figure 1.5e and f also present the similar TEM and PCR images of the $PE_{50}PS_{50}$ composite with graphene. The PCR image clearly confirmed the distribution of graphene platelets inside the PE phase with a fraction also stabilizing the interface between PE and PS phases.

Morphological analysis of the composites generated from the blends with other compositions was also carried out to confirm the findings reported in Figure 1.4 and 1.5 for $PE_{50}PS_{50}$ blend based composites. Figure 1.6 demonstrates the TEM, PCR and AFM images of the $PE_{75}PS_{25}$ blend based composites with silica and graphene. The $PE_{50}PS_{50}$ blend had a co-continuous morphology, whereas the blend $PE_{75}PS_{25}$ exhibited dispersed morphology of the PS phase inside PE. Similar filler distribution was observed in the composites. The filler particles were localized in the PE phase, with a fraction of filler present at the interface, thus, confirming no change in the nature of component dispersion with blend composition.

Table 1.2 demonstrates the tensile properties of the blends and nanocomposites. The blend with 75% PE exhibited the tensile modulus of 1284 MPa, which was enhanced to 1435 and 1593 MPa respectively for the blend with 50% and 25% PE. It indicated that the modulus of the blend was correspondingly enhanced as the extent of the stiffer phase was increased. The addition of filler in the blends increased the tensile modulus significantly. Comparing among all the fillers, graphene resulted in the largest extent of enhancement in the

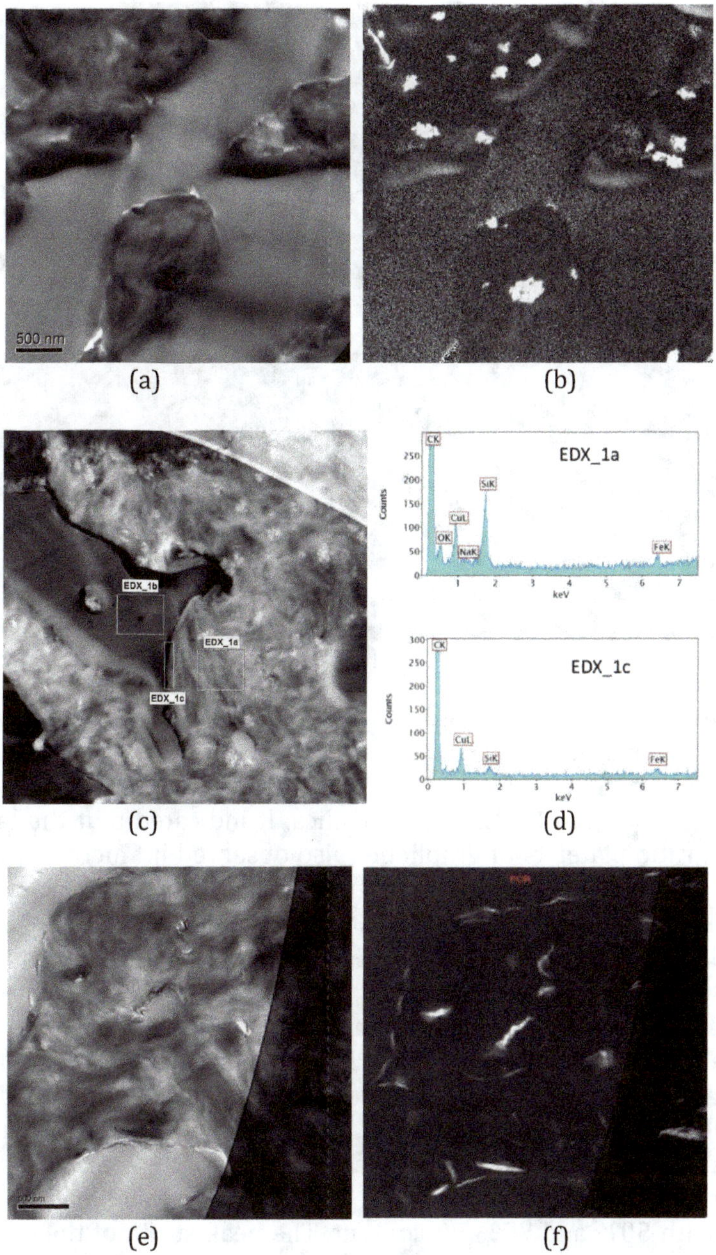

Figure 1.5 (a) TEM and (b) corresponding PCR image; (c) TEM and (d) corresponding EDX analysis for determining the distribution of silica particles in the composite for $PE_{50}PS_{50}$/Silica composite; (e) TEM and (f) corresponding PCR image for $PE_{50}PS_{50}$/Graphene composite.

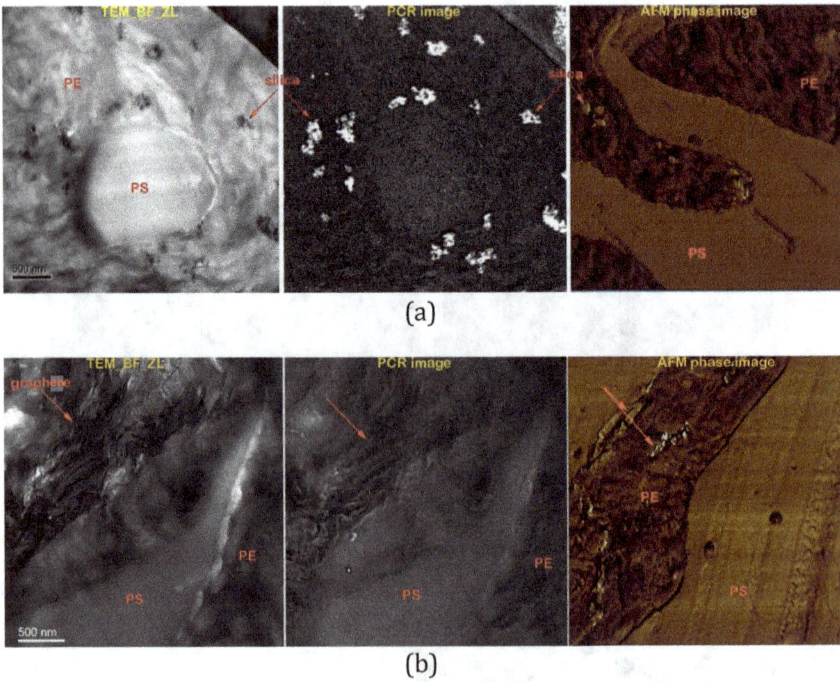

Figure 1.6 Morphological analysis of (a) $PE_{75}PS_{25}$/Silica and (b) $PE_{75}PS_{25}$/Graphene composites.

modulus, followed by silicate and silica. It indicated that the larger aspect ratio platelets of graphene (also observed in microscopy images) contributed to effective stress transfer from the polymer chains. Overall, an increase of 17% in the modulus was observed for the blend composites with 25% PE, whereas the increment was 29% and 36% for the blend composites with 50% and 75% PE content. It has also to be noted that the maximum enhancement in the modulus of the blend composite with 75% PE occurred even after reduction of crystallinity in the composites on addition of filler. This indicated that the filler may have higher extent of dispersion in this case due to larger fraction of available PE phase. Similarly, the peak stress and stress at break was also observed to enhance for the blend composites with 50% and 75% PE content. The peak strain of the composites was also largely retained when the PE content in the blend was high. It also indicated that the specific localization of the filler in the PE phase did not largely impact the flexibility. Thus, the blend composite with 75% PE exhibited more optimal enhancement in the mechanical properties, followed by the blend composites with 50% PE.

Table 1.2 Mechanical properties of the PE/PS blends and composites

	Tensile modulus[a], MPa	Peak stress[b], MPa	Peak strain[c], %	Stress at break[d], MPa
$PE_{25}PS_{75}$	1593	26	1.5	26
$PE_{50}PS_{50}$	1435	17	1.6	15
$PE_{75}PS_{25}$	1284	15	2.5	12
$PE_{25}PS_{75}$/Silica	1719	21	1.4	20
$PE_{25}PS_{75}$/Silicate	1760	21	1.1	20
$PE_{25}PS_{75}$/Graphene	1864	25	1.4	24
$PE_{50}PS_{50}$/Silica	1561	20	1.2	19
$PE_{50}PS_{50}$/Silicate	1682	20	1.1	18
$PE_{50}PS_{50}$/Graphene	1853	19	1.2	18
$PE_{75}PS_{25}$/Silica	1272	13	2.5	13
$PE_{75}PS_{25}$/Silicate	1423	14	2.3	14
$PE_{75}PS_{25}$/Graphene	1749	17	2.0	17

[a]Relative probable error 3%; [b]relative probable error 3%; [c]relative probable error 5%; [d]relative probable error 10%.

Figure 1.7 shows the dynamic mechanical performance of the blends and nanocomposites as a function of temperature (room temperature till 120 °C). Similar to the tensile modulus, the storage modulus at room temperature was observed to be higher for the blends with higher extent of PS. On increasing the temperature, the storage modulus decreased gradually, though the trend between the blends was retained. The blends exhibited a sharp decrease in the modulus around 100 °C, which is the glass transition temperature of PS. The reduction was most prominent in the $PE_{25}PS_{75}$ blend owing to the large fraction of PS in the blend. Thus, beyond this temperature, both polymer phases were in rubbery state, thereby, resulting in lower storage modulus. Due to this, no difference in the performance of different blends was observed beyond this temperature. The storage modulus was observed to increase in the blend composites with 50% and 75% PE content, with maximum extent of increment achieved on the incorporation of graphene, similar to the findings from tensile analysis. However, the blend composite with 25% PE did not exhibit any increment in the modulus on filler addition, in fact, the modulus was observed to decrease to some extent. Tensile analysis also exhibited least extent of modulus enhancement for this case. As the filler was contained in the 25% PE fraction in this case, probably some filler aggregation may have taken place. As the blends, all the composites exhibited sharp decline in the modulus around 100 °C, though this transition was observed to shift to higher temperatures in the case of graphene containing composites. This transition, owing to the glass transition temperature of PS, is also depicted in Figure 1.7e. The transition in the blend with 75% PS was sharp and well defined. In the graphene based composite generated with this blend, the transition was observed to shift to a higher temperature, as observed earlier. Thus, it confirmed that the addition of graphene enhanced the thermal resistance of the polymer due to effective heat transfer and conductivity of graphene platelets. The transition was observed to become broad as the amount of PS was decreased in the blends. Similar to the case with 75% PS, the blend composites of graphene with 50% PS and 25% PS also exhibited the shift of transition temperature

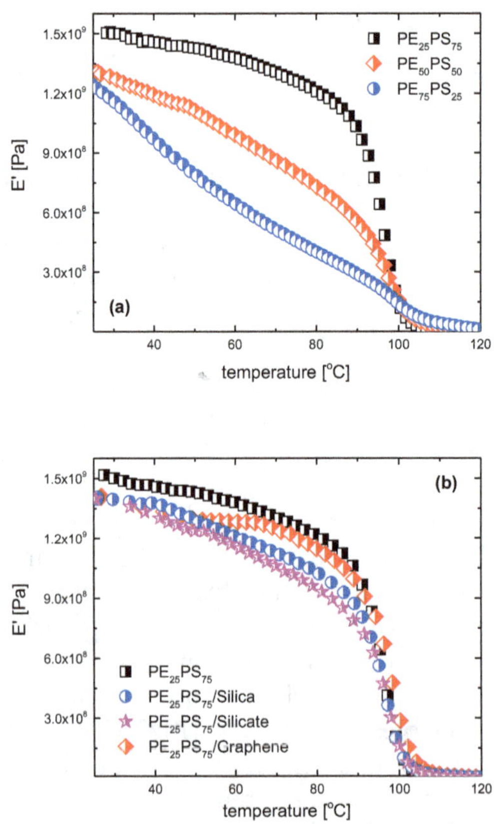

Figure 1.7 (continued on next page)

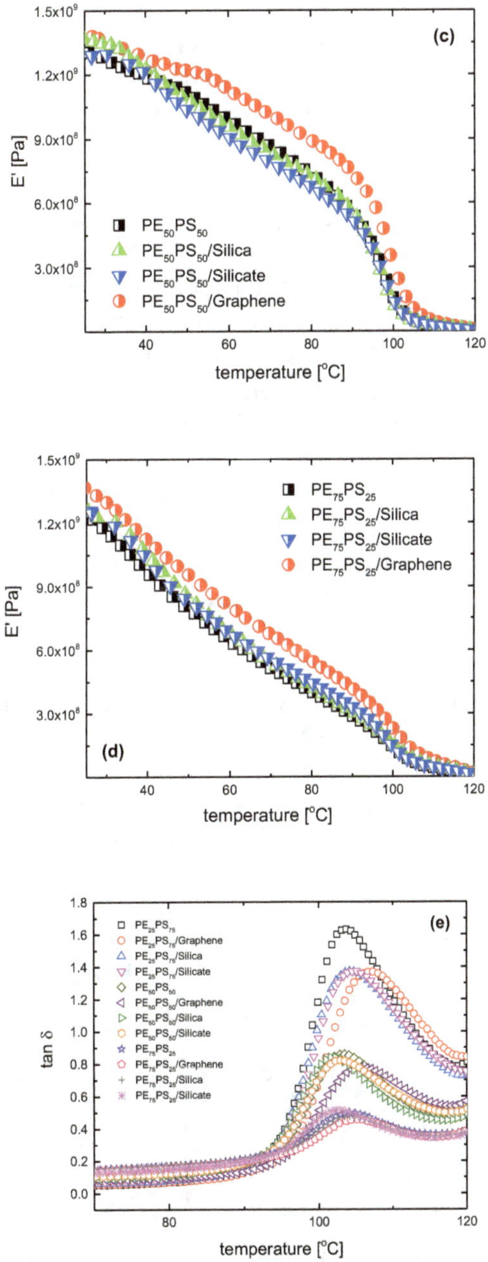

Figure 1.7 E' vs T curves of polymer blends (a) and silica, silicate and graphene composites with polymer blends with compositions of (b) $PE_{25}PS_{75}$, (c) $PE_{50}PS_{50}$ and (d) $PE_{75}PS_{25}$; (d) tan delta vs. T curves.

of PS to a higher temperature, and the extent of enhancement was also a function of PS content in the composites. Therefore, both mechanical and thermal properties were significantly affected by large aspect ratio graphene platelets, as compared to silicate and silica fillers. Even though the silicate filler had also large aspect ratio platelets, however, reinforcement effect was not as high as graphene platelets. Higher polarity of silicate platelets may have led to lower degree of their dispersion in the nanocomposites as compared to graphene platelets.

Figure 1.8 also demonstrates the rheological performance of the blends and nanocomposites as a function of angular frequency at a temperature of 205 °C. All the samples exhibited gradual decline in the extent of modulus enhancement as a function of angular frequency due to shear thinning effect. At the melt shearing conditions, the blend with 75% PE exhibited largest complex modulus, probably due to the more viscous nature of the PE fraction at these conditions. The blend with 50% PE also exhibited slightly higher complex modulus due to this reason. Even at higher angular frequency, the differences between the samples were maintained indicating composition dependent behavior of the blends. In the composites, the addition of filler resulted in enhancement of the modulus, with graphene exhibiting highest extent of increment, similar to earlier findings. The complex viscosity of the composites also increased with the addition of filler, as shown in Figure 1.9. Similar to the modulus, addition of graphene resulted in maximum extent of viscosity increase. However, the composites were still processable using the same conditions as pure blends.

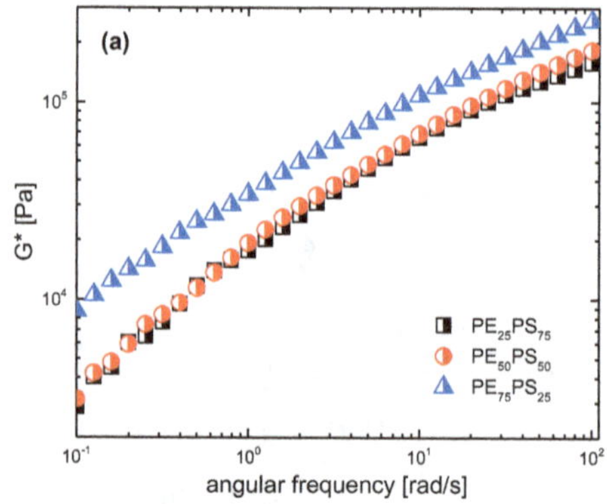

Figure 1.8 (continued on next page)

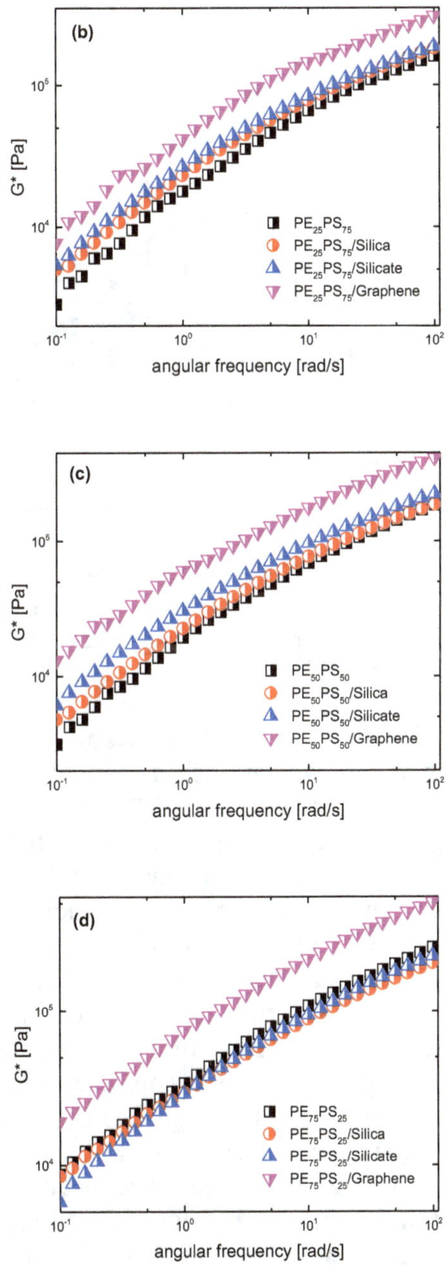

Figure 1.8 Complex modulus of polymer blends (a) and silica, silicate and graphene composites with polymer blends with compositions of (b) $PE_{25}PS_{75}$, (c) $PE_{50}PS_{50}$ and (d) $PE_{75}PS_{25}$.

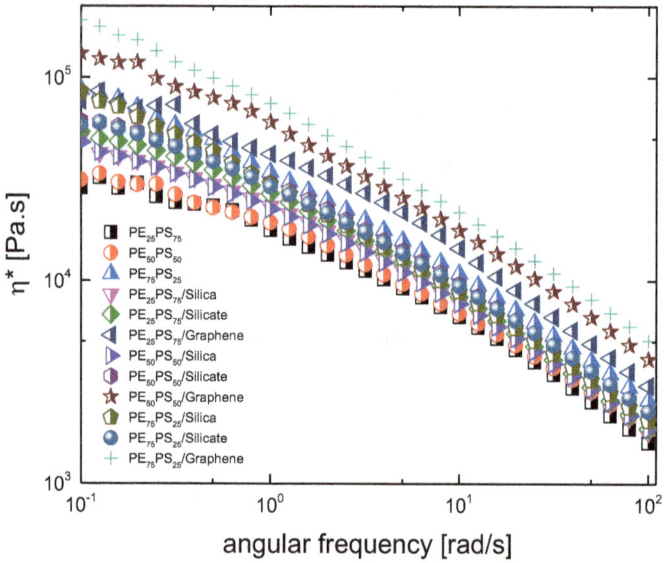

Figure 1.9 Complex viscosity of polymer blends and composites as a function of angular frequency.

The miscibility of the blends was further studied using Cole-Cole viscosity plot which develops relationships between real (η') and imaginary (η") parts of complex viscosity [51-53]. A smooth, semi-circular shape of the graph would suggest miscible systems with homogenous phase. The deviation from this behavior indicates phase segregation due to immiscibility of the components. Though these analyses have been traditionally developed for polymer blends, however, some insights can also be gained about the phase miscibility in blend composites. Figure 1.10 shows the Cole-Cole plots for the materials. In the case of blend with 25% PE content, the viscosity plot did not exhibit semi-circular shape. In comparison, the plots in the composites exhibited nearly semi-circular shape indicating good degree of miscibility, though mixing at molecular level was absent. Thus, it indicated that the presence of filler would have resulted in phase stabilization and reduction of phase domains, as suggested earlier. In the case of blend composites with 50% and 75% PE, the silicate-based composites exhibited specifically semi-circular shape of the viscosity plot. It also corroborated with the earlier findings from the TGA analysis, where a single degradation pattern was observed in the thermo-

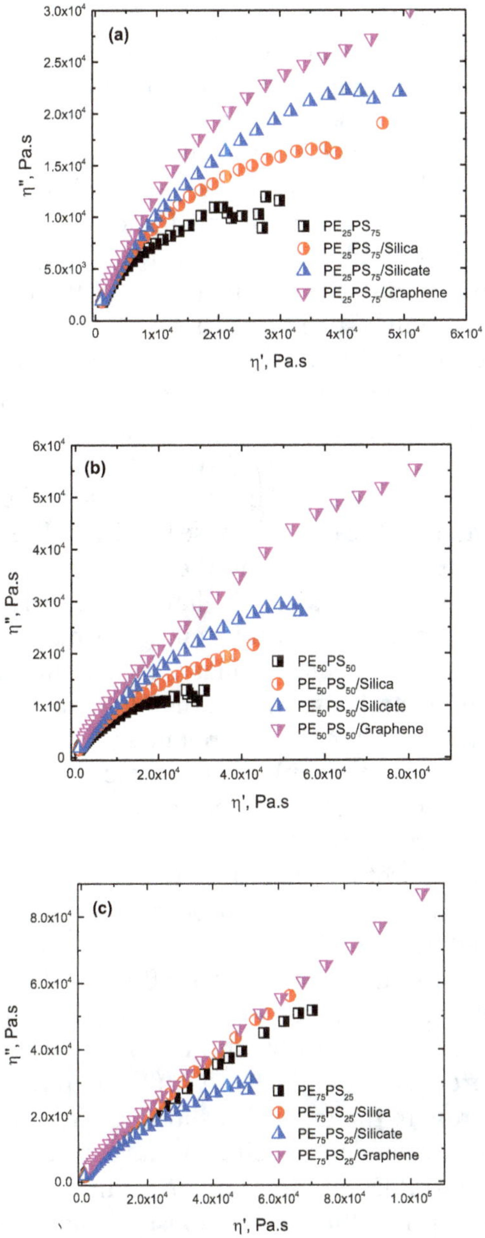

Figure 1.10 Analysis of phase miscibility thorough Cole-Cole plots (viscosity plots between η' and η") plots for the silica, silicate and graphene hybrids with polymer blends with compositions of (a) $PE_{25}PS_{75}$, (b) $PE_{50}PS_{50}$ and (c) $PE_{75}PS_{25}$.

grams, thereby, confirming enhancement in the phase intermixing and reduction of domain sizes.

To gain further insights into the development of morphology of the composites during compounding, different processing protocols were also employed. As observed earlier, the fillers were observed to localize in the polymer phase forming the melt earlier (i.e. PE). Thus, to confirm this further, the composites were also prepared by (i) melting PE and PS, followed by filler addition, (ii) melting PE, followed by addition of graphene and PS, (iii) melting PS, followed by addition of graphene and PE, (iv) melt mixing the components together using a rotational speed of 120 rpm (7 min mixing time) or using a mixing time of 15 min (rotational speed of 80 rpm). For instance, Figure 1.11 shows the light microscopy micrographs of the $PE_{50}PS_{50}$/Graphene composite when (a) PE and PS were melted first followed by graphene addition and (b) PS was melted first followed by addition of graphene and PE. The bright phase in the micrographs represents PS, whereas the darker phase is PE. In the first case, when the polymer phases were already in the melt form before the filler was added, the filler particles did not have any specific localization, as these were observed to be uniformly distributed (observed as black domains) in both PE and PS phases in the micrograph. On the other hand, in the second case, where the filler was added to the PS melt before PE was added to the compounder, the filler thus had opportunity to mix specifically with the PS phase. This resulted in predominant localization of the filler in PS. As suggested earlier, this confirmed that the observed morphology in the composites evolved due to the kinetic mixing protocol employed to generate the nanocomposites. In other words, the filler was observed to localize in the polymer which formed the melt first.

In addition to the morphology, the effect of the different processing protocols was also studied on the composite properties. Table 1.3 demonstrates the effect of different processing conditions on the mechanical properties of $PE_{50}PS_{50}$/Graphene composite. For comparison, bottom row in the table also shows the mechanical properties of the composite generated using the earlier processing method of one step mixing at 205 °C at 80 rpm for 7 min.

As evident from the table, the composite generated with earlier processing method exhibited the best mechanical performance. Mixing all the components together at 120 rpm as well as for 15 min probably led to some degree of thermal degradation of the polymer, thus, resulting in reduced modulus. On the other hand, localizing filler

(a)

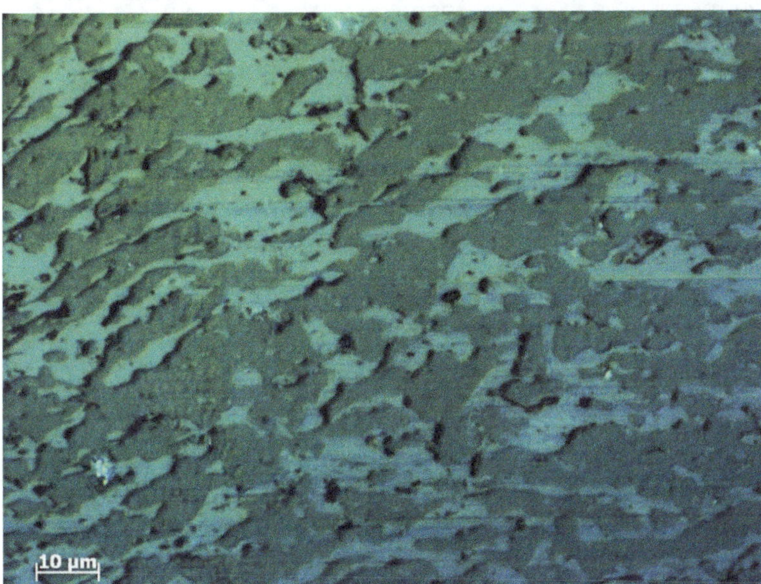

(b)

Figure 1.11 Light microscopy images of $PE_{50}PS_{50}$/Graphene composites when (a) PE and PS were melted followed by graphene addition and (b) PS was melted followed by addition of graphene and PE.

in PS was also not optimal. Even the trial where PE was melted and added with graphene before PS did not lead to similar mechanical performance as the composite where all the components are added together, which will also result in similar morphology. Longer duration of compounding in the former case would have caused some polymer degradation. Localizing filler in both PS and PS phases led to probably some aggregation or phase separation as the mechanical properties were observed to deteriorate significantly. Thus, the findings from the mechanical performance indicated that the earlier

Table 1.3 Effect of processing conditions on the mechanical performance of $PE_{50}PS_{50}$/Graphene composite

Composite	Tensile modulus[a], MPa	Peak stress[b], MPa	Peak strain[c], %	Stress at break[d], MPa
melt PE+PS followed by graphene addition and mixing for 7 min at 205 °C and 80 rpm	1034	3	0.4	3
melt PE, add graphene, followed by PS addition and mixing for 7 min at 205 °C and 80 rpm	1778	17	1.3	16
melt PS, add graphene, followed by PE addition and mixing for 7 min at 205 °C and 80 rpm	1654	13	1.1	13
add PE, PS and graphene together followed by mixing for 7 min at 205 °C and 120 rpm	1642	11	1.0	11
add PE, PS and graphene together followed by mixing for 15 min at 205 °C and 80 rpm	1746	18	1.3	18
add PE, PS and graphene together followed by mixing for 7 min at 205 °C and 80 rpm	1853	19	1.2	18

[a]Relative probable error 3%; [b]relative probable error 3%; [c]relative probable error 5%; [d]relative probable error 10%.

processing method resulted in optimum morphology with significant extent of property enhancement. In addition, the method was also efficient due to single step addition of the components, thus, reducing the mixing time as well as exposure of the polymer to high temperature for longer durations.

Table 1.4 Effect of processing conditions on the calorimetric performance of $PE_{50}PS_{50}$/Graphene composite

	Peak melting temperature T_m, °C	Melt enthalpy ΔH_m, J/g	Peak crystallization temperature T_c, °C	Crystallization enthalpy ΔH_c, J/g
melt PE+PS followed by graphene addition and mixing for 7 min at 205 °C and 80 rpm	130.2	129	111.9	112
melt PE, add graphene, followed by PS addition and mixing for 7 min at 205 °C and 80 rpm	131.9	138	112.1	133
melt PS, add graphene, followed by PE addition and mixing for 7 min at 205 °C and 80 rpm	133.7	140	111.8	138
add PE, PS and graphene together followed by mixing for 7 min at 205 °C and 120 rpm	131.5	140	111.1	133
add PE, PS and graphene together followed by mixing for 15 min at 205 °C and 80 rpm	132.2	146	114.4	138
add PE, PS and graphene together followed by mixing for 7 min at 205 °C and 80 rpm	133.1	144	116.0	148

Table 1.4 also shows the effect of processing conditions on the calorimetric performance of the $PE_{50}PS_{50}$/Graphene composite. The bottom row depicts the properties of the composites generated using earlier processing method. Similar to the mechanical performance, the composite generated using earlier processing protocol also exhibited optimal calorimetric response with high melt enthalpy and peak melting point. The peak crystallization point of the composite was also the highest among the composites, thus, indicating earlier initiation of crystallization, which might be because of the localization of filler in the semi-crystalline PE phase. The thermal degradation behavior of the composites was also similar, with no significant difference in the onset and peak degradation temperatures of the composites generated using different processing methods.

Figure 1.12 also demonstrates the evolution of complex modulus of the $PE_{50}PS_{50}$/Graphene composite as a function of different processing conditions. Similar to the mechanical and calorimetric properties, the complex modulus of the earlier developed composite was either same or higher than the other composites. Thus, these findings confirmed that the earlier followed processing methodology of one step addition of the polymers and filler phase was efficient due to faster compounding (lower energy consumption) and lower risk of polymer degradation, which thus led to the development of stable morphology with optimal composite properties.

Overall, it was observed that the morphology of the composites was significantly affected by processing method employed to generate these composites. The methodology involving single step mixing of the components was suitable to generate optimal morphology of filler stabilization of the phases and thus corresponding extent of property enhancement. In addition, the longer mixing time as well as higher rotational speed also deteriorated the composite performance.

1.4 Conclusions

Blend composites have evolved as materials of significant interest due to the potential of synergistic combination of component properties. The blends are generally created to achieve toughness, whereas the filler phase imparts stiffness to the system, thereby leading to simultaneous enhancement of these properties, which otherwise is challenging to achieve in ether blends or composites alone. These combinations of blends and fillers also lead to occasional cost saving

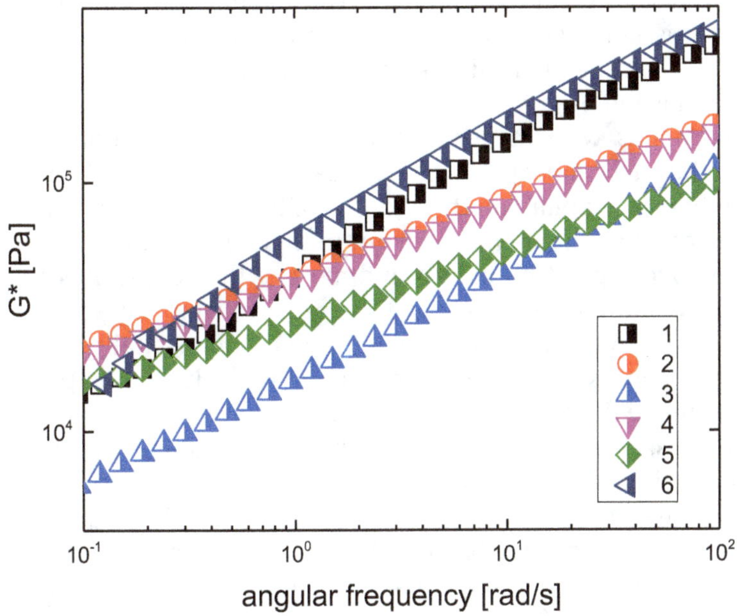

Figure 1.12 Complex modulus of the $PE_{50}PS_{50}$/Graphene composites as a function of different processing methodologies.
1: melt PE+PS followed by graphene addition and mixing for 7 min at 205 °C and 80 rpm
2: melt PE, add graphene, followed by PS addition and mixing for 7 min at 205 °C and 80 rpm
3: melt PS, add graphene, followed by PE addition and mixing for 7 min at 205 °C and 80 rpm
4: add PE, PS and graphene together followed by mixing for 7 min at 205 °C and 120 rpm
5: add PE, PS and graphene together followed by mixing for 15 min at 205 °C and 80 rpm
6: add PE, PS and graphene together followed by mixing for 7 min at 205 °C and 80 rpm

by reducing the overall cost of the composites, without sacrificing the properties. In the current study, a functional PE/PS blend system has been successfully used to provide optimal enhancement in the properties when incorporated with different fillers. One step melt mixing process could be used for generating the composites, which resulted in the kinetic embedding of the filler particles in PE phase due to its earlier melting during compounding. A fraction of the filler was also observed at the interface between PE and PS phases. In addition, the filler particles were also observed to cause reduction in the phase domains, thus, enhancing the phase intermixing between immiscible PE and PS. Different fillers like silica, silicate and graphene exhibited similar morphology.

The blends with higher fraction of PE were observed to have higher thermal stability. The composites also exhibited enhancement in the thermal stability on the addition of filler, with silicate filler imparting the maximum enhancement in the onset and peak degradation temperatures. The blends and most of the composites (except silicate containing composites) exhibited two step degradation pattern in the TGA thermograms owing to phase immiscibility. The calorimetric analysis also revealed reduction in peak melting point and crystallinity of PE on enhancing the amount of PS in the blends. In the composites, especially generated with blends containing 25 and 50% PE, the melt enthalpy of PE was observed to increase, though there was no difference in the peak meeting and crystallization temperatures, as compared to the blends. The blend and composites with 75% PS content also exhibited the glass transition of PS at slightly higher temperature of 110 °C.

The tensile modulus in the blend composites increased with filler with maximum extent of increase provided by graphene platelets as well as in the blends containing 75% PE. High aspect ratio platelets of graphene thus were beneficial in generating good degree of stress transfer. Though the silicate also contained large aspect ratio platelets, however, its high polarity may have resulted in lower degree of filler dispersion in the polymer matrix. The storage modulus of the composites also increased with the addition of filler, especially in the composite generated using the blends with 50% and 75% PE. The transition temperature of PS was also observed to shift to higher temperatures with filler as well as PS content. The complex modulus also increased with filler addition to the composites as a function of angular frequency, with maximum enhancement exhibited by graphene. Variation in the processing methodologies led to significant changes in the composite morphology. In the case of pre-melting of PS phase, the filler was also observed to localize in PS, thus, confirming the kinetic mixing of filler in the compounder. The morphology generated earlier by one step mixing of the components at optimal processing parameters (80 rpm rotational speed and 7 min mixing time) resulted in the highest extent of enhancement in the composite performance, as modulus, melt enthalpy as well as shear viscosity were observed to be maximum in this case. The generated composites, thus, open the opportunities for the development of phase stabilized polymer blend composites using a variety of fillers.

References

1. Fekete, E., Foldes, E., and Pukanszky, B. (2005) Effect of molecular interactions on the miscibility and structure of polymer blends. *European Polymer Journal*, **41**, 727-736.
2. Lipatov, Y. S. (2002) Polymer blends and interpenetrating polymer networks at the interface with solids. *Progress in Polymer Science*, **27**, 1721-1801.
3. Litmanovich, A. D., Plate, N. A., and Kudryavtsev, Y. V. (2002) Reactions in polymer blends: interchain effects and theoretical problems. *Progress in Polymer Science*, **27**, 915-970.
4. Sundararaj, U., and Macosko, C. W. (1995) Drop breakup and coalescence in polymer blends: The effects of concentration and compatibilization. *Macromolecules*, **28**, 2647-2657.
5. Katz, H. S., and Milewski, J. V. (1987) *Handbook of Fillers for Plastics*, Van Nostrand, USA.
6. Rothon, R. (1995) *Particulate-Filled Polymer Composites*, Longman, UK.
7. Brydson, J. A. (1975) *Plastic Materials*, Newnes-Butterworths, UK.
8. Karger-Kocsis, J. (1995) *Polypropylene Structure, Blends and Composites*, Chapman & Hall, UK.
9. Elias, H.-G. (2003) *Anwendungen von Polymeren*, Wiley-VCH, Germany.
10. Robertson, G. L. (1993) *Food Packaging - Principles and Practice*, volume 6 of Packaging and Converting Technology, Marcel Dekker, Inc, USA.
11. Diaz, M. F., Barbosa, S. E., and Capiati, N. J. (2007) Reactive compatibilization of PE/PS blends. Effect of copolymer chain length on interfacial adhesion and mechanical behavior. *Polymer*, **48**, 1058-1065.
12. Xu, S.-A., and Cham, C.-M. (1998) Polystyrene/high density polyethylene blends compatibilized by a tri-block copolymer I. Properties and morphology. *Polymer Journal*, **30**, 552-558.
13. Ismail, H. H., and Nasir, M. (2002) The effect of various compatibilizers on mechanical properties of polystyrene/polypropylene blend. *Polymer Testing*, **21**(2), 163-170.
14. D'orazio, L., Guarino, R., Mancarella, C., Martuscelli, E., and Cecchin, G. (1997) Effects of the addition of a graft copolymer of propylene with styrene. *Journal of Applied Polymer Science*, **65**(8), 1539-1553.
15. Aziz, A. A., Akil, H. M., Jamaludin, S. M. S., and Ramli, N. A. M. (2011)

The effect of multiple compatibilizers on the impact properties of polypropylene/polystyrene (PP/PS) blend. *Polymer-Plastics Technology and Engineering*, **50**(8), 768-775.

16. Kallel, T., Massardier-Nageotte, V., Jaziri, M., Gerard, J.-F., and Elleuch, B. (2003) Compatibilization of PE/PS and PE/PP blends. I. Effect of processing conditions and formulation. *Journal of Applied Polymer Science*, **90**(9), 2475-2484.
17. Galloway, J. A., Jeon, H. K., Bell, J. R., and Macosko, C. W. (2005) Block copolymer compatibilization of cocontinuous polymer blends. *Polymer*, **46**, 183-191.
18. Osman, M. A., Mittal, V., and Suter, U. W. (2007) Poly(propylene)-layered silicate nanocomposites: Gas permeation properties and clay exfoliation. *Macromolecular Chemistry and Physics*, **208**, 68-75.
19. van Ballegooie, P., and Rudin, A. (1988) Reactive extrusion of polystyrene/polyethylene blends. *Polymer Engineering & Science*, **28**(21), 1434-1442.
20. Lan, T., Kaviratna, P. D., and Pinnavaia, T. J. (1994) On the nature of polyimide-clay hybrid composites. *Chemistry of Materials*, **6**, 573-575.
21. Chin, I.-J., Thurn-Albrecht, T., Kim, H.-C., Russell, T. P., and Wang, J. (2001) On exfoliation of montmorillonite in epoxy. *Polymer*, **42**, 5947-5952.
22. Lim, S. K., Kim, J. W., Chin, I.-J., Kwon, Y. K., and Choi, H. J. (2002) Preparation and interaction characteristics of organically modified montmorillonite nanocomposite with miscible polymer blend of poly(Ethylene Oxide) and poly(Methyl Methacrylate). *Chemistry of Materials*, **14**, 1989-1994.
23. Wang, Z., and Pinnavaia, T. J. (1998) Nanolayer reinforcement of elastomeric polyurethane. *Chemistry of Materials*, **10**, 3769-3771.
24. Yano, K., Usuki, A., and Okada, A. (1997) Synthesis and properties of polyimide-clay hybrid films. *Journal of Polymer Science, Part A: Polymer Chemistry*, **35**, 2289-2294.
25. Shi, H., Lan, T., and Pinnavaia, T. J. (1996) Interfacial effects on the reinforcement properties of polymer-organoclay nanocomposites. *Chemistry of Materials*, **8**, 1584-1587.
26. Kuilla, T., Bhadra, S., Yao, D., Kim, N. H., Bose, S., and Lee, J. H. (2010) Recent advances in graphene based polymer composites. *Progress in Polymer Science*, **35**, 1350-1375.
27. Cai, D., and Song, M. (2010) Recent advance in functionalized graphene/polymer nanocomposites. *Journal of Materials Chemistry*, **20**, 7906-7915.
28. Steurer, P., Wissert, R., Thomann, R., and Muelhaupt, R. (2009) Functionalized graphenes and thermoplastic nanocomposites based upon expanded graphite oxide. *Macromolecular Rapid Communications*, **30**, 316-327.

29. Nguyen, D. A., Lee, Y. R., Raghu, A. V., Jeong, H. M., Shin, C. M., and Kim, B. K. (2009) Morphological and physical properties of a thermoplastic polyurethane reinforced with functionalized graphene sheet. *Polymer International*, **58**, 412-417.
30. Fang, M., Wang, K., Lu, H., Yang, Y., and Nutt, S. (2009) Covalent polymer functionalization of graphene nanosheets and mechanical properties of composites. *Journal of Materials Chemistry*, **19**, 7098-7105.
31. Chaudhry, A. U., and Mittal, V. (2013) High-density polyethylene nanocomposites using masterbatches of chlorinated polyethylene/graphene oxide. *Polymer Engineering and Science*, **53**, 78-88.
32. Sinha Ray, S., Pouliot, S., Bousmina, M., and Utracki, L. A. (2004) Role of organically modified layered silicate as an active interfacial modifier in immiscible polystyrene/polypropylene blends. *Polymer*, **45**(25), 8403-8413.
33. Wahit, M. U., Hassan, A., Rahmat, A. R., Lim, J. W., and Ishak, Z. A. M. (2006) Effect of organoclay and ethylene-octene copolymer inclusion on the morphology and mechanical properties of polyamide/polypropylene blends. *Journal of Reinforced Plastics and Composites*, **25**(9), 933-955.
34. Austin, J. R., and Kontopoulou, M. (2006) Effect of organoclay content on the rheology, morphology, and physical properties of polyolefin elastomers and their blends with polypropylene. *Polymer Engineering and Science*, **46**(11), 1491-1501.
35. Yang, H., Zhang, X., Qu, C., Li, B., Zhang, L., Zhang, Q., and Fu, Q. (2007) Largely improved toughness of PP/EPDM blends by adding nano-SiO_2 particles. *Polymer*, **48**(3), 860-869.
36. Si, M., Araki, T., Ade, H., Kilcoyne, A. L. D., Fisher, R., Sokolov, J. C., and Rafailovich, M. H. (2006) Compatibilizing bulk polymer blends by using organoclays. *Macromolecules*, **39**(14), 4793-4801.
37. Kim, J., Cote, L. J., Kim, F., Yuan, W., Shull, K. R., and Huang, J. (2010) Graphene oxide sheets at interfaces. *Journal of the American Chemical Society*, **132**(23), 8180-8186.
38. Kim, H., Abdala, A. A., and Macosko, C. W. (2010) Graphene/polymer nanocomposites. *Macromolecules*, **43**(16), 6515-6530.
39. Cao, Y., Zhang, J., Feng, J., and Wu, P. (2011) Compatibilization of immiscible polymer blends using graphene oxide sheets. *ACS Nano*, **5**(7), 5920-5927.
40. Elias, L., Fenouillot, F., Majeste, J. C., Alcouffe, P., and Cassagnau, P. (2008) Immiscible polymer blends stabilized with nano-silica particles: Rheology and effective interfacial tension. *Polymer*, **49**(20), 4378-4385.
41. Thareja, P., and Velankar, S. (2007) Particle-induced bridging in immiscible polymer blends. *Rheologica Acta*, **46**(3), 405-412.
42. Pawar, S. P., Misra, A., Bose, S., Chatterjee, K., and Mittal, V. (2015)

Enzymatically degradable and flexible bio-nanocomposites derived from PHBV and PBAT blend: Assessing thermal, morphological, mechanical, and biodegradation properties. *Colloid and Polymer Science*, **293**, 2921-2930.

43. Tanaka, H., Lovinger, A. J., and Davis, D. D. (1994) Pattern evolution caused by dynamic coupling between wetting and phase separation in binary liquid mixture containing glass particles. *Physical Review Letters*, **72**(16), 2581-2584.
44. Ginzburg, V. V., Qiu, F., Paniconi, M., Peng, G., Jasnow, D., and Balazs, A. C. (1999) Simulation of hard particles in a phase-separating binary mixture. *Physical Review Letters*, **82**(20), 4026-4029.
45. Nesterov, A. E. (1999) Compatibilizing effect of a filler in binary polymer mixtures. *Polymer*, **40**(5), 1347-1349.
46. Nesterov, A. E. (2001) Effect of an interface with solid on the component distribution in separated phases of binary polymer mixtures. *European Polymer Journal*, **37**(2), 281-285.
47. Hong, J. S., Namkung, H., Ahn, K. H., Lee, S. J., and Kim, C. (2006) The role of organically modified layered silicate in the breakup and coalescence of droplets in PBT/PE blends. *Polymer*, **47**(11), 3967-3975.
48. McAllister, M. J., Li, J. L., Adamson, D. H., Schniepp, H. C., Abdala, A. A., Liu, J., Herrera-Alonso, M., Milius, D. L., Car, R., Prud'homme, R. K., and Aksay, I. A. (2007) Single sheet functionalized graphene by oxidation and thermal expansion of graphite. *Chemistry of Materials*, **19**, 4396.
49. Mittal, V., and Chaudhry, A. U. (2015) Polymer-graphene nanocomposites: Effect of polymer matrix and filler amount on properties. *Macromolecular Materials and Engineering*, **300**(5), 510-521.
50. Joshi, M., Butola, B. S., Simon, G., and Kukaleva, N. (2006) Rheological and viscoelastic behavior of HDPE/octamethyl-POSS nanocomposites. *Macromolecules*, **39**, 1839-1849.
51. Cho, K., Lee, B. H., Hwang, K. M., Lee, H., and Choe, S. (1998) Rheological and mechanical properties in polyethylene blends. *Polymer Engineering and Science*, **38**, 1969-1975.
52. Kim, H. K., Rana, D., Kwag, H., and Choe, S. (2001) Melt rheology of ethylene 1-octene copolymer blends synthesized by Ziegler-Natta and metallocene catalysts. *Korea Polymer Journal*, **8**, 34-43.
53. Kwag, H., Rana, D., Cho, K., Rhee, J., Woo, T., Lee, B. H., and Choe, S. (2000) Binary blends of metallocene polyethylene with conventional polyolefins: Rheological and morphological properties. *Polymer Engineering and Science*, **40**, 1672-1681.

2

Effect of MOF-Graphene based Hybrid Fillers on the Properties of High Density Polyethylene Nanocomposites

2.1 Introduction

Polymer nanocomposites have generated a substantial research interest owing to their promising properties as compared to conventional micro-scale nanocomposites. Polymer nanocomposites are advanced materials in which at least one of the components has a dimension in the nanoscale range (less than 100 nm). Nanofillers influence the properties of the polymers even at very low volume fractions due to their high surface area and aspect ratio [1]. Nanofillers can be divided into three different classes depending upon their dimensions. The filler materials with all dimensions in the nanometer range are known as 0-D nanofillers (e.g., silica nanoparticles) [2,3]. When two dimensions are in the nanometer range and the third is in the micrometer range, the materials are termed as 1-D nanofillers (e.g., carbon nanotubes (CNTs)) [4,5]. In case, one of the dimensions is in the nanometer range, whereas the other two are in the micrometer scale, the materials are called 2-D nanofillers (e.g., layered silicates) [6]. Overall, a large number of studies on the polymer nanocomposites with a variety of fillers have reported a significant improvement in the mechanical, thermal, gas-barrier, flame retardation and rheological properties [7,8].

Recently, hybrid nanofillers have received an extensive research attention in the field of polymer nanocomposites due to the synergetic combination of the properties of the two or more different fillers. The hybrid nanofillers are made up of two or more components prepared using different physical and/or chemical methods. The hybridization of nanomaterials possessing different dimensions (e.g., 0-D, 1-D and 2-D) results in the development of the high performance hybrid materials as compared to the individual nanofillers, thus, leading to the superior enhancement of properties in the polymer nanocomposites.

Swati Singh, Muthukumaraswamy R. Vengatesan, Georgios Karanikolos and Vikas Mittal
Khalifa University of Science and Technology, Abu Dhabi, UAE

Graphene contains sp^2 hybridized carbon atoms in a honeycomb structure and has been used as a nanofiller in the polymer matrices due to its superior thermal, mechanical and electrical properties [9-13]. Graphene based 2-D nanomaterials have also been commonly used with other materials for the preparation of hybrid nanofillers. Recently, Chen *et al.* [14] developed a grafted hybrid filler from CNTs and graphene platelets and incorporated it in the polyurethane matrix. It was reported that the hybrid filler displayed better dispersion than the individual fillers, which resulted in an improved electric-induced strain in the nanocomposites. Sharmila *et al.* [15] prepared epoxy nanocomposites by employing the hybrid filler generated using reduced graphene oxide and iron nanoparticles. The improved tensile strength, dielectric properties and thermal stability as compared to the pure polymer were attributed to the homogenous dispersion and effective interfacial adhesion owing to the hybrid filler incorporated in the epoxy matrix.

Metal organic frameworks (MOFs) possess many superior properties, such as uniform porosity, large surface area [16] and pore volume, high thermal and chemical stability, non-toxic nature [17,18] as well as easily tunable structure, which make them efficient candidates for different applications in gas storage and separation, drug delivery [19], chemical sensing [20], catalysis [21,22], among others. MOFs, also known as coordination polymers, are composed of inorganic metal ions or metal clusters and organic linkers held together via coordination bond. Moreover, owing to the array of metal centers and functional ligands, the MOF based materials provide the opportunity of developing new types of composite materials with enhanced performance [23-32]. In this respect, it is of significant interest to study the preparation of the hybrid nanofillers of MOFs with 2-D (e.g., graphene) nanomaterials by employing both *in-situ* and *ex-situ* methods.

In the present work, the effect of the hybrid nanofillers (based on MOF and graphene) on the thermal, mechanical and rheological properties of the high density polyethylene (HDPE) matrix has been analyzed. Two different types of nanocomposites were developed: (i) composites generated with MOF coated graphene which was prepared via an *in-situ* method and (ii) composites generated using MOF and graphene via *ex-situ* method, i.e. MOF and graphene fillers were separately incorporated in the polyethylene matrix. In order to achieve homogenous dispersion of nanofillers, polyethylene-*alt*-maleic anhydride (PE-*alt*-MA) was used as a compatibilizer.

2.2 Experimental

2.2.1 Materials

HDPE, PE-*alt*-MA, xylene, dimethyl formamide (DMF), methanol, acetone, copper sulfate, trimesic acid and deionized water were procured from Sigma Aldrich. Graphite oxide (GO) was synthesized by using improved Tour's method [33].

2.2.2 Preparation of HKUST-MOF, Thermally Reduced Graphene (TRG) and HKUST-MOF/Graphene Hybrid

HKUST-MOF was prepared using the method reported in literature [34].

For the preparation of TRG, the dried GO was thermally reduced at 900° C for 30 s, using a tube furnace with nitrogen flow gas medium. The black powdered material was collected from the tube after cooling down to room temperature.

The hybrid of HKUST-MOF and graphene was prepared by one step pot method. 0.5 g of GO was mixed with 450 mL of DMF and sonicated for 1 h (solution A). Separately, 0.6 g of copper sulfate was mixed with 0.3 g of trimesic acid in 15 mL of DMF. The solution was subsequently sonicated for 5 min. 15 mL of de-ionized water (DI) was subsequently added to the solution, followed by ultra-sonication (solution B). Both solutions were subsequently mixed and kept under reflux condensation with constant stirring for 24 h. Subsequently, the solution was filtered and effectively washed with methanol and acetone.

2.2.3 Preparation of HDPE Nanocomposites using *Ex-situ* Fillers

0.1 g of MOF and 0.1 g of TRG were mixed using 400 mL of xylene in a round bottom flask and kept for 1 h under ultrasonication. 1 g of PE-*alt*-MA and 8.8 g of PE were then added into the solution and kept under reflux condensation for 2 h at 130° C. The solution was subsequently dried under vacuum at 100 °C so as to remove the solvent molecules.

The nanocomposite test specimens were molded using mini-injection molding machine (HAAKE MiniJet PRO, Thermo Scientific) with a cylinder temperature of 180 °C, mold temperature of 125 °C, injection pressure of 430 bar for 10 s and post pressure of 500 bar for 6 s.

The resultant nanocomposite was termed as PE-*ex*-MG1 (1 wt%MOF+1wt%graphene). Similarly, other composites were prepared by varying the weight percentage of the fillers as 2.5wt%MOF+2.5wt%graphene and 5wt%MOF+5wt%graphene, which were subsequently termed as PE-*ex*-MG2 and PE-*ex*-MG3 respectively.

2.2.4 Preparation of Nanocomposites using *In-situ* Prepared Hybrid Filler

0.2 g of the *in-situ* synthesized hybrid nanofiller was dispersed in 400 mL of xylene using ultrasonication for 1 h. Subsequently, 1 g of PE-*alt*-MA and 8.8 g of HDPE were added. The mixture was kept under reflux condensation for 2 h at 130° C. The solution was then dried under vacuum atmosphere at 100° C.

The test specimens were prepared by using the same procedure as composites with *ex-situ* fillers. The sample with 2 wt% MOF/graphene hybrid was termed as PE-*in*-MG1. Similarly, other composites were prepared by incorporating 5 wt% and 10 wt% of *in-situ* synthesized hybrid nanofiller and were termed as PE-*in*-MG2 and PE-*in*-MG3 respectively.

2.2.5 Characterization

Calorimetric properties of the materials were studied using dynamic scanning calorimetry (Discovery DSC, TA Instruments) in nitrogen atmosphere. A sample size of 3-7 mg was used for the DSC analysis. The samples were heated from 35 to 200 °C followed by cooling from 200 to -60 °C using heating and cooling rates of 10 °C/min. The second heating and cooling cycles, performed in the same temperature range, were used for the analysis of the calorimetric properties of the materials.

The thermal properties of the materials were evaluated using thermogravimetric analyzer (Discovery TGA, TA instruments) with nitrogen as carrier gas. The samples (10-20 mg weight) were heated from 35 to 700 °C by employing a heating rate of 10 °C/min.

Tensile testing of the pure polymer, polymer-compatibilizer blend and composites was performed on universal testing machine (Testometric, UK). The sample dimensions for tensile test were: sample length 73 mm, gauge length 30 mm, width 4 mm and thickness 2 mm. A loading rate of 4 mm/min was used, and the tests were carried out at room temperature. Tensile modulus and yield stress were calculated using built in software Win Test Analysis. An average of five values was recorded.

The melt rheological properties (storage modulus (G'), loss modulus (G"), tan δ and complex viscosity (η*) as a function of angular frequency (ω)) of the materials were studied using AR 2000 shear rheometer from TA Instruments (ASTM D4440). Injection molded disc shaped samples of diameter 25 mm and thickness 1.5 mm were used, and the analysis was carried out at 180 °C in parallel plate geometry of diameter 25 mm with a geometric gap of 1.4 mm. Dynamic frequency scans were performed from 0.1 to 100 rad/s at 2% strain.

X-ray diffraction (XRD) patterns of the MOF, TRG, MOF/graphene hybrid and HDPE nanocomposites were measured on a Panalytical powder diffractometer in reflection mode with CuKα radiation (λ = 1.5406 Å). The samples were scanned at room temperature from 2θ = 5-60° with a step size of 0.02° and 10 s as step time. Zero-background holder was used to minimize the noise.

The microstructure of the samples was analyzed using transmission electron microscope (TEM) Philips CM 20 (Philips/FEI, Eindhoven) at 120 and 200 kV accelerating voltages. Thin sections of 70-90 nm thickness were microtomed from the block of the nanocomposite specimens using diamond knife at -60 °C. The thin sections were subsequently supported on 100-mesh grids sputter coated with a 3 nm thick carbon layer.

2.3 Results and Discussion

Calorimetric properties of the pure HDPE, HDPE-compatibilizer blend (10 wt%) and nanocomposites are plotted in Figure 2.1. The degree of crystallinity (X_c), peak melting temperature (T_m), peak crystallization temperature (T_c) and melting (ΔH_m)/crystallization (ΔHc) enthalpy are also listed in Table 2.1. The degree of crystallinity was determined by taking the melt enthalpy of pure crystalline HDPE as 293 J/g [28]. For pure HDPE, T_m and T_c were observed to be 135 °C and 117 °C respectively, whereas the degree of crystallinity was calculated as 56%. The melting and crystallization temperatures of HDPE did not exhibit significant change in the blend and nanocomposites. On the other hand, X_c decreased to 52% on the addition of compatibilizer (10 wt%), which might be due to the plasticization effect of the compatibilizer. Also, the incorporation of both *in-situ* prepared MOF/graphene hybrid and *ex-situ* added nanofillers to the

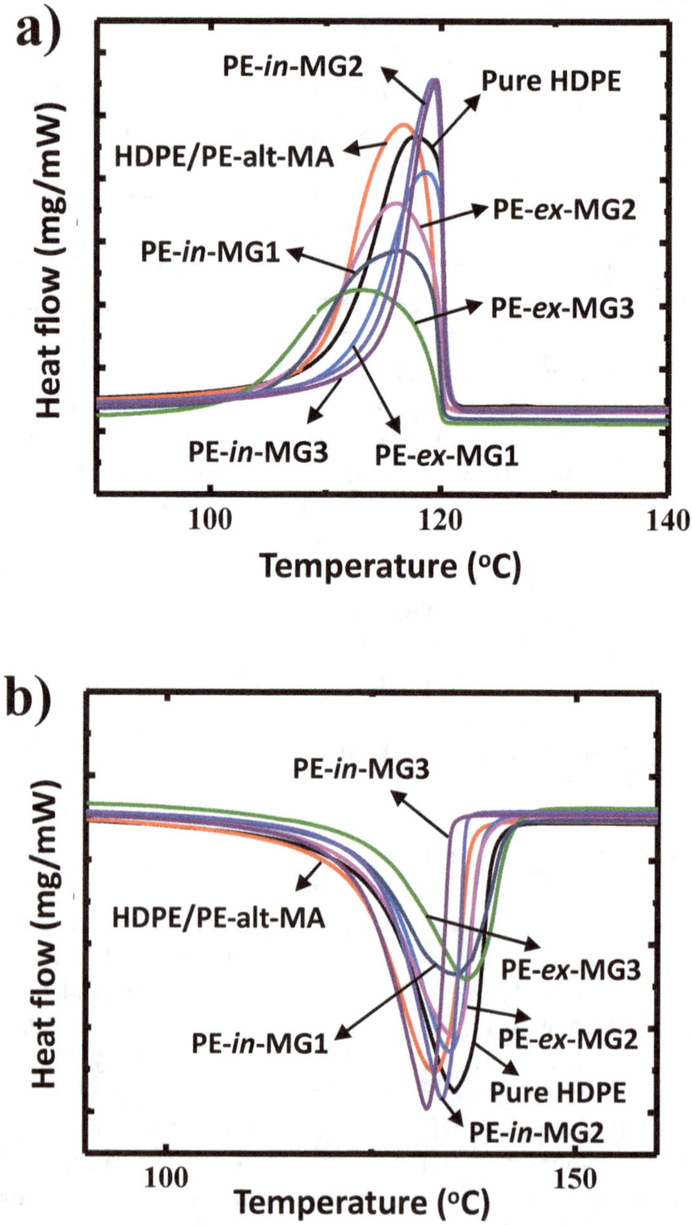

Figure 2.1 DSC (a) melting and (b) crystallization thermograms of HDPE, HDPE-compatibilizer blend and HDPE/hybrid filler nanocomposites.

Table 2.1 Calorimetric properties of HDPE, HDPE-compatibilizer blend and HDPE/hybrid filler nanocomposites

Sample	T_m(°C)	Tc(°C)	ΔH_m	ΔH_c	X_c(%)
HDPE	135	117	39.3	43.2	56
HDPE/PE-*alt*-MA	133	116	37.5	39.7	53
PE-*ex*-MG1	134	117	31.4	34.8	53
PE-*ex*-MG2	135	115	31.3	33.4	51
PE-*ex*-MG3	137	112	26.9	30.0	48
PE-*in*-MG1	136	115	28.9	32.8	47
PE-*in*-MG2	132	118	32.9	32.9	56
PE-*in*-MG3	132	119	33.5	34.0	59

HDPE matrix impacted the X_c values. In the case of *ex-situ* added nanofiller composites, X_c was observed to gradually decrease with increasing filler content. The observed decrease in the crystallinity might have resulted from the non-uniform distribution of the nanofillers. Also, as the fillers were only physically mixed, one filler might have impacted the crystallinity of the polymer more than the other. Similarly, one of the fillers might have poorer distribution than the other, thus, leading to the overall decrease in crystallinity. In the case of *in-situ* generated MOF/graphene hybrid filled nanocomposites, X_c was observed to decrease to 46% for the composite with 2 wt% filler content. However, the crystallinity increased with enhancing the filler content in the composites. For instance, for the composite with 10 wt% filler content, the crystallinity value was observed to be 59%. The initial decrease in the X_c might be due to the interference by the hybrid nanofiller and compatibilizer with the arrangement of chain molecules in the crystal lattice of the polymer. Further addition of the hybrid filler in the HDPE matrix exhibited a gradual nucleating effect which resulted in improved X_c values.

The thermal properties of the materials were examined under nitrogen atmosphere to ascertain the onset degradation temperature and char yield. Table 2.2 lists T^d_{10} (°C) and T^d_{40} (°C) values (temperatures at which 10 and 40% weight loss has taken place, respectively) as well as char yield at 700 °C (% of non-volatile material present at 700 °C). Figure 2.2 also presents the TGA thermograms of the materials. As observed from the table, the initiation of degradation (10% weight loss) temperature was observed to be higher for pure HDPE as compared to the blend and nanocomposites, probably due to the presence of the compatibilizer. On the other hand, the temperatures for 40 wt% degradation were observed to be slightly higher in the nanocomposites. It indicated that the nanofillers did not significantly impact the thermal behavior of the polymer, and the polymer still retained high temperature stability. The char yield was observed to increase with the amount of filler due to the increment in the fraction of the non-volatile material.

Table 2.2 Thermal stability and char yield of HDPE, HDPE-compatibilizer blend and HDPE/hybrid filler nanocomposites

Sample	T^{d}_{10}(°C)	T^{d}_{40}(°C)	Char yield at 700°C (%)
Pure HDPE	445	472	0.00
HDPE/PE-alt-MA	443	477	0.25
PE-*ex*-MG1	446	479	1.77
PE-*ex*-MG2	430	479	3.96
PE-*ex*-MG3	431	477	7.97
PE-*in*-MG1	426	477	1.89
PE-*in*-MG2	427	479	2.83
PE-*in*-MG3	423	479	5.54

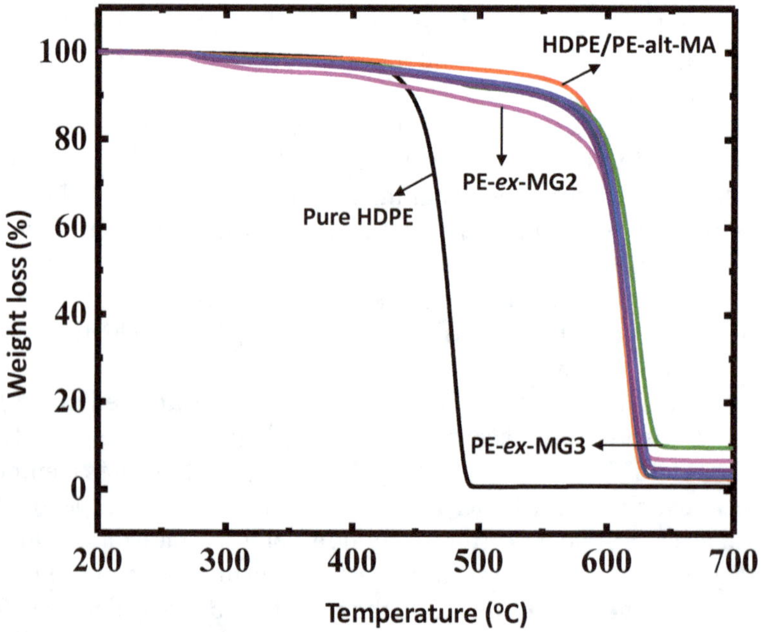

Figure 2.2 TGA curves of HDPE, HDPE-compatibilizer blend and HDPE/hybrid filler nanocomposites.

The effect of the *in-situ* hybrid nanofiller and *ex-situ* mixed fillers on the mechanical properties of the materials has been depicted in Table 2.3. The addition of 10 wt% compatibilizer to the HDPE matrix resulted in a reduction in the tensile modulus due to the plasticization effect. The *in-situ* hybrid MOF/graphene incorporated nanocomposites showed a gradual increase in the modulus with enhancement in the hybrid filler content. In case of *ex-situ* mixed filler system, the addition of nanofillers also increased the modulus for the composite with 2 wt% filler content. However, the modulus was observed to gradually decrease with increase in the filler content. It indicated that the *in-situ* hybrid fillers imparted superior mechanical performance to the polymer. In the case of *in-situ* hybrid filler, MOF molecules are expected to be uniformly stacked on the surface of the graphene sheets which reduces sheet stacking, whereas in the case of *ex-situ* method, the individual filler materials may have distributed in different regions in the polymer matrix, thus, resulting in less optimal enhancement in the mechanical performance (Figure 2.3). The superior performance of *in-situ* hybrid filler system can also be gauged from the simultaneous increment in the tensile strength of the polymer, which was observed to decrease in the case of *ex-situ* added fillers. For instance, the *in-situ* hybrid filler based polymer nanocomposite with 10 wt% filler fraction exhibited higher tensile strength (36 MPa) than the pure polymer (34 MPa), whereas the same nanocomposite with *ex-situ* filler addition exhibited a much lower value of 24 MPa. The strain at break was observed to decrease in both *in-situ* and *ex-situ* systems, however, the reduction was more severe in the case of *ex-situ* added filler, thus, exhibiting higher extent of brittleness.

Table 2.3 Mechanical properties of HDPE, HDPE-compatibilizer blend and HDPE/hybrid filler nanocomposites

Sample	UTS (MPa)	Modulus (MPa)	Strain at Break (%)
Pure HDPE	34	598	31
HDPE/PE-alt-MA	31	567	41
PE-*ex*-MG1	32	730	11
PE-*ex*-MG2	27	707	10
PE-*ex*-MG3	24	634	4
PE-*in*-MG1	32	710	17
PE-*in*-MG2	32	756	15
PE-*in*-MG3	36	798	9

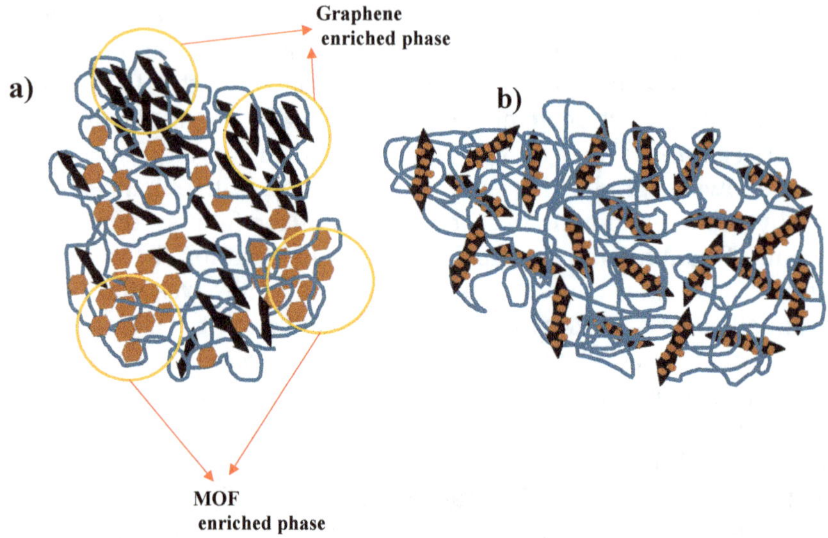

Figure 2.3 Schematic representation of a) *ex-situ* hybrid filler nanocomposites and b) *in-situ* hybrid filler nanocomposites.

The effect of hybrid fillers on the polymer chain motion, filler dispersion and polymer-filler interactions has been studied by the melt rheological analysis [35,36]. The shape of nanofiller has been reported to influence the melt rheological properties of the nanocomposites [37]. Figure 2.4a,b shows the plots of storage (G') and loss modulus (G") as a function of angular frequency. The G' and G" values of the HDPE matrix and nanocomposites increased with frequency. The nanofillers had an insignificant impact on the storage and loss modulus of the polymer. Also, for both types of filler systems, a well-defined plateau of G' was observed at lower frequency in the composites, indicating a transition from liquid-like to pseudo solid-like viscoelastic behavior leading to a percolation network structure [38,39]. The miscibility analysis using the rheological data also indicated higher extent of compatibility in the case of *in-situ* filler system, thus, confirming the higher extent of property enhancement as compared to *ex-situ* system. Figure 2.4c also presents the complex viscosity of the materials as a function of angular frequency. The nanocomposites showed a frequency independent Newtonian flow like behavior at lower frequencies (less than 1 rad/s), followed by a shear thinning behavior [40,41]. As the viscosity of the polymer remained largely unaffected by the addition of nanofillers, the composites could be processed using the same processing conditions as the pure polymer.

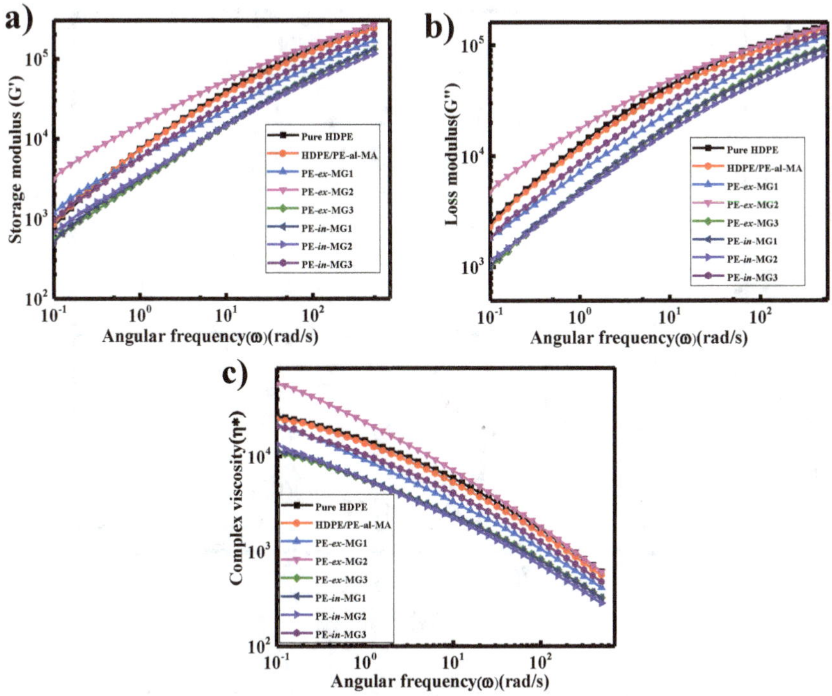

Figure 2.4 Melt rheology of HDPE, HDPE-compatibilizer blend and HDPE/hybrid filler nanocomposites: a) storage modulus, b) loss modulus and c) complex viscosity as a function of angular frequency.

XRD was performed on the pristine polyethylene and nanocomposites to study the changes in the crystalline structure with filler incorporation, as shown in Figure 2.5. High density polyethylene exhibited two diffraction peaks at 2θ = 22.4° and 24.7° corresponding to (110) and (200) planes. It indicated the orthorhombic crystal structure of polyethylene, as reported in literature [25]. Other weak reflections were observed at 30° and 37.1° for pristine polyethylene corresponding to (210) and (020) reflection planes respectively [26,27]. No noticeable change in the characteristic diffraction peaks of (110) and (200) planes was observed in the nanocomposites, thus, indicating that the incorporation of nanofillers did not affect the crystal structure of the polymer.

The TEM micrographs of the *in-situ* generated filler and composites with 5 wt% *in-situ* and *ex-si*tu filler systems are presented in Figure 2.6. The TEM micrograph of the *in-situ* filler system in Figure 2.6a showed uniform stacking of MOF on the graphene platelets, which

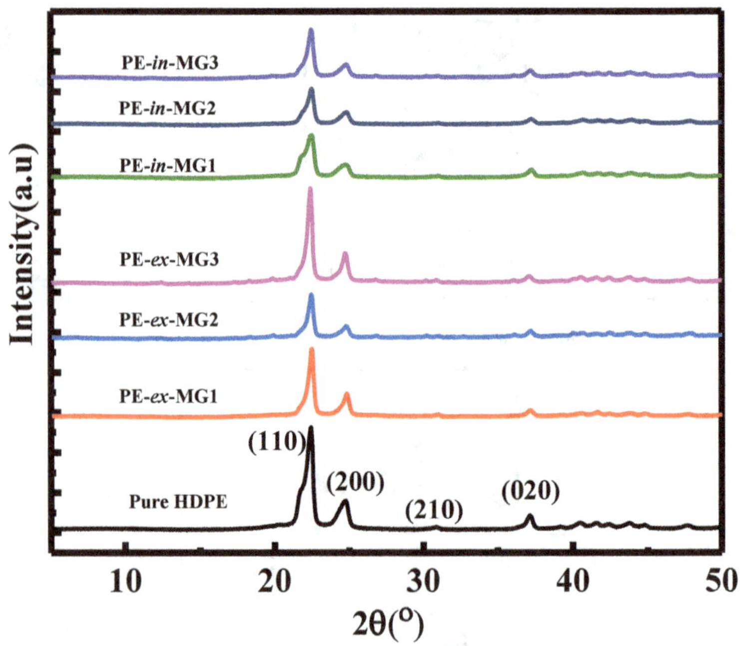

Figure 2.5 XRD of HDPE and HDPE/hybrid filler nanocomposites.

helped to enhance sheet delamination and reduce filler aggregation in the polymer matrix. The TEM image of PE-*ex*-MG2 composite in Figure 2.6b presents the enriched phase of MOF in the polymer region, along with a small degree of agglomeration. It confirmed the

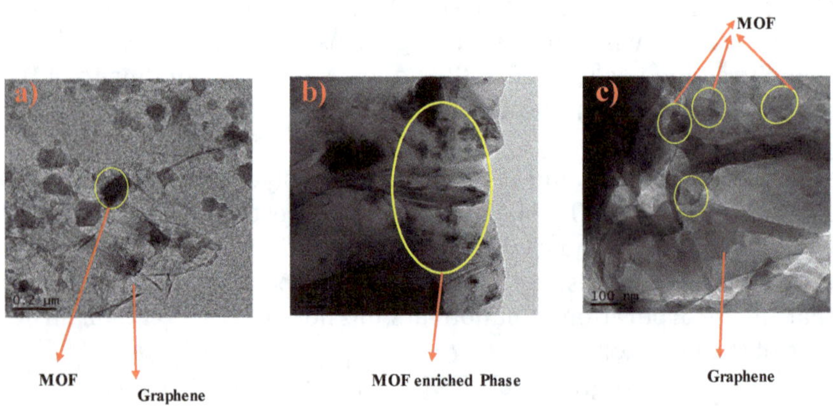

Figure 2.6 TEM images of a) *in-situ* MOF coated graphene, b) PE-*ex*-MG2 and c) PE-*in*-MG2.

earlier notion that the filler phases in the *ex-situ* system may be dispersed at different locations in the polymer matrix. In the TEM micrograph of the *in-situ* hybrid filler nanocomposite (PE-*in*-MG2) (Figure 2.6c), a uniform dispersion of nanofiller in the polymer phase was observed, which can be attributed to the relatively better property profiles as compared to the composites with *ex-situ* added fillers.

2.4 Conclusion

In this work, the effect of hybrid nanofillers composed of MOF and graphene was analyzed on the properties of the polyethylene matrix. The hybrid fillers were generated either *in-situ* prior to the incorporation in the polymer matrix or were added individually to the polymer matrix (*ex-situ*). *In-situ* hybrid nanofiller based nanocomposites exhibited higher modulus and strength than the *ex-situ* hybrid nanofiller based composites. The degree of crystallinity of the *in-situ* hybrid nanofiller based nanocomposites also showed an improvement with filler content. The thermal and rheological properties of the polymer were not significantly affected by the presence of the hybrid nanofillers. The morphological analysis confirmed the uniform filler distribution in the case of *in-situ* hybrid filler based composites, thus, indicating the advantage of hybridizing the fillers prior to the incorporation in the polymer matrix.

References

1. Brechet, Y., Cavaille, J. Y., Chabert, E., Chazeau, L., Dendievel, R., Flandin, L., and Gauthier, C. (2001) Polymer based nanocomposites: effect of filler-filler and filler-matrix interactions. *Advanced Engineering Materials*, **3**(8), 571-577.
2. Mark, J. E. (1996) Ceramic-reinforced polymers and polymer-modified ceramics. *Polymer Engineering and Science*, **36**(24), 2905-2920.
3. Chang, W. Y., Lee, K. J., and Nam, J. D. (2003) Reactive dispersion and mechanical property of dicyanate/montmorillonite nanocomposite. *Polymer Korea*, **27**, 75-83.
4. Calvert, P. (1997) Potential applications of nanotubes. In: *Carbon Nanotubes: Preparation and Properties*, Ebbesen, T. W. (ed.), CRC Press, USA, pp. 277-292.
5. Favier, V., Canova, G. R., Shrivastava, S. C., and Cavaille, J. Y. (1997) Mechanical percolation in cellulose whisker nanocomposites. *Polymer Engineering and Science*, **37**, 1732-1739.

6. Mittal, V. (2006) *Organic Modifications of Clay and Polymer Surfaces for Specialty Applications*, PhD Thesis, ETH Zurich, Switzerland.
7. Mittal, V. (2007) polypropylene-layered silicate nanocomposites: filler matrix interactions and mechanical properties. *Journal of Thermoplastic Composite Materials,* **20**, 575-599.
8. Chaudhry, A. U., and Mittal, V. (2013) High-density polyethylene nanocomposites using masterbatches of chlorinated polyethylene/graphene oxide. *Polymer Engineering and Science,* **53**, 78-88.
9. Neto, A. H. C., Guinea, F., Peres, N. M. R., Novoselov, K. S., and Geim, A. K. (2009) The electronic properties of graphene. *Reviews of Modern Physics,* **81**, 109-162.
10. Katsnelson, M. I., Novoselov, K. S., and Geim A. K. (2006) Chiral tunneling and the Klein paradox in graphene. *Nature Physics,* **2**, 620-625.
11. Zhang, Y. B., Tan, Y. W., Stormer, H. L., and Kim P. (2005) Experimental observation of the quantum Hall effect and Berry's phase in graphene. *Nature,* **438**, 201-204.
12. Reiner-Rozman, C., Larisika, M., Nowak, C., and Knoll, W. (2015) Graphene-based liquid-gated field effect transistor for biosensing: theory and experiments. *Biosensors and Bioelectronics,* **70**, 21-27.
13. Novoselov, K. S., Geim, A. K., Morozov, S. V., Jiang, D., Zhang, Y., Dubonos, S. V., Grigorieva I. V., and Firsov, A. A. (2004) Electric field effect in atomically thin carbon films. *Science,* **306**, 666-669.
14. Chen , T., Pan, L., Lin, M., Wang, B., Liu, L., Li, Y., Qiu, J., and Zhu, K. (2015) Dielectric, mechanical and electro-stimulus response properties studies of polyurethane dielectric elastomer modified by carbon nanotube-graphene nanosheet hybrid fillers. *Polymer Testing,* **40**, 4-11.
15. Sharmila, T. K. B., Antony, J. V., Jayakrishnan, M. P., Beegum, P. M. S., and Thachil, E. T. (2015) Mechanical, thermal and dielectric properties of hybrid composites of epoxy and reduced graphene oxide/iron oxide. *Materials and Design,* **90**, 66-75.
16. Chae, H. K., Siberio-Pérez, D. Y., Kim, J., Go, Y., Eddaoudi, M., Matzger, A. J., O'Keeffe, M., and Yaghi, O. M. (2004) A route to high surface area porosity and inclusion of large molecules in crystals. *Nature,* **427**, 523-527.
17. Furukawa, H., Cordova, K. E., O'Keeffe, M., and Yaghi, O. M. (2013) The chemistry and applications of metal-organic frameworks. *Science,* **341**(6149), 1230444.
18. Tanabe, K. K., and Cohen, S. M. (2011) Post synthetic modification of metal-organic frameworks-a progress report. *Chemical Society Review,* **40**, 498-519.
19. Cook, T. R., Zheng, Y. R., and Stang, P. J. (2013) Metal-organic frameworks and self-assembled supramolecular coordination complexes: Comparing and contrasting the design, synthesis, and functionality

of metal-organic materials. *Chemical Reviews,* **113**, 734-777.

20. Cheetham, A. K., Rao, C. N., and Feller, R. K. (2006) Structural diversity and chemical trends in hybrid inorganic-organic framework materials. *Chemical Communications,* **46**, 4780-4795.
21. Vaesen, S., Guillerm, V., Yang, Q., Wiersum, A. D., Marszalek, B., Gil, B., Vimont, A., Daturi, M., Devic, T., Llewellyn, P. L., Serre, C., Maurin, G., and Weireld, G. D. (2013) A robust amino-functionalized titanium(IV) based MOF for improved separation of acid gases. *Chemical Communications,* **49**, 10082-10084.
22. Hamilton, T. D., Papaefstathiou, G. S., and MacGillivray, L. R. (2005) Template-controlled reactivity: Following nature's way to design and construct metal-organic polyhedra and polygons. *Journal of Solid State Chemistry,* **178**, 2409-2413.
23. Anbia, M., and Hoseini, V. (2012) Development of MWCNT@MIL-101 hybrid composite with enhanced adsorption capacity for carbon dioxide. *Chemical Engineering Journal,* **191**, 326-330.
24. Buso, D., Jasieniak, J., Lay, M. D. H., Schiavuta, P., Scopece, P., Laird, J., Amenitsch, H., Hill, A. J., and Falcaro, P. (2012) Highly luminescent metal-organic frameworks through quantum dot doping. *Small,* **8**, 80-88.
25. Jahan, M., Bao, Q., and Loh, K. P. (2012) Electrocatalytically active graphene-porphyrin MOF composite for oxygen reduction reaction. *Journal of the American Chemical Society,* **134**, 6707-6713.
26. Xu, Y., Wen, Y., Zhu, W., Wu, Y.-n., Lin, C., and Li, G. (2012) Electrospun nanofibrous mats as skeletons to produce MOF membranes for the detection of explosives. *Materials Letters,* **87**, 20-23.
27. Prasanth, K. P., Rallapalli, P., Raj, M. C., Bajaj, H. C., and Jasra, R. V. (2011) Enhanced hydrogen sorption in single walled carbon nanotube incorporated MIL-101 composite metal-organic framework. *International Journal of Hydrogen Energy,* **36**, 7594-7601.
28. Yang, Y., Ge, L., Rudolph, V., and Zhu, Z. (2014) In situ synthesis of zeolitic imidazolate frameworks/carbon nanotube composites with enhanced CO_2 adsorption. *Dalton Transactions,* **43**, 7028-7036.
29. Lee, H., Choi, Y. N., Choi, S. B., Seo, J. H., Kim, J., Cho, I. H., Gang, S., and Jeon, C. H. (2014) In situ neutron powder diffraction and X-ray photoelectron spectroscopy analyses on the hydrogenation of MOF-5 by Pt-doped multiwalled carbon nanotubes. *Journal of Physical Chemistry C,* **118**, 5691-5699.
30. Yang, S. J., Choi, J. Y., Chae H. K., Cho, J. H., Nahm, K. S., and Park, C. R. (2009) Preparation and enhanced hydrostability and hydrogen storage capacity of CNT@MOF-5 hybrid composite. *Chemistry of Materials,* **21**, 1893-1897.
31. Lu, G., Li, S., Guo, Z., Farha, O. K., Hauser, B. G., Qi, X., Wang, Y., Wang, X., Han, S., Liu, X., DuChene, J. S., Zhang, H., Zhang, Q., Chen, X., Ma, J., Loo, S. C. J., Wei, W. D., Yang, Y., Hupp, J. T., and Huo, F. (2012) Imp-

arting functionality to a metal-organic framework material by controlled nanoparticle encapsulation. *Nature Chemistry,* **4**, 310-316.

32. Mao, Y., Li, J., Cao, W., Ying, Y., Hu, P., Liu, Y., Sun, L., Wang, H., Jin, C., and Peng, X. (2014) General incorporation of diverse components inside metal-organic framework thin films at room temperature. *Nature Communications,* **5**, 5532.
33. Marcano, D. C., Kosynkin, D. V., Berlin, J. M., Sinitskii, A., Sun, Z., Slesarev, A., Alemany, L. B., Lu, W., and Tour, J. M. (2010) Improved Synthesis of Graphene Oxide. *ACS Nano,* **4**, 4806-4814.
34. Ke, F., Qiu, L. G., Yuan, Y. P., Peng, F. M., Jiang, X., Xie, A. J., Shen Y. H., and Zhu, J. F. (2011) Thiol-functionalization of metal-organic framework by a facile coordination-based post synthetic strategy and enhanced removal of Hg^{2+} from water. *Journal of Hazardous Materials,* **196,** 36-43.
35. Bousmina, M. (2006) Study of intercalation and exfoliation processes in polymer nanocomposites. *Macromolecules,* **39**, 4259-4263.
36. Cassagnau, P. (2008) Melt rheology of organoclay and fumed silica nanocomposites. *Polymer,* **49**, 2183-2196.
37. Knauert, S. T., Douglas, J. F., and Starr, F. W. (2007) The effect of nanoparticle shape on polymer-nanocomposite rheology and tensile strength. *Journal of Polymer Science, Part B: Polymer Physics,* **45**, 1882-1897.
38. Achaby, M. E., and Qaiss, A. (2013) Processing and properties of polyethylene reinforced by graphene nanosheets and carbon nanotubes. *Materials and Design,* **44**, 81-89.
39. Horst, M. F., Quinzani, L. M., and Failla, M. D (2014) Rheological and barrier properties of nanocomposites of HDPE and exfoliated montmorillonite. *Journal of Thermoplastic Composite Materials,* **27**, 106-125.
40. Li, Y., Zhu, J., Wei, S., Ryu, J., Sun, L., and Guo, Z. (2011) Poly (propylene)/graphene nanoplatelet nanocomposites: Melt rheological behavior and thermal, electrical, and electronic properties. *Macromolecular Chemistry and Physics,* **212**, 1951-1959.
41. Essabir, H., Rodrigue, D., Bouhfid, R., and Qaiss, A. K. (2018) Effect of nylon 6 (PA6) addition on the properties of glass fiber reinforced acrylonitrile-butadiene-styrene. *Polymer Composites,* **39**(1), 14-21.

3

PE-CPE Blends and PE-CPE-Graphene Oxide Nanocomposites: Morphology, Phase Miscibility, Interfacial Interactions and Reduced Low Temperature Brittleness

3.1 Introduction

Processes like polymer blending or incorporation of specific fillers to the polymer matrices are commonly performed to acquire new polymeric materials with altered properties [1,2]. Polymer blends have advantages such as property profile combinations, aid in processing, low capital costs as compared to development of new monomers and polymers of required properties. On the other hand, composites fulfill requirements for different applications where cost and weight reduction, dimensional stability, opacity, heat stability, and processability are required. The properties of the resulting materials depend on a number of factors like interaction between the individual phases, dispersion of one phase into the other, compatibilization, processing conditions, filler volume fraction, dimensional characteristics (size, aspect ratio, etc.) of the reinforcement phase, etc.

Polyolefins, for example polyethylene (PE), have wide range of properties including low cost, ease of recycling, good processability, non-toxicity, bio-compatibility, and good chemical resistance. The number of individual polyolefin members is limited and the number of applications based on such materials is constantly increasing. Various polyolefin based blends and composites are thus required to meet the increasing demand of materials with required properties and processing suitable for attaining a wide range of specific applications [1,3].

Chlorinated polyethylene (CPE) of different chlorination levels is generally blended with polymers for different purposes such as toughening of matrix polymers [4], enhancing the interfacial interactions between two polymers [5,6], improving processing [7] and

Vikas Mittal[a], Gisha E. Luckachan[a] and Nadejda Matsko[b]
[a]Khalifa University of Science and Technology, Abu Dhabi, UAE; [b]Graz Centre of Electron Microscopy, Graz, Austria

adhesion properties [8], enhancing ignition resistance [9] and modifying gas transport properties [10]. In reported studies on PE+CPE systems, CPE with 36% chlorination content was blended with high density polyethylene and low density polyethylene [11] and the mechanical properties were observed to depend on morphological changes occurring at higher concentration of CPE36 in the blends [11,12]. Blends of CPE have also been reported with other polymer systems, however, studies on the miscibility, structural properties and performance of commercially important blends of different types of chlorinated polyethylene with polyolefins like polyethylene are still missing.

Graphene has been a filler of choice in studies on polymer nanocomposites due to its excellent electrical and mechanical properties, which are significantly better than other filler materials [13-18]. Graphene (and graphene oxide (GO)) can be produced from graphite by different methods such as thermal expansion of chemically intercalated graphite, micromechanical exfoliation of graphite, chemical vapor decomposition and chemical reduction method [15]. The compatibility of the graphene derivatives with the polymer matrices is of utmost importance to achieve optimum enhancement in the polymer properties. Graphene has very low surface energy owing to the presence of less number of functional groups (such as carboxyl, epoxide, and hydroxyl, etc.) on the surface which leads to lower compatibility with polar polymer matrices [13]. On the other hand, dispersion of polar graphene oxide in non-polar polymers like polyethylene also suffers due to absence of positive interactions between them. Addition of amphiphilic compatibilizer having polar and non-polar groups is used to compatibilize filler and host polymer, thus, resulting in improved filler dispersion. For example, maleic anhydride was reported to compatibilize nanocomposites of low density polyethylene and exfoliated graphite nanoplatelets [19]. Kim *et al.* [20] also reported the importance of functionalized polyethylene as well as solution mixing method for exfoliation of graphene in the polyethylene matrix. CPE has also been reported to be efficient compatibilizer in composites. For example, CPE was used to comaptibilize poly(vinyl chloride) and wood flour composites (chlorine content ranged from 25% to 42%), and improvements in mechanical and rheological properties were reported [21]. In a recent study by the authors of this study, CPE polymers of 25% and 35% chlorination levels were also used to comaptibilize PE-graphene oxide system and rheological and mechanical properties of the nanocomposites were observed to depend on the degree of chlorination as well as CPE content [22].

Due to the immense potential of graphene in enhancing the material characteristics, further analysis of its nanocomposites with commercially important polymers is required in terms of characterization of miscibility of the phases and changes in the crystalline structure of the matrix polymer. In addition, it is also needed to study physical and chemical interactions between the compatibilizer and filler in comparison to interactions between polymer and compatibilizer phases as well as polymer and filler phases. To accomplish this, two types of chlorinated polyethylene polymers (with 25% and 35% chlorination content, i.e. from the thermoplastic and elastomer range respectively [23]) were used to generate blends with PE as well as PE+CPE nanocomposites with graphene oxide. A recent study on these nanocomposites reported their thermal, mechanical and rheological properties [22], however, more fundamental aspects of morphology as well as interfacial interactions have been studied in the current work in order to optimize the filler exfoliation to achieve optimum composite design. To generate masterbatches of CPE compatibilizer and graphene oxide, solution blending technique was used. The melt mixing method was used to obtain PE+CPE blends as well as composites of masterbatches with PE. The morphology of the blends and composites was specifically characterized using microscopy (electron and atomic force microscopy), shear rheology, thermogravimetric analysis (TGA) coupled with mass spectrometer, as well as Raman and infra-red (IR) spectroscopy.

3.2 Experimental

3.2.1 Materials

High density polyethylene BB2581 was kindly supplied by Abu Dhabi Polymers Company Limited (Borouge), UAE. Chlorinated polyethylene grades Weipren® 6025 (25% chlorine content, named as CPE25) and CPE 135A (35% chorine content, named as CPE35) were obtained from Lianda Corporation, USA and Weifang Xuran Chemicals, China respectively. The CPE25 was in the form of white granules (specific gravity 1.1-1.3), whereas the appearance of CPE35 was white powder (specific gravity 1.1-1.16). The melt flow index (190 °C, 2.16 kg) for CPE25 and CPE35 was measured to be

1.8 and 1.9 g/10 min respectively, whereas the pure polymer had a melt flow index of 0.35 g/10 min (190 °C, 2.16 kg). The heat of fusion of CPE25 and CPE35 polymers was 45 and 2 J/g respectively. The polymer materials were used as obtained. Graphene oxide was synthesized by thermal exfoliation of precursor graphite oxide using modified Hummer's method as reported earlier [22,24,25]. The density of the obtained graphene oxide was measured by tapped density tester to be 0.0161 g/ml.

3.2.2 PE+CPE Blends

Melt mixing was used to generate polymer blends of CPE25 and CPE35 with PE using mini twin conical screw extruder (MiniLab HAAKE Rheomex CTW5, Germany). A mixing temperature of 190 °C for three minutes at 60 rpm with batch size of 5 g was used. The screw length and screw diameter were 109.5 mm and 5/14 mm conical respectively. The percentage of the CPE phase was fixed at 5 wt%. Disc shaped test samples were prepared by mini injection molding machine (HAAKE MiniJet, Germany) at a processing temperature of 190 °C. The injection pressure was 700 bar for 6 s whereas holding pressure was 400 bar for 3 s. The temperature of the mold was kept at 50 °C.

3.2.3 Nanocomposite Generation

The nanocomposites were prepared by either solution mixing followed by melt mixing or direct melt mixing, as described earlier [22]. Solution mixing of CPE in graphene oxide was achieved by dissolving CPE polymer in *p*-xylene solvent (3 wt% solid content) at 100 °C under reflux and dispersing graphene oxide in *p*-xylene for 1 h at room temperature followed by mixing of the two at 100 °C. The slurry was sonicated using ultrasonic horn (70% amplitude). The mixture was cooled and kept overnight at room temperature followed by 40 °C for 24 h in order to remove any solvent residues. The so-formed CPE-graphene oxide masterbatches were melt mixed with PE at 190 °C using mini twin screw extruder using the same processing conditions used to form blends. The overall amount of graphene oxide was fixed at 0.5 wt%. Melt mixing of pure PE as well as PE-graphene oxide was also similarly performed. Disc shaped test specimens for melt rheological analysis were similarly injection molded at 190 °C, as described above.

3.2.4 Characterization of Blends and Nanocomposites

AR 2000 rheometer from TA Instruments was used to characterize the rheological properties such as storage modulus (G′), loss modulus (G″), viscosity (η′) and elasticity (η″). Disc shaped samples of 25 mm diameter and 2 mm thickness were measured at 185 °C using a gap opening of 1.2 mm. Strain sweep scans were recorded at ω = 1 rad/s from 0.1 to 100% strain and the samples were observed to be shear stable up to 10% strain. For the comparison, frequency sweep scans (dynamic testing) of all the samples were recorded at 4% strain from ω = 0.1 to 100 rad/s.

For the microscopy analysis, the composite samples were mounted in special holders which at the same time fit in the microtome and were suitable for the examination of the block face by atomic force microscopy. Ultrathin sections (30-90 nm) as well as block faces of the composite samples were obtained using a Leica Ultracut E microtome (Leica, Austria) equipped with a diamond knife (Diatome, Switzerland) at -120 °C. Sections for transmission electron microscopy (TEM) analysis were collected on 400 mesh formvar electron microscopy grids (coated with holey carbon film), and were examined in a Philips CM 20 (Philips/FEI, Germany) electron microscope at 200 kV at room temperature without staining. TEM image processing was performed using DigitalMicrograph software (Gatan, USA). All energy dispersive X-ray spectroscopy (EDXS) experiments were performed in STEM mode with a probe current of 20 nA and a beam diameter of 100 nm. EDXS spectra were collected using a HPGe detector (solid angle 0.13 sr). For the calculation of elemental distribution images, jump ratio method was used and EDXS spectra have been visualized using Digital Micrograph software (Gatan, USA). The block faces of specimens after cryo-ultramicrotomy were investigated at ambient conditions using Dimension 3100 AFM/SPM (Veeco, USA) atomic force microscope. AFM images were collected in tapping mode using silicon nitride cantilevers with natural frequencies in the 300 kHz range (force constant 20 N/m, tip radius 10 nm (NT-MDT, Russia)). AFM image processing was performed using Nanoscope v720 software (Veeco, USA).

IR spectra were collected on Nicolet iS10 spectrometer equipped with SmartiTR diamond ATR accessory (angle of incidence of 45°), DTGS KBr detector and KBr beam splitter. It had a diamond ATR crystal (index of refraction 2.4 at 1000 cm^{-1}) and a depth penetration of 2 μm at 1000 cm^{-1} for sample with refractive index of 1.5. Spectra were recorded by the OMNIC software in the 4000-525 cm^{-1} region with a resolution of 4 cm^{-1} from 16 scans. Raman spectra were recorded on a Jobin Yvon Horiba LabRAM spectrometer with back-scattered confocal configuration using a HeNe laser (633 nm). A long working distance objective with magnification 50x was used both to collect the scattered light and to focus the laser beam on the sample surface. CCD (charge-coupled device) detector exposure time was 10 s and an average of 2 cycles was used to increase S/N ratio. Raman shifts were calibrated using Si wafer signal at 520.7 cm^{-1}.

LINSEIS STA PT1600 simultaneous thermogravimetric analysis - differential thermal analysis (TGA-DTA) system was used to measure the weight change during heating. This system is coupled with a Phipher mass spectrometer (MS) which allowed to determine HCl (m/z=36), H_2O (m/z=18) and CO_2 (m/z=44) elimination during heating. All measurements are realized with a heating rate of 10 °C/min, under air atmosphere (flow rate 20 ml/min).

Wide angle X-ray diffraction analysis of the composites was performed on Panalytical Powder Diffractometer (X'Pert PRO) using CuKα radiation (λ = 1.5406 A°) in reflection mode. Zero-background holder was used to minimize the noise. The samples were step-scanned from 10-70° 2Θ at room temperature using a step size of 0.02° 2Θ and a step time of 10 s.

3.3 Results and Discussion

In the current study, the morphological features of PE+CPE blends and PE+GO+CPE nanocomposites, generated using chlorinated polyethylene with two different chlorination levels, were characterized. The chlorination contents were differed in order to study the impact on miscibility with polyethylene as well as dispersion of graphene oxide, thus, affecting the final morphology of these systems. For example, CPE is expected to disperse graphene oxide in PE, however, the overall miscibility in the ternary system may be affected. It was also known from earlier reported results, CPE25 was semi-crystalline in nature with a peak melting temperature of 130 °C, whereas CPE35 was amorphous as no crystalline melting peak was observed in its DSC thermogram [22]. Hence, due to different structural characteristics of the CPE polymers, different morphology was expected in blends and nanocomposites containing these polymers.

The oscillatory torque increased for all the samples as a function of angular frequency during the rheological testing due to enhanced resistance of the melt at higher rates of deformation (Figure 3.1). The build-up of torque as a function of angular frequency for PE+GO, PE+CPE25 and PE+CPE35 was higher than the PE, PE+GO+CPE25 and PE+GO+CPE35 samples. It indicated that the CPE copolymers in the blends also generated similar torque as graphene oxide platelets in the un-compatibilized composite. Thus, stiff CPE chains acted as filler particles in the matrix, however, it should also be noted that the amount of graphene in the PE+GO composites was very low as compared to CPE amounts in PE+CPE blends. Apart from that, graphene oxide platelets were also observed to be very flexible and could be bent and folded under shear. The observed torque was much lower in PE+GO+CPE nanocomposites and the least values were obtained for PE+GO+CPE35 composite. As PE+GO as well as PE+CPE systems individually had higher torque values, the observed decrease in the composites, thus, can be traced to the interactions between CPE and graphene oxide phases. As graphene oxide

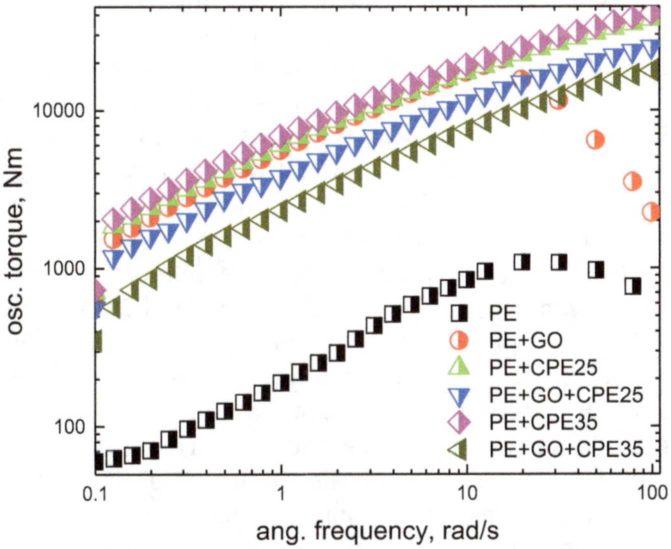

Figure 3.1 Buildup of oscillatory torque during the rheological analysis as a function of angular frequency.

has a polar surface, therefore, it is expected to interact positively with CPE chains. CPE35 due to its higher chlorination content can have even enhanced extent of these interactions. As a result, the dispersion of graphene oxide platelets under shear in the PE matrix would improve as extent of polarity of compatibilizer was increased, leading to lower torque generation due to the well dispersed graphene oxide platelets. However, such selective partitioning of CPE phase near to the graphene surface may also lead to immiscibility with the polymer matrix. PE and PE+GO were also observed to have sudden shear thinning at ~20 rad/s angular frequency. In the case of PE+GO nanocomposite, such a behavior has been reported to be due to the rupture of interface between the polymer and graphene oxide at higher frequencies [22].

Studies on the miscibility of the blends and nanocomposites were performed using Cole-Cole [26-28], van Gurp [29,30] and Han Chuang [31,32] analyses, as shown in Figure 3.2. Cole-Cole analysis develops relationships between real (η') and imaginary (η'') parts of complex viscosity. A smooth, semi-circular shape of the graph would suggest miscible phases, whereas, the deviation from this behavior indicates phase segregation due to immiscibility of the components. As observed in Figure 3.2a, the plots of PE+GO, PE+CPE25 and PE+CPE35 followed a semi-circular shape indicating the miscible phase morphology in these samples. However, the PE+GO+CPE nanocomposites showed deviation from the semi-circular path at higher viscosity (or lower frequency) values indicating presence of immiscibility in the phases. However, it is necessary to further support the findings from Cole-Cole analysis as these findings could sometimes be misleading [33]. Figure 3.2b shows the van Gurp plots representing the relationship between complex modulus (G^*) and phase angle delta (δ). The findings from the Cole-Cole analysis were confirmed by van Gurp plots. In the case of PE, PE+GO, PE+CPE25 and PE+CPE35 samples, the time-temperature superposition principle was observed to hold as indicated by the merging of the curves into a common curve. On the other hand, the PE+GO+CPE samples deviated from the common curve, thus, confirming the presence of immiscibility in these samples. The nanocomposite with CPE35 compatibilizer exhibited most notable deviation indicating that the extent of immiscibility in this case was higher than the PE+GO+CPE25 composite. The curves of both PE+GO+CPE composites almost converged with each other at higher frequency (lower delta value), however, they still did not merge with the common curve. The rheological behavior of the samples was also analyzed by using the criteria reported by Han and Chuang using a plot between

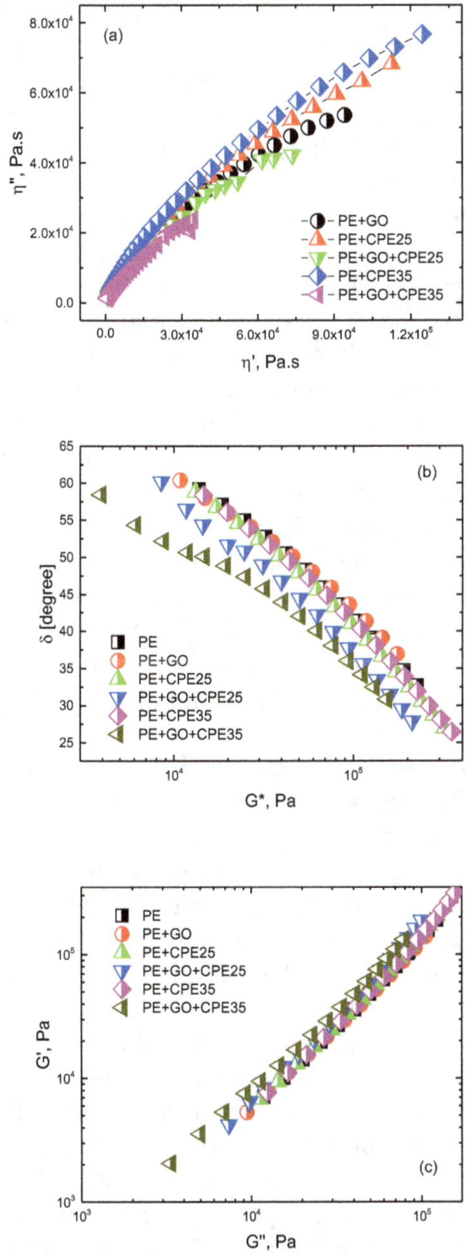

Figure 3.2 Miscibility studies on PE+CPE blends and PE+GO+CPE composites using a) Cole-Cole analysis, b) van Gurp analysis, and c) Han Chuang analysis.

G' and G'', as shown in Figure 3.2c. The curves for the ternary composite systems did not merge with others indicating the presence of immiscibility in these systems. Similar to the van Gurp analysis, the nanocomposite with CPE35 compatibilizer had maximum deviation. The deviation between PE+GO+CPE25 and PE+GO+CPE35 was similarly the maximum at lower frequency, but the curves merged with each other at higher frequency, however, they retained their deviation from the other samples. The miscibility studies, thus, confirmed that the polymer blends as well as PE+GO nanocomposite were miscible with uniform phase (slope of curves ~2), however, the PE+GO+CPE nanocomposites had immiscible phase [34,35]. These also signal towards the earlier suggested notion that due to the good interaction of CPE with graphene surface, the concentration of CPE phase near to interface with graphene oxide increased leading to immiscibility with the PE matrix. CPE35 containing composite, thus, expectedly had higher extent of immiscibility as compared to the CPE25 containing composite. This also explains the different mechanical properties obtained for these composites in an earlier study [22]. CPE25 containing composite had higher modulus than PE due to less immiscibility as compared to much lower modulus in the case of CPE35 containing composite due to immiscible phases. However, it should also be noted that the CPE25 is semi-crystalline in nature, while CPE35 was amorphous. Therefore, resulting changes in the crystalline structure of the PE chains would also contribute to the changes in the mechanical response.

In the TEM micrographs of the PE polymer and PE+CPE blends in Figure 3.3, the morphology of the pure polymer was observed to change on incorporating the CPE polymer. The pure PE due to its semi-crystalline nature showed distinct morphological features, which were also present in the PE+CPE25 blend though these were modified. These features owing to crystalline morphology were significantly lost in the case of PE+CPE35 blend indicating that the two CPE polymers had noticeable differences in their effect on polymer morphology. In an earlier study, the morphology of these systems was also studied using differential scanning calorimetry and it was observed that the melt enthalpy in the case of PE+CPE35 blend was 130 J/g as compared to 151 J/g for pure polymer, whereas the PE+CPE25 blend had a melt enthalpy of 140 J/g, which further confirmed the findings from TEM micrographs [34]. Insights into the morphology of blends and nanocomposites samples were also achieved using atomic force microscopy performed in tapping mode, as shown in Figure 3.3. The scales in the TEM and AFM images were kept constant in order to achieve comparison of the morphological features observed in the samples. Similar to TEM, significant differences in the sample morphology were observed. The PE sample was uniform with bright areas (in the phase image) corresponding to crystal structure and dark areas related to the amorphous content. The PE+CPE25 sample also had uniform distribution of bright phase inside the dark areas indicating that the CPE25 polymer still maintained the semi-crystalline nature of the pure polymer. However, its morphology was quite different from pure PE as in this case the chains seemed to be shorter in length. Also, the addition of CPE25 generated polycrystallinity in the sample. PE+CPE35 blend had the most significant effect on the polymer microstructure, which was also confirmed earlier by significant reduction in polymer crystallinity on its addition [22]. The crystalline phases were still present but the order or periodicity was disturbed. Also, the crystalline phases were randomly distributed due to the hindrance of CPE35 phase to

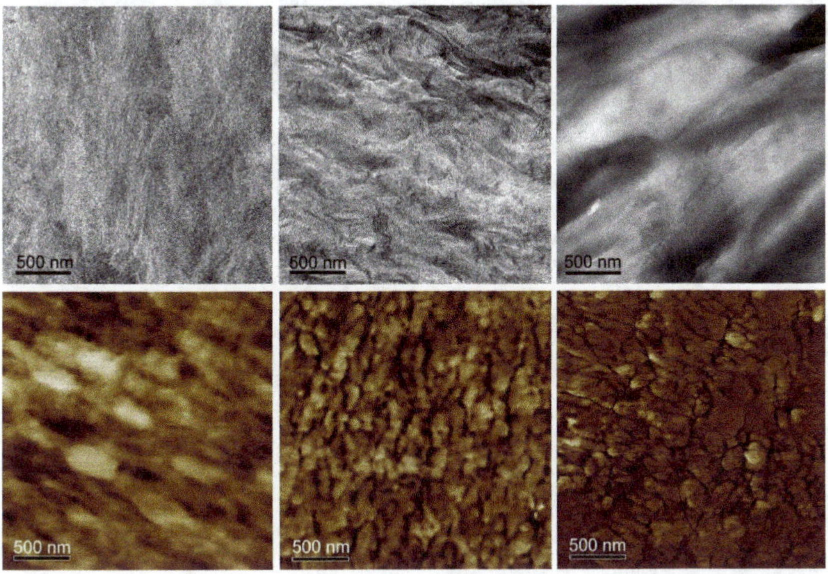

Figure 3.3 TEM and AFM images of (left) PE, (middle) PE+CPE25, and (right) PE+CPE35 samples.

PE crystallization. However, the system was still miscible as confirmed by the rheological analysis. It is difficult to comment on the distribution of CPE chains in the matrix, as the CPE25 is semi-crystalline in nature, so it is expected to align with the crystalline phase as well as amorphous phase of PE. However, in the case of CPE35 polymer, it would be expected to be in the dark regions due to its amorphous nature. It is also possible that the chlorination leads to local stiffness/hardness in CPE polymer chains, in this case, their concentration could be higher in brighter areas of phase images. This was also confirmed by the fact that CPE polymers in the blends generated similar torque as graphene oxide platelets in the composite without CPE.

Figure 3.4 also depicts the TEM characterization of nanocomposite samples. The graphene oxide platelets in the powder form in an earlier study were shown to stack with each other and were also observed to be flexible [22]. The incorporation of graphene oxide to PE without any additional compatibilizer did not lead to optimal filler dispersion and graphene oxide particles with high stack thickness were observed in the micrograph (Figure 3.4a). The PE+GO+CPE25

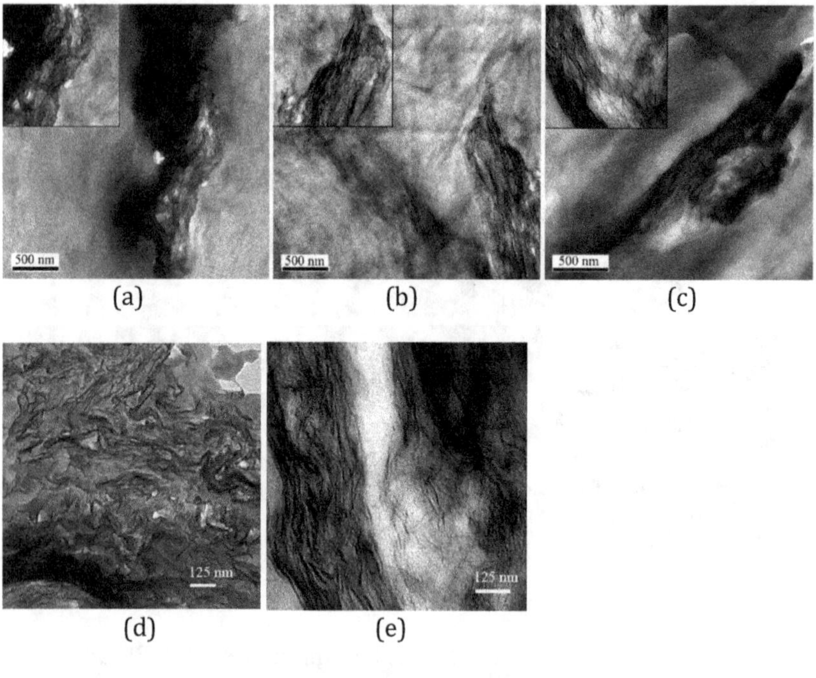

Figure 3.4 Transmission electron micrographs of a) PE+GO hybrid, b) PE+GO+CPE25 composite, and c) PE+GO+CPE35 composite. Figures d) and e), respectively, show the high-magnification micrographs of PE+GO+CPE25 and PE+GO+CPE35 nanocomposites. The black lines represent the cross-section of graphene platelets.

improved the dispersion of graphene oxide in the polymer matrix, but varying degrees of stack thicknesses could still be observed in the micrographs indicating less optimal filler dispersion (Figure 3.4b). In the other case, PE+GO+CPE35 nanocomposite showed significant extent of polymer intercalation inside the stacks of graphene oxide platelets and in the process also led to exfoliation of large extent of these stacks (Figure 3.4c). This, thus, confirmed that the higher extent of chlorination in the system resulted in better interactions between the CPE compatibilizer and graphene oxide leading to much better filler dispersion in the composite. Also, the blend as well as nanocomposite with CPE35 were observed to become extremely flexible and could not be sectioned even at -140 °C indicating the significant reduction of the T_g of polymer along with degree of crystallinity.

Figure 3.5 demonstrates the WAXRD analysis of the blends and nanocomposites. The diffraction peak for pure graphene oxide was observed at 26.4° 2Θ. However, this peak was not visible in the nanocomposites due its very low concentration in the polymer. The polymer crystalline structure was also observed to change in the presence of CPE and graphene. The 110 (21.8° 2Θ) and 200 (24.2° 2Θ) diffraction peaks of pure polymer were observed to have shoulders at higher angle indicating the presence of different crystal structures in the polymer. The shoulder at 110 peak was observed to develop into well-defined peaks in the nanocomposites, indicating possible nucleating effect of graphene oxide. This was also confirmed by the absence of such a phenomenon in PE+CPE blends. In the earlier study [22], the impact of better filler dispersion due to the addition of CPE35 in the nanocomposites was reported to enhance the mechanical properties, but the plasticization effect of the CPE35 reduced them.

In order to confirm the interactions between the CPE compatibilizer and graphene oxide, EDX studies were also performed on PE+GO+CPE25 and PE+GO+CPE35 nanocomposites. As shown in Figure 3.6 for PE+GO+CPE25 composite, the EDX analysis of matrix area (marked as EDX b in TEM image, i.e. area without graphene oxide) did not show any signal for Cl, whereas the area with graphene oxide platelets (marked as EDX a in TEM image) showed only a weak signal of Cl. It has to be noted that amount of CPE was only 5 wt% in the composites and, thus, effective amount of Cl atoms was further very low due to only 25-35% chlorination levels, thus, resulting in the weak signal. It is also possible that some amount of the

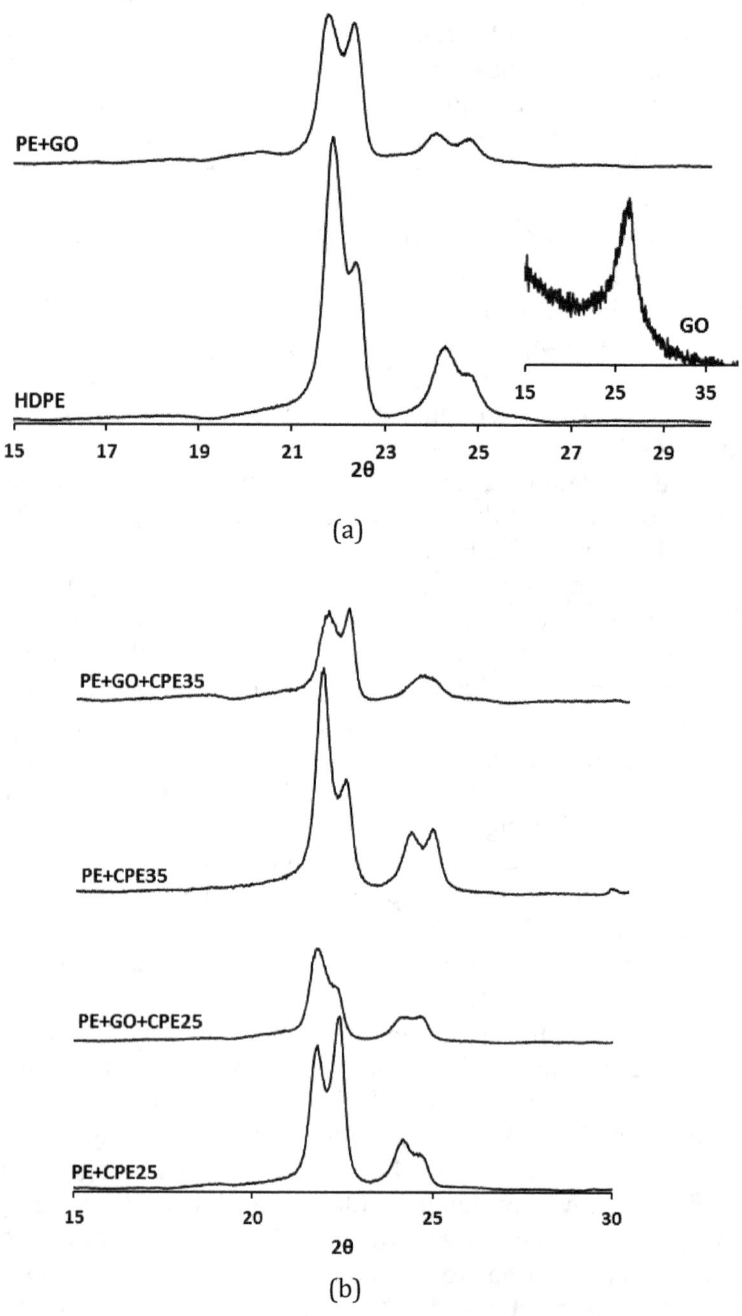

Figure 3.5 X-ray diffraction analysis of a) PE and PE+GO composite, b) PE+CPE blends and nanocomposites.

CPE phase was dispersed in the polymer, but its concentration was very low to observe its signal in EDX spectrum. However, these findings did not eliminate the possibility of a chemical interaction between the graphene and CPE phases thus leading to depletion of Cl from the CPE. Thermal degradation of CPE during melt compounding could also result in the loss of Cl atoms from the system. Figure 3.7 also shows the EDX results for PE+GO+CPE35 nanocomposite. In the selected area representing the matrix phase (marked as EDX_matr in TEM image), no signal for Cl was observed, whereas in the selected area on polymer-graphene surface (marked as EDX_gr in TEM image), a weak Cl signal (though stronger than the signal observed in PE+GO+CPE25) was observed. These studies thus necessitated further analysis to understand the interactions between the graphene and CPE phases.

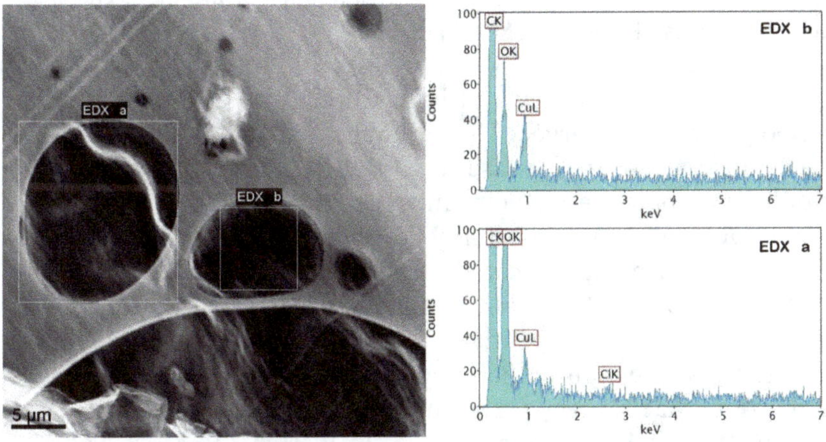

Figure 3.6 TEM image of PE+GO+CPE25 sample; EDX signal from matrix (area marked as EDX b in TEM image) and polymer-graphene surface (area marked as EDX a in TEM image).

Figure 3.8 shows higher magnification height and phase images of the nanocomposite samples PE+GO, PE+GO+CPE25 and PE+GO+CPE35 nanocomposites. The graphene oxide platelets or clusters were observed to be surrounded (or covered) by bright phase indicating hardness. The distribution of graphene oxide in PE was very poor as indicated by a large cluster of graphene oxide particles. The matrix in the background was similarly semi- crystalline in nature as observed in Figure 3.3. The distribution of graphene oxide improved on adding the compatibilizer as smaller cluster sizes

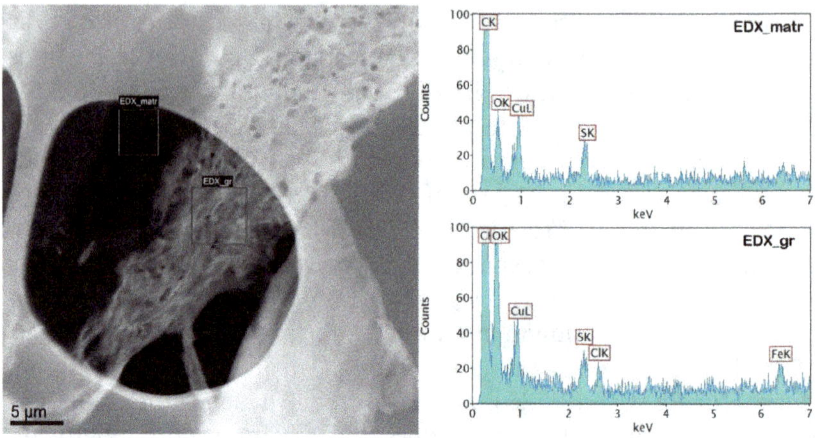

Figure 3.7 TEM image of PE+GO+CPE35 sample; EDX signal from matrix (area marked as EDX_matr in TEM image) and polymer-graphene surface (area marked as EDX_gr in TEM image).

of graphene oxide particles in both CPE25 and CPE35 containing composites were observed. The graphene platelets were occasionally also observed as dark phase (e.g. PE+GO+CPE25 composite in the middle row) because of their embedding below the surface of the sample or due to their flexibility. The distribution of harder areas was also uniform and indicated the semi-crystalline polymer phase. In the case of PE+GO+CPE35 composite, the distribution of the harder areas was significantly affected in the matrix. It should also be noted that the melt enthalpy in the case of nanocomposites was higher than that of blends [22]. In the case of PE+GO, it was 142 J/g as compared to 151 J/g for pure PE. In PE+GO+CPE25, it was 151 J/g, whereas in the case of PE+GO+CPE35 composite, a marginal decrease to 147 J/g in the melt enthalpy was observed. This may indicate that the graphene platelets act as nucleating agents and this effect also enhanced as the dispersion of graphene oxide platelets was improved in PE. The overall crystallinity is also thus accordingly affected; for PE+GO due to very poor distribution of graphene in PE, the effect was not observed. For the composite with semi-crystalline CPE25 compatibilizer, the crystallinity of pure polymer was retained due to much better nucleating effect of the filler as in the PE+CPE25 blend system, the melt enthalpy was only 140 J/g. Similarly, in the composite with amorphous CPE35 polymer, the overall crystallinity observed was result of competing factors of nucleating effect of graphene as well as decrease in crystallinity of the polymer because of CPE35 addition. As areas around the graphene oxide platelets are bright depicting the harder or crystalline phase, it would also confirm the notion of graphene oxide acting as nucleating agent. Moreover, as it was earlier observed that these areas also have higher concentration of CPE chains, it would thus indicate that the harder regions may also form due to ordering/alignment of CPE chains around the particles. However, as such regions are also

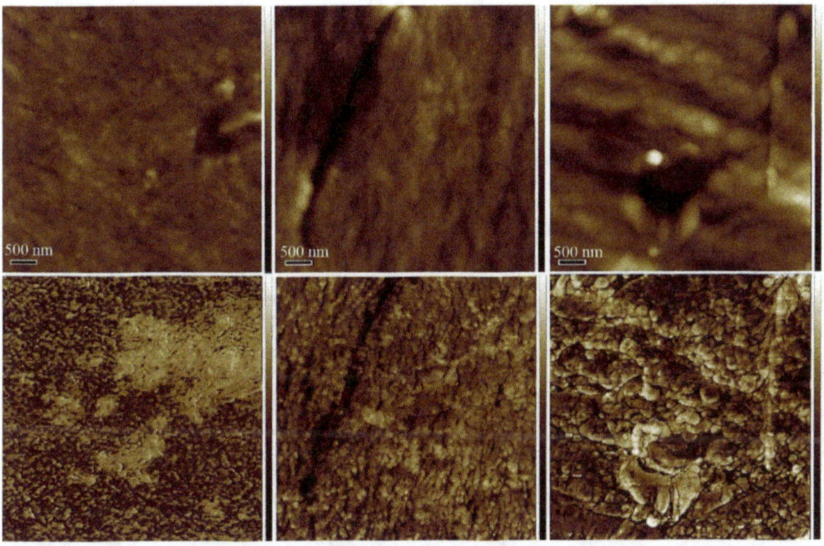

Figure 3.8 High-magnification AFM height (top) and phase (bottom) images of the samples; (left) PE+GO, (middle) PE+GO+CPE25, and (right) PE+GO+CPE35.

formed in the composite without CPE polymers, therefore, it can be concluded that both the factors would have contributed in the formation of these hard phases around the graphene platelets. Figure 3.9 also shows lower magnification height and phase images of the nanocomposites. The concentration of bright phase around the graphene oxide platelets was further confirmed. Also, Figure 3.9c demonstrates much better dispersion of the graphene oxide filler for the PE+GO+CPE35 nanocomposite.

Interactions between CPE compatibilizer and graphene oxide in PE+GO+CPE nanocomposites were also studied by IR as shown in Figures 3.10 and 3.11. In addition to the characteristic $V_{C\text{-}Cl}$ vibrations of CPE35 at 654 cm^{-1} and 607 cm^{-1}, IR spectrum of CPE35+GO showed a new peak at 1735 cm^{-1} which was attributed to $V_{C=O}$ vibrations of ester linkages (Figure 3.10 and Table 3.1). This new peak along with the reduced intensity of $V_{C\text{-}Cl}$ vibrations indicated that C-Cl groups on CPE were involved in a chemical reaction with polar functional groups on graphene oxide (-COOH, -OH) under the mild mixing conditions (100°C, 30 min sonication). Melt mixing of this composite with PE at 170 °C under shear (60rpm) facilitated more reaction between CPE and graphene oxide, that resulting in an intense $V_{C=O}$ vibration peak of ester linkages at 1715 cm^{-1} with a shoulder at 1740 cm^{-1} and $V_{C\text{-}O}$ vibration peak of ester and ether linkages at 1156 cm^{-1} (Figure 10). Chances of presence of carboxylic acid in the final composite (carboxylic acid also gives signal at 1715 cm^{-1}) could be safely ignored as the IR spectrum of PE+GO+CPE35 didn't show any hydroxyl contribution at ~3500 cm^{-1}. Utilization of chlorine for chemical bonding (ester, ether etc.) between CPE and graphene oxide in the PE+GO+CPE35 nanocomposite could be further

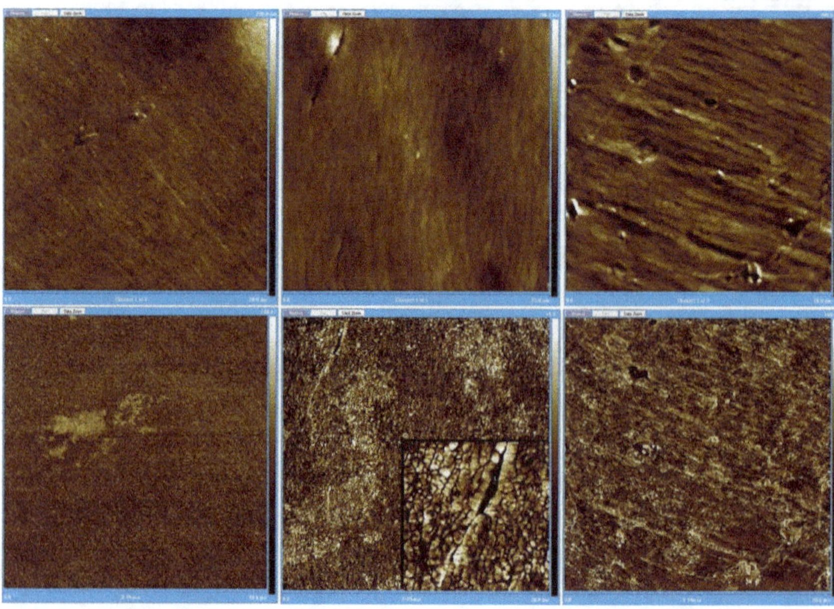

Figure 3.9 Low-resolution AFM height (top) and phase (bottom) images of the samples; (left) PE+GO, (middle) PE+GO+CPE25, and (right) PE+GO+CPE35.

confirmed from the absence of $V_{C\text{-}Cl}$ vibrations at 654 cm^{-1} and 607 cm^{-1}. CPE25 exhibited exactly similar behavior to that of CPE35 in its composite with graphene oxide and PE (Figure 3.11). However, the extent of reaction between the functional groups on graphene oxide and CPE25 was less due to the low percentage of chlorine content that could be observed in the IR spectrum of PE+GO+CPE25 as a low intense $V_{C=O}$ vibration at 1737 cm^{-1} (Figure 3.11). Interestingly, IR spectra of PE+CPE35 and PE+CPE25 didn't show any features from CPE part (Figure 3.12) which indicated that C-Cl bonds would have been destroyed at higher temperature used during compounding. Overall lower concentration of CPE in the blend might also partly contribute to this effect. Since the contribution of graphene oxide in nanocomposites was difficult to detect in IR, Raman spectra of

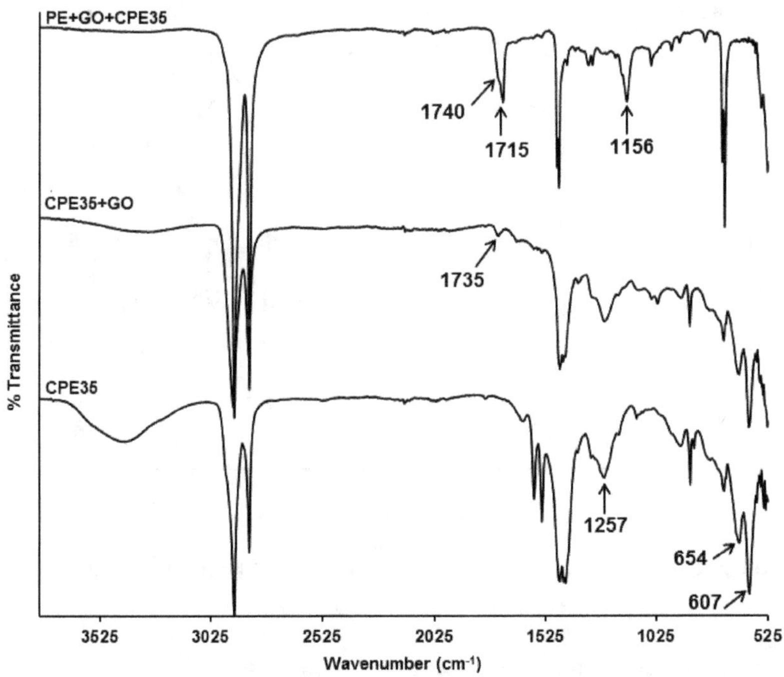

Figure 3.10 IR spectra of CPE35 and its composites with graphene oxide and PE.

PE+GO, PE+GO+CPE35 & PE+GO+CPE25 are given in Figure 3.13. Characteristic G band of graphene oxide was observed at ~1580 cm^{-1} and the D band was merged with PE signals in the 1250 cm^{-1} to 1350 cm^{-1} region. It could be concluded from above IR and Raman results that CPE compatibilizer helped graphene oxide to be dispersed well in the PE matrix through the formation of strong chemical bonds like ester and ether between CPE and graphene oxide layers, and the effect increased with the increase of chlorine content in CPE. Moreover, since the Cl atoms were sacrificed from CPE due to chemical reaction and possible thermal degradation, the amount of chlorine content in the final mixture would be negligible. The schematic representation of these findings is shown in Scheme 3.1.

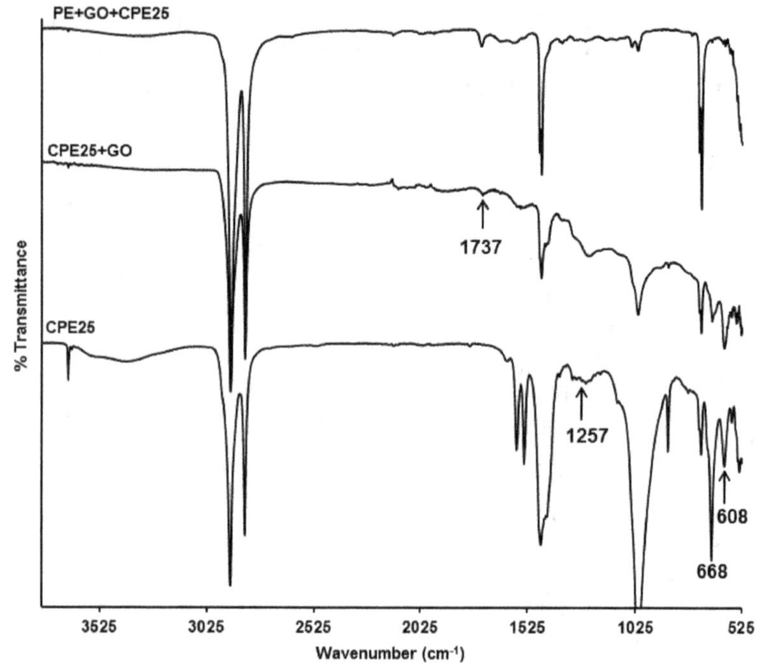

Figure 3.11 IR spectra of CPE25 and its composites with graphene oxide and PE.

Confirmation into the content of chlorine in the final product could also be obtained from the TGA /DTA/MS analysis of CPE35 and its nanocomposite with PE and graphene oxide. Since the nanocomposite contained only 5 wt% CPE35, weight loss analysis of chlorine from TGA plots was within the range of instrumental error. However, the elimination of chlorine could be clearly monitored from DTA curves when compared with corresponding mass fragments in mass spectrometry analysis. DTA plot of CPE35 given in Figure 3.14 showed three endothermic peaks corresponding to the first thermal degradation step in TGA (150 °C-360 °C). The first endothermic peak at T_{max} at 211 °C was due to sample melting. Melting

Table 3.1 FTIR band assignment of PE, CPE and its composites with graphene oxide [36]

Sample	Wavenumber (cm^{-1})	assignment
HDPE	2914	V_a CH_2
	2847	V_s CH_2
	1472, 1462	δ CH_2
	730, 719	γ_r CH_2
CPE25 & CPE35	2916,2849	V_{C-H} of methylene (-C**H**$_2$-CH_2-)
	1464	δ_{C-H} of methylene (-C**H**$_2$-CH_2-)
	1430	δ_{C-H} of methylene with adjacent chlorine atom (-C**H**$_2$-CHCl-)
	1257	δ_{C-H} of (-CH_2-C**H**Cl-)
	719	γ_r CH_2
	668 (654 in the case of CPE35), 608	V_{C-Cl} of CPE
	1576, 1540,1013, 874	carbontetrachloride contamination [37]
	3406, 1620	H_2O
CPE25+GO	1737	$V_{C=O}$ of (-COO-) all other peaks are same as in CPE25
CPE35+GO	1735	$V_{C=O}$ of (-COO-)
	1156	V_{C-O} of (-COO-), (C-O-C) all other peaks are same as in CPE35
PE+GO+CPE25	~3400	$C=O_{str}$ overtone of (-COO-)
	2915, 2847	V_a CH_2 & V_s CH_2
	1737	$\delta_{C=O}$ vibrations of (-COO-)
	1472,1462	δ CH_2
	730,719	γ_r CH_2
PE+GO+CPE35	~3400	$C=O_{str}$ overtone of (-COO-)
	2915, 2847	V_a CH_2 & V_s CH_2
	1740	$C=O_{str}$ of saturated esters
	1715	$C=O_{str}$ of aryl esters or α,β-unsaturated esters
	1472,1462	δ CH_2
	1329	δ_{C-H} of (-CH_2-C**H**OCO-)
	1156	V_{C-O} of (-COO-)
	730,719	γ_r CH_2

(V=stretching, δ=bending, γ_r= rocking)

phenomena seemed to result in the loss of water and some amorphous carbon as it showed small humps at 211 °C in the mass spectrometer profiles of H_2O (m/z=18) and CO_2 (m/z=44) given in Figure 3.15. The second endothermic peak (T_{max}=292 °C) in the DTA plot of CPE35 could be assigned to the loss of chlorine because the

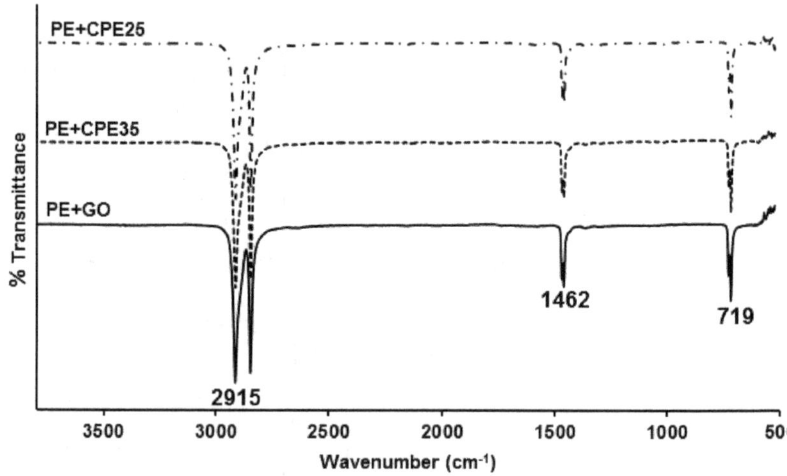

Figure 3.12 IR spectra of PE+GO composite and PE blends with CPE35 and CPE25.

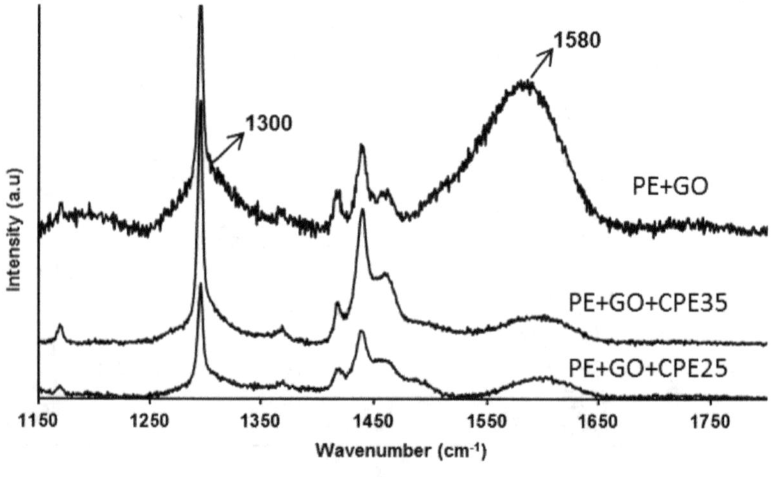

Figure 3.13 Raman spectra of PE+GO, PE+GO+CPE35, and PE+GO+CPE25 composites.

mass spectra contains HCl (m/z=36) profile in the temperature range of 250 °C to 350 °C. Comparison of DTA plot of CPE35 with those of PE as well as PE+CPE35 revealed that all plots had first endothermic peak, but for PE, the second endothermic peak was absent (Figure 3.14). It, thus, confirmed that the second endothermic peak in the DTA plot of PE+CPE35 was from the loss of HCl which appeared as smaller but clearly visible contribution at 273 °C (T_{max}). Even though the amount of CPE in PE+CPE35 was only 5 wt%, the

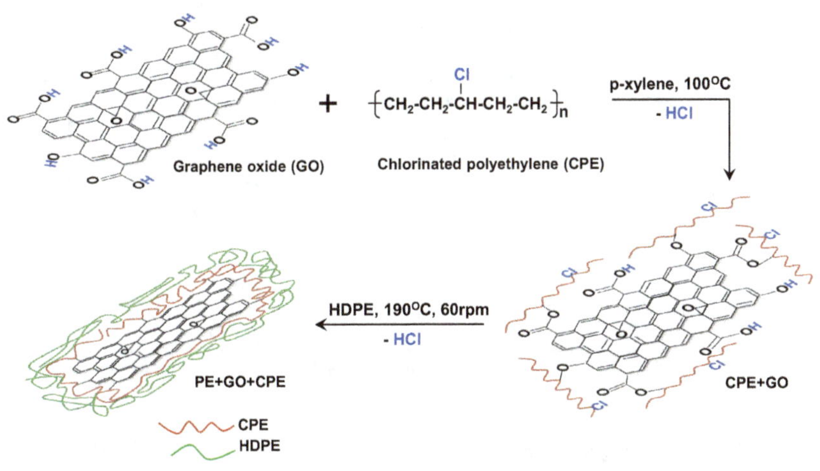

Scheme 3.1 Scheme representing the chemical interaction between the CPE and graphene oxide.

evolution of HCl was clearly detected in the DTA plot. The final composite PE+GO+CPE35 showed a significant decrease in the intensity of second endothermic peak and/or it appeared as a small shoulder around 300 °C. The presence of chlorine contamination in the final composite could not be eliminated as the IR spectra contained weak signals at 1576 cm^{-1}, 1540 cm^{-1} and 874 cm^{-1} (Figure 3.10, Table 3.1). Thus, it could be concluded from TGA/DTA /MS analysis that the amount of chlorine in PE+CPE35 and PE+GO+CPE35 decreased significantly due to the destruction of C-Cl bond during melt compounding. The amount of remaining Cl in the PE+GO+CPE35 was even lower than the PE+CPE35 blend due to significant depletion of Cl atoms during their reactions with functional groups on graphene oxide which resulted in strong chemical bonds such as ester, ether etc. between CPE chains and graphene oxide as confirmed from FTIR.

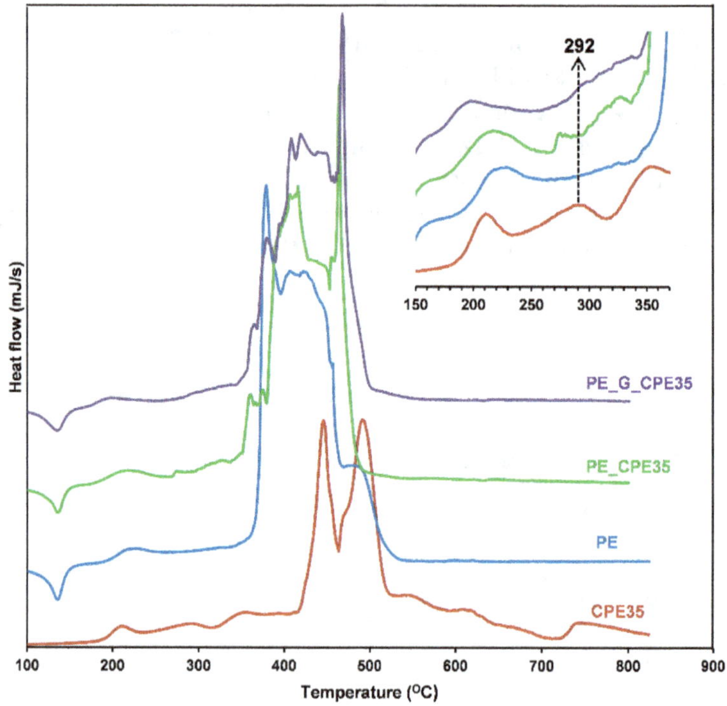

Figure 3.14 DTA plots of PE, CPE35, PE+CPE35, and PE+GO+CPE35.

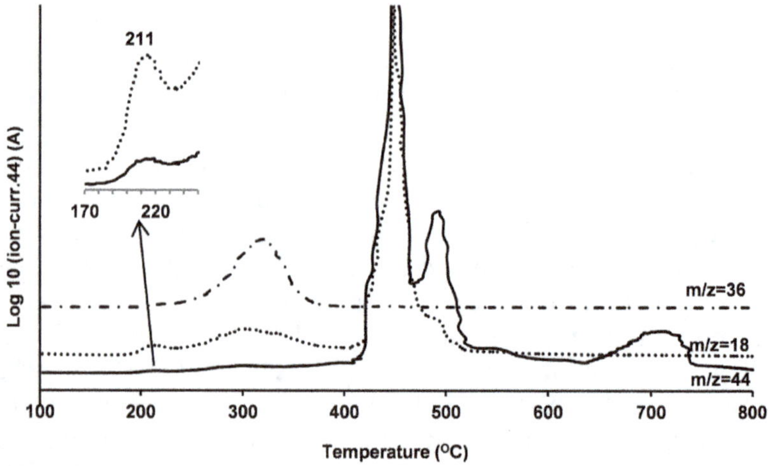

Figure 3.15 Mass spectrometer profile of H_2O (m/z = 18), CO_2 (m/z = 44), and HCl (m/z = 36) released from CPE35.

3.4 Use of Raman Mapping for Component Distribution

Luckachan and Mittal [38] developed further the use of Raman mapping technique to characterize the phase distribution of individual components in polymer blends and polymer nanocomposites systems (0.5 wt% filler and 5 wt% CPE), as shown in Figures 3.16 and 3.17.

Two visualizations shown in Figure 3.16 A and B for PE+GO nanocomposite system were obtained by selecting G band of graphene at 1580 cm^{-1} and V_{as}C-C band of PE at 1060 cm^{-1} respectively. In both cases, colour signal varying from black to yellow in the intensity bar indicated increasing intensity of peak chosen as contrast parameter. Three representative spectra of the different colour zone from the map in Figure 3.16A are shown in Figures 3.16(C)-(E) to demonstrate the variation of graphene content in the mapping area. Spectra showed maximum concentration of graphene in the red region. The concentration decreased in the green region, whereas the presence of graphene was almost nil in the black region. Moreover, the red and green coloured regions were observed to exist in the mapping image (Figure 3.16A) as bounded in discrete domains which indicated the immiscibility of graphene phase in the PE matrix. High stack thickness of graphene platelets observed in TEM micrograph of the PE+G nanocomposite (Figure 3.16F) further supported the absence of nano-scale graphene dispersion in PE+GO nanocomposite.

Similarly, mapping of PE+GO+CPE35 nanocomposite given in Figure 3.17 also showed two representative visualizations corresponding to the intensity of G band of graphene and V_{as}C-C band of PE. For instance, in Figure 3.17A, majority of the mapping areas exhibited red and green colour signal, though a small yellow domain was also observed to be confined in the red region. The representative spectra of the different colour domains exhibited the majority of the yellow domain to contain graphene, however, a significant amount of graphene was also observed in the red and green colour domains. Thus, it indicated that the distribution of graphene signal in various colour domains was an indicative of uniform graphene dispersion in the PE matrix. A significant extent of polymer intercalation inside the stacks of graphene platelets observed in the TEM micrograph of PE+GO+CPE35 nanocomposite (Figure 3.17E) also supported the findings of uniform dispersion of graphene in the PE matrix with the addition of CPE35 compatibilizer.

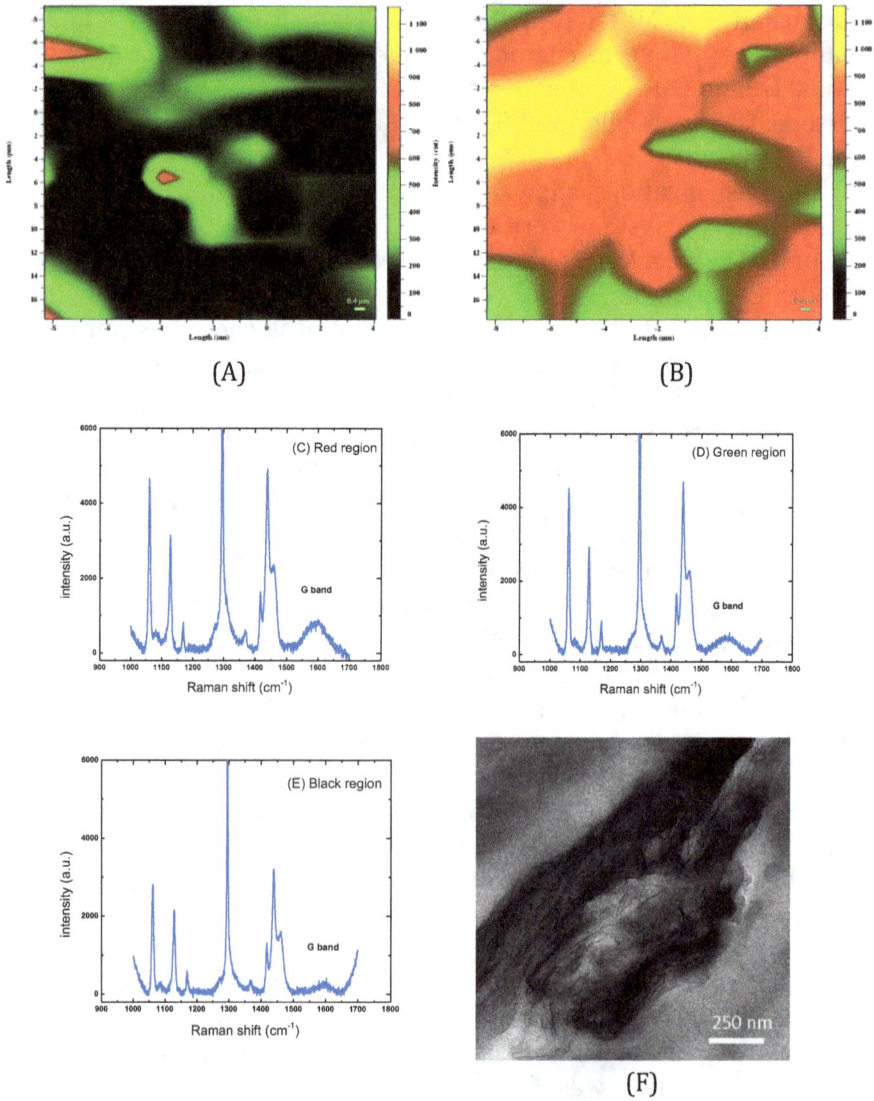

Figure 3.16 Raman mapping of PE+GO nanocomposite: (A) and (B) represent the maps of graphene and PE separately. Spectra were collected from (C) red, (D) green and (E) black regions of the map in (A). Figure 3.16F shows the TEM micrograph of PE+GO nanocomposite. Reproduced from Reference 38 with permission from Springer.

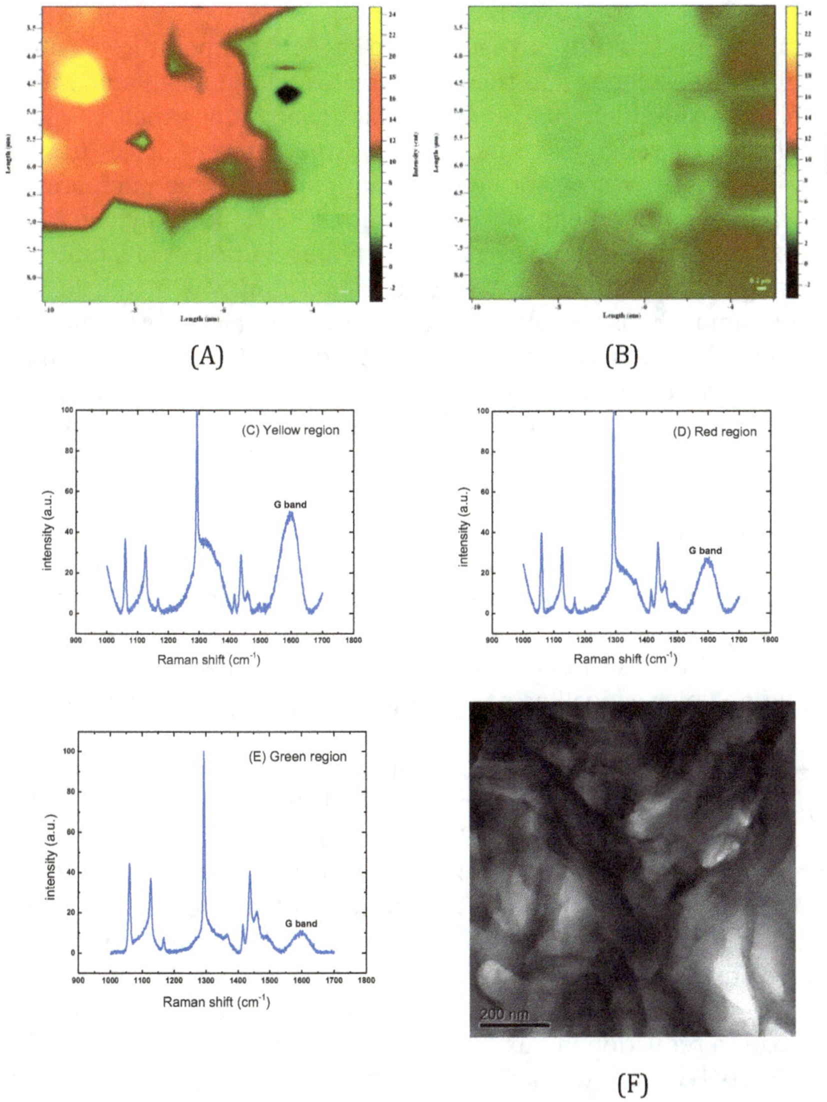

Figure 3.17 Raman mapping of PE+GO+CPE35 nanocomposite: (A) and (B) represent maps of graphene and PE separately. Spectra were collected from (C) yellow, (D) red and (E) green regions of (A). Figure 3.17F shows the TEM micrograph of PE+GO+CPE35 nanocomposite. Reproduced from Reference 38 with permission from Springer.

3.5 Reduced Low Temperature Brittleness

Mittal and Matsko [39] further studied the low temperature flexibility of PE+CPE blends and their composites with a small amount of graphene oxide filler (0.1 wt%). Surface topography analysis (surface height or roughness variation) at three different places on the sample surface was performed thorough atomic force microscopy. The height variation enhanced on adding CPE to polyethylene indicating the plasticization of the matrix on the addition of low molecular weight polymer. The extent of plasticization was higher in the case of amorphous CPE35 polymer where the height variation was significantly large as compared to PE blend with CPE25. The composites followed similar trend as the polymer blends, however, the height variation was even higher in the composite samples (Figure 3.18). For example, for PE+GO sample, a value of 113 nm for surface variation was calculated which was enhanced markedly to 198 nm for PE+GO+CPE25 nanocomposite and 564 nm for PE+GO+CPE35 nanocomposites. It indicated that the addition of small amount of graphene oxide did not negatively impact the flexibility of the samples.

Figure 3.19 also demonstrates the fracture analysis performed on the polymer as well as nanocomposite samples at two temperatures: -140 °C and -195 °C. The PE as well as PE+CPE25 and PE+CPE35 blends (with 5% CPE content) were flexible at -140 °C, whereas only PE+CPE35 sample was still flexible at -195 °C. The other two samples exhibited a brittle failure at -195 °C. It confirmed the findings from the AFM analysis about the flexibility of CPE35 containing samples at -180 °C and indicated that the samples were still flexible at -195 °C. Similar to the PE+CPE35 blend, the nanocomposite PE+GO+CPE35 was also observed to be fully flexible at -195 °C indicating that addition of graphene oxide to this system did not lead to reduction in flexibility. The other two composites PE+GO and PE+GO+CPE25 were flexible only near to the glass transition temperature of PE, whereas failed in a brittle manner at -195 °C. These findings are of substantial importance as these can ensure the application of polyethylene based materials at very low temperature conditions, which otherwise is not feasible due to low temperature brittleness. In addition, though these findings relate specifically to polyethylene based materials, however, other commercially important polyolefins such as polypropylene can also benefit from these insights.

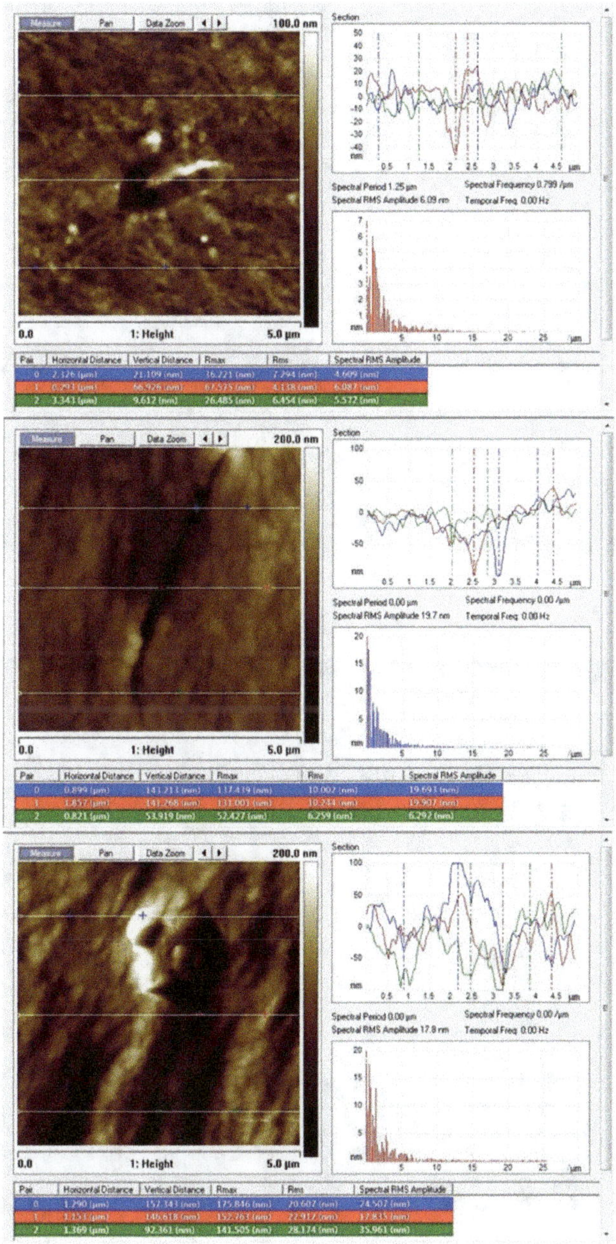

Figure 3.18 AFM height images and corresponding height corrugation analysis for PE+GO (top image), PE+GO+5%CPE25 (middle row) and PE+GO+5%CPE35 (bottom row) when sectioned at -140 °C. Reproduced from Reference 39 with permission from Springer.

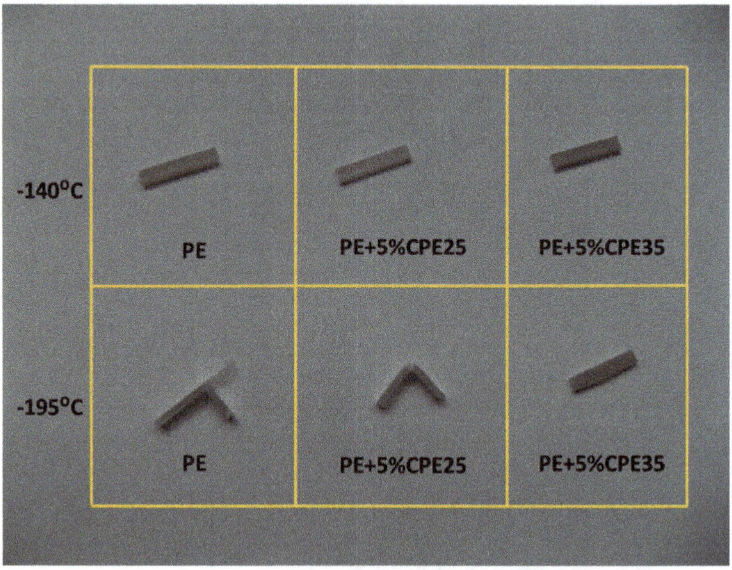

(a)

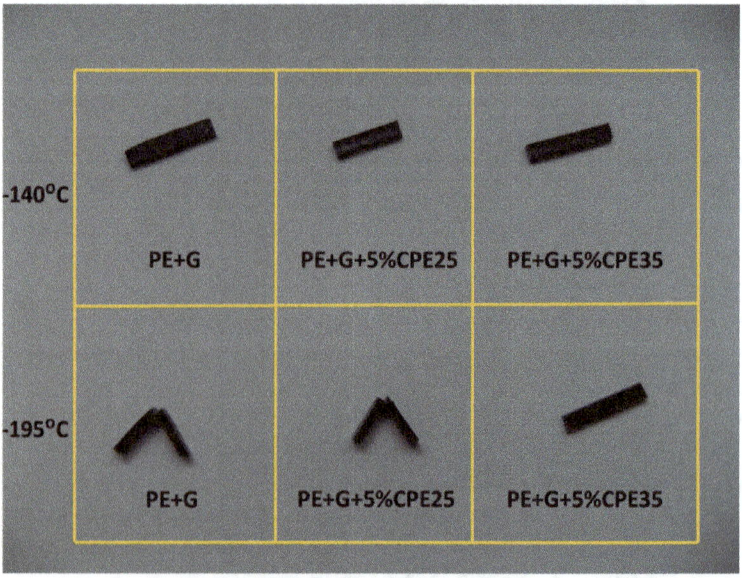

(b)

Figure 3.19 Fracture analysis of samples; (a) PE and PE+CPE blends, (b) PE+GO+CPE nanocomposites. Reproduced from Reference 39 with permission from Springer.

3.6 Conclusions

In the current study, morphological characterization of the PE+CPE blends and PE+GO+CPE nanocomposites, generated using CPE of two different chlorination contents, was studied. The rheological analysis using Cole-Cole, van Gurp and Han Chuang plots showed miscibility for the PE+GO, PE+CPE25 and PE+CPE35 samples, whereas the nanocomposite materials PE+GO+CPE25 and PE+GO+CPE35 were concluded to have immiscible phase. The extent of immiscibility was observed to be highest in PE+GO+CPE35 nanocomposite. The TEM analysis of the samples confirmed the best distribution of graphene oxide platelets in the composite containing CPE35 compatibilizer, which also resulted in least value of oscillatory torque for this composite. Apart from that, the polymer microstructure was also affected by the incorporation of CPE polymers, the most significant change was observed for CPE35 system due to its amorphous nature. The atomic force microscopy analysis also complemented the TEM findings that the crystalline microstructure of the PE polymer was drastically affected by CPE35 addition (as confirmed with lowest extent of crystallinity measured for this sample using DSC). The filler surface was also observed to be surrounded with hard phase which could have been resulted from the enhanced nucleation action of the graphene platelets as well as ordering of CPE chains near the graphene surface due to their positive interactions. This was also confirmed by higher extent of overall crystallinity in the composite samples as compared to PE+CPE blends. IR and Raman analysis confirmed a chemical reaction between the CPE and graphene oxide platelets. The blends of PE and CPE did not show C-Cl peaks thus indicating that the Cl was depleted not only due to the chemical reaction, but also by thermal degradation, which was also confirmed by DTA-MS analysis.

Role of compatibilizer in the dispersion of graphene filler in the polyethylene matrix was also successfully visualized using Raman mapping technique in PE+GO and PE+GO+CPE35 nanocomposites. The dispersion of graphene was confirmed to be less optimum in the PE+GO composites, whereas in the PE+GO+CPE35 composites, the addition of comapatibilizer resulted in improved degree of filler dispersion.

The blend and composite samples with CPE35 content could both be sectioned at -140 °C and -180 °C without brittle fracture, thus, confirming the flexibility at lower temperature, whereas the other systems with CPE25 or composite without CPE exhibited a brittle fracture at -180 °C. The fracture analysis carried out with samples further confirmed the low temperature flexibility of PE+CPE35 and PE+GO+CPE35 systems even at -195 °C.

Acknowledgements

The work (except Sections 3.4 and 3.5) presented in this study has been published earlier in the journal Macromolecular Chemistry and Physics, 2014, 215, 255-268, doi:10.1002/macp.201300613. The work has been reproduced in this chapter with permission from Wiley.

References

1. Jancar, J. (1999) *Mineral Fillers in Thermoplastics I: Raw Materials and Processing*, Springer-Verlag, Germany.
2. Mittal, V. (2012) *Functional Polymer Blends: Synthesis, Properties and Performance*, CRC Press, USA.
3. Hippi, U. (2008) *Polyolefin Composites*, John Wiley & Sons, USA.
4. Zhou, L., Wang, X., Lin, Y., Yang, J., and Wu, Q. (2003) Comparison of the toughening mechanisms of poly(vinyl chloride)/chlorinated polyethylene and poly(vinyl chloride)/acrylonitrile-butadiene-styrene copolymer blends. *Journal of Applied Polymer Science*, **90**, 916-924.
5. Eastwood, E. A., and Dadmun, M. D. (2002) Compatibilization of poly(vinyl chloride) and polyolefin elastomer blends with multi-block/blocky chlorinated polyethylenes. *Polymer*, **43**, 6707-6717
6. He, P., Huang, H., Xiao, W., Huang, S., and Cheng, S. (1997) A study on PVC/LLDPE blends with solid-state-chlorinated polyethylene as compatibilizer. *Journal of Applied Polymer Science*, **64**, 2535-2541.
7. Stoeva, S. (2006) Study of poly(vinyl chloride) blends with solid-state chlorinated polyethylene. *Journal of Applied Polymer Science*, **101**, 2602-2613.
8. Waddington, S., and Briggs, D. (1991) Adhesion mechanisms between polymer coatings and polypropylene studied by XPS and SIMS. *Polymer Communications*, **32**, 506-508.
9. Ogoe, S. A., Liou, D. W., Ainsworth, O. C., Bethea, J. R., Flores, D. R., Ellebracht, S. R., Muskopf, J. W., Werling, C. L., Wilson, E. J., Cocup, J. W., and Allison, J. L. (1996) Blends of Polycarbonate and Chlorinated Polyethylene, US patent 5,525,651.
10. Barrie, J. A., and Webb, W. D. (1989) Gas transport in miscible ble-

nds of ethylene-vinyl acetate copolymer and chlorinated polyethylene. *Polymer*, **30**, 327-332.

11. Maksimov, R. D., Ivanova, T., Zicans, J., and Kalkis, V. (2005) Mechanical properties of blends of low-density polyethylene with chlorinated polyethylene. *Mechanics of Composite Materials*, **41**, 267-276.
12. Maksimov, R. D., Ivanova, T., Kalnins, M., and Zicans, J. (2004) *Mechanics of Composite Materials*, **40**, 331-340.
13. Mukhopadhyay, P., and Gupta, R. K. (2011) Trends and frontiers in graphene-based polymer nanocomposites. *Plastics Engineering*, **67**, 32-42.
14. Kim, H., Abdala, A. A., and Macosko, C. W. (2010) Graphene/polymer nanocomposites. *Macromolecules*, **43**, 6515-6530.
15. Kuilla, T., Bhadra, S., Yao, D., Kim, N. H., Bose, S., and Lee, J. H. (2010) Recent advances in graphene based polymer composites. *Progress in Polymer Science*, **35**, 1350-1375.
16. Rafiee, M. A., Rafiee, J., Wang, Z., Song, H., Yu, Z. Z., and Koratkar, N. (2009) Enhanced mechanical properties of nanocomposites at low graphene content. *ACS Nano*, **3**, 3884-3890.
17. Kim, H., Miura, Y., and Macosko, C. W. (2010) Graphene/polyurethane nanocomposites for improved gas barrier and electrical conductivity. *Chemistry of Materials*, **22**, 3441-3450.
18. Nguyen, D. A., Lee, Y. R., Raghu, A. V., Jeong, H. M., Shin, C. M., and Kim, B. K. (2009) Morphological and physical properties of a thermoplastic polyurethane reinforced with functionalized graphene sheet. *Polymer International*, **58**, 412-417.
19. Schniepp, H. C., Li, J.-L., McAllister, M. J., Sai, H., Herrera-Alonso, M., Adamson, D. H., Prud'homme, R. K., Car, R., Saville, D. A., and Aksay, I. A. (2006) Functionalized single graphene sheets derived from splitting graphite oxide. *Journal of Physical Chemistry B*, **110**, 8535-8539.
20. Kim, H., Kobayashi, S., AbdurRahim, M. A., Zhang, M. L. J., Khusainova, A., Hillmyer, M. A., Abdala, A. A., and Macosko, C. W. (2011) Graphene/polyethylene nanocomposites: Effect of polyethylene functionalization and blending methods. *Polymer*, **52**, 1837-1846.
21. Guffey, V. O., and Sabbagh, A. B. (2002) PVC/wood-flour composites compatibilized with chlorinated polyethylene. *Journal of Vinyl & Additive Technology*, **8**, 259-263.
22. Chaudhry, A. U., and Mittal, V. (2013) High-density polyethylene nanocomposites using masterbatches of chlorinated polyethylene/graphene oxide. *Polymer Engineering and Science*, **53**, 78-88.
23. Donskoi, A. A., M. A. Shashkina, and G. E. Zaikov. (2003) *Fire Resistant and Thermally Stable Materials Derived from Chlorinated Polyethylene*, VSP, The Netherlands.

24. McAllister, M. J., Li, J. L., Adamson, D. H., Schniepp, H. C., Abdala, A. A., Liu, J., Herrera-Alonso, M., Milius, D. L., Car, R., Prud'homme, R. K., and Aksay, I. A. (2007) Single sheet functionalized graphene by oxidation and thermal expansion of graphite. *Chemistry of Materials*, **19**, 4396-4404.
25. Hummers W. S., and Offeman, R. E. (1958) Preparation of Graphitic Oxide. *Journal of American Chemical Society*, **80**, 1339-1339.
26. Cho, K., Lee, B. H., Hwang, K. M., Lee, H., and Choe, S. (1998) Rheological and mechanical properties in polyethylene blends. *Polymer Engineering and Science*, **38**(12), 1969-1975.
27. Kim, H. K., Rana, D., Kwag, H., and Choe, S. (2001) Melt rheology of ethylene 1-octene copolymer blends synthesized by Ziegler-Natta and metallocene catalysts. *Korea Polymer Journal*, **8**, 34-43.
28. Kwag, H., Rana, D., Choe, K., Rhee, J., Woo, T., Lee, B. H., and Choe, S. (2000) Binary blends of metallocene polyethylene with conventional polyolefins: Rheological and morphological properties. *Polymer Engineering and Science*, **40**(7), 1672-1681.
29. Joshi, M., Butola, B. S., Simon, G., and Kukaleva, N. (2006) Rheological and viscoelastic behavior of HDPE/octamethyl-POSS nanocomposites. *Macromolecules*, **39**(5), 1839-1849.
30. van Gurp, M., and Palmen, J. (1998) Time-temperature superposition for polymeric blends. *Rheology Bulletin*, **67**(1), 5-8.
31. Chuang, H. K., and Han, C. D. (1984) Rheological behavior of polymer blends. *Journal of Applied Polymer Science*, **29**, 2205-2229.
32. Han, C. D., and Chuang, H. K. (1985) Criteria for rheological compatibility of polymer blends. *Journal of Applied Polymer Science*, **30**(11), 4431-4454.
33. Utracki, L. A. (1989) *Rheology of Polymer Alloys and Blends*, Hanser Publishers, Germany.
34. Ko, S. W., Gupta, R. K., Bhattacharya, S. N., Choi, H. J. (2010) Rheology and physical characteristics of synthetic biodegradable aliphatic polymer blends dispersed with MWNTs. *Macromolecular Materials and Engineering*, **295**, 320-328.
35. Choi, H. J., and Ray, S. S. (2011) A review on melt-state viscoelastic properties of polymer nanocomposites. *Journal of Nanoscience and Nanotechnology*, **11**, 8421-8449.
36. Krimm, S., Liang, C. Y., and Sutherland, G. B. B. M. (1956) Infrared Spectra of High Polymers. II. Polyethylene. *Journal of Chemical Physics*, **25**(3), 549-562.
37. Smook, M. A., Pieski, E. T., and Hammer, C. F. (1953) Derivatives of chlorosulfonated polyethylene and their infrared spectra. Industrial & Engineering Chemistry, **45**(12), 2731-2737.
38. Luckachan, G. E., and Mittal, V. (2015) Evaluation of crystallinity variation and phase dispersion in polymer blends and nanocomposites by Raman mapping. *Journal of Polymer Research*, **22**:237,

doi:10.1007/s10965-015-0878-0.

39. Mittal, V., and Matsko, N. (2013) PE-CPE blends and their graphene oxide nanocomposites with reduced low temperature brittleness. *Colloid and Polymer Science*, **291**(8), 1949-1961.

4

CNTs Reinforced Polyethylene/Polyamide Blend Nanocomposites: Structural Characterization and Properties

4.1 Introduction

Polymer blends have received significant research interest in the past few years because of the ideal balance of cost and properties. For obtaining the blends possessing improved properties, the advantageous selection of the blend components and the control of the microstructure developed at the time of blending are required [1]. In addition to the blending of distinct polymers, an alternate method for enhancing the polymer performance is the incorporation of solid particles by melt mixing the components. Utilization of the nanosized particles permits an exceptional reduction in the quantity of filler necessary to enhance the properties as compared to the typical micron-sized particles [2]. Among the different immiscible polymer blends of technological significance, several research studies have concentrated on the blends of polyamides (PA) and polyethylene (PE), the polymers broadly utilized in a wide variety of automotive and packaging products due to their effective mechanical properties, barrier performance and processability. Polyamides act as a superior barrier to oxygen and offer excellent mechanical and thermal properties, whereas polyethylene performs as a good barrier to humidity and contributes better impact resistance, thus, leading to the durable blends.

Recently, the biopolymers have received research attention due to the increasing consciousness towards the usage of environmentally beneficial materials [3]. The biopolymers have replaced the conventional polymers in various industrial, automotive, packaging, electronic/electrical and consumer good applications. Presently, the greatest necessity for the bio-based polymers is in the packaging sector, where there are remarkable opportunities for the biopolymers as engineering resins. Nevertheless, the biopolymers are mostly expensive and possess poor property profiles, as compared to

Haleema Saleem and Vikas Mittal
Khalifa University of Science and Technology, Abu Dhabi, UAE

several commercial thermoplastic polymers. Therefore, the biopolymers need improvement in their properties for effectively competing with the conventional polymers. Polyamides are a category of synthetic engineering polymers, comprising of the amide group repeating units in the polymer chains [4,5]. Most of the commercial PAs are based on the petroleum-based monomers. Nonetheless, during past several decades, a developing interest in the development and commercialization of PAs based on bio-based monomers has been noticed. The most important bio-based commercial PAs are PA1010, PA11, PA610 and PA1012. PA1010 or Nylon 1010 is a commercially developed engineering PA [6] possessing several unique characteristics like high toughness, abrasion resistance, strength and elasticity [7]. Nevertheless, it has a few disadvantages such as significant moisture adsorption, which can generally be overcome by developing the blends as well as nanocomposites of PA1010 [8].

As mentioned earlier, there has been a developing interest in the blends of polyolefins and PAs in the recent years. When appropriately blended, the polyolefin and PA blends can conceivably contribute an extensive variety of desirable qualities, for example, low cost, lesser water sorption and greater chemical resistance. The mechanical characteristics of the aforementioned blends are strongly associated with the ratio of PA to polyolefin [9]. The blending of polyolefin with PA also enhances the impact resistance. As also mentioned earlier, the inclusion of a comprehensive range of fillers in the conventional polymers has been noted to result in significant enhancements in mechanical, gas barrier, thermal and rheological characteristics [10]. Nanofillers such as graphene, carbon nanotubes (CNTs) and nanoclays can develop the filler networks inside the polymer matrices at lower loadings due to the greater aspect ratios [11-13]. In order to generate high value materials from the biopolymers, corresponding nanocomposites employing the fillers such as layered silicates, nanotubes, etc., have been reported [14]. Polyamide nanocomposites have also obtained significant attention due to exceptional qualities such as effective barrier performance, modulus, electrical conductivity as well as non-flammability [15,16].

The CNTs, a one-dimensional (1-D) carbonaceous material, have brought about remarkable activities in various territories of engineering and science due to their exclusive physico-chemical properties. The diameter in CNTs is at the nanometer scale, and these can be either multi-walled (MWCNT) or single-walled (SWCNT). Due to their high aspect ratio, stiffness and strength at lower thickness, the CNTs have drawn the attention as reinforcing fillers for the development of low-weight and superior performance polymer based materials [17]. The methods for the dispersion and distribution of the CNTs inside the polymer materials play a crucial role in achieving the properties of the final product. For the preparation of CNTs filled polymer nanocomposites, different methods such as solution casting, melt mixing and *in-situ* polymerization have been utilized.

As reported earlier, the polyolefin/PA blend systems display excellent impact strength [18-20], along with high mechanical, thermal and oil resistance properties [21,22]. The polyolefin/PA blends have remarkable importance for the use in food packaging because of the superior moisture barrier characteristics of the polyolefin component and the excellent oxygen barrier properties of the PA phase [23]. The influence of CNTs on the morphology, electrical and mechanical properties of the immiscible polyethylene/polypropylene (PE/PP) blends was also examined by Al-Saleh [24]. The authors noted significant increase in the tensile strength with an increment in the CNT concentration. Also, the electrical conductivity and mechanical properties of the CNTs reinforced polyamide 6 and its blends with acrylonitrile/butadiene/styrene(ABS) were reported by Meincke *et al.* [25]. It was noted that the CNTs-filled polymer blends displayed excellent properties in Izod notched impact and tensile tests. The nanocomposites exhibited an enhancement of 27% in the Young's modulus, nevertheless, the elongation at break of the aforementioned nanocomposites significantly reduced due to the embrittlement of polyamide 6.

In this study, the thermo-mechanical properties of high density polyethylene (HDPE)/PA1010 blend nanocomposites with CNTs have been studied, at varying amounts of CNTs (1, 2, 3 and 5 wt%), so as to further develop this important class of blend composites for various applications.

4.2 Experimental

4.2.1 Materials

Renewably sourced biodegradable polyamide PA1010, having the trade name Zytel® RS LC1200 BK385 (70 wt% renewably sourced ingredient), was provided by DuPont. It had a melt viscosity of 500 Pa.s at 220 °C and density of 1.03 g/cm^3. This material was UV and heat stabilized and was suitable for multiple extrusion applications. HDPE grade PE BB2581 was provided by Abu Dhabi Polymers Company Limited. The MWCNTs utilized in the study were commercially procured and had outside diameter <8 nm, inside diameter 2-5 nm, purity >95%, length 10-30 um and SSA >500 m^2/g.

4.2.2 Preparation of HDPE/PA1010 Blend

HDPE/PA1010 blend (1: 1 ratio) was generated by the melt mixing technique. HDPE (2 g) and PA1010 (2 g) were added to a twin conical screw extruder (Thermo Scientific Haake Minilab II). A mixing temperature of 220 °C was utilized, and the mixing was carried out for 5 min at 100 rpm in a co-rotating mode, with a batch size of 4 g.

4.2.3 Preparation of HDPE/PA1010/CNTs Nanocomposites

HDPE/PA1010/CNT nanocomposites were prepared by the melt mixing technique, using suitable amount of HDPE, PA1010 and MWCNTs nanofiller employing a twin conical screw extruder (Thermo Scientific Haake Minilab II). A mixing temperature of 220 °C was utilized, and the mixing was carried out for 5 min at 100 rpm in a co-rotating mode, with a batch size of 4 g. A series of HDPE/PA-1010/CNTs nanocomposites were prepared by varying the content of MWCNTs as 1, 2, 3 and 5 wt%, as presented in Table 4.1.

Table 4.1 Composition of CNTs based HDPE/PA1010 nanocomposites

Blend/Nanocomposites	HDPE (%)	PA1010 (%)	CNTs (%)
HDPE+ PA1010 blend	50.0	50.0	0.0
HDPE/PA1010+1% CNTs	49.5	49.5	1.0
HDPE/PA1010+2% CNTs	49.0	49.0	2.0
HDPE/PA1010+3% CNTs	48.5	48.5	3.0
HDPE/PA1010+5% CNTs	47.5	47.5	5.0

After the extrusion process, the injection molding of the specimens was carried out in an injection molding machine (Thermo Scientific Haake Minijet Pro). The cylinder and mold temperatures were fixed as 235 °C and 125 °C, respectively. Also, the post and injection pressures were set as 470 and 570 bars, respectively, for a time period of 10 s. Various sample shapes, such as dumbbells and circular discs, were generated for each composition. The blend and nanocomposites were designated as PEPA (HDPE/PA1010 blend),

PEPACNT1 (HDPE/PA1010+1 wt% CNTs), PEPACNT2 (HDPE/PA1010+2 wt% CNTs), PEPACNT3 (HDPE/PA1010+3 wt% CNTs) and PEPACNT5 (HDPE/PA1010+5 wt% CNTs).

4.2.4 Characterization

Crystallization behavior of PEPA blend and PEPA blend nanocomposites was analyzed using differential scanning calorimetry (DSC) under N_2 atmosphere. The DSC thermograms were obtained utilizing a TA Discovery DSC with the help of TRIOS software. The extruded specimens of almost 3-7 mg weight were employed for the characterization. The tests were performed under non-isothermal conditions. Two heating and cooling cycles were carried out. During the first cycle, the specimens were heated from 25 °C to 275 °C, and subsequently cooled to -30 °C at a rate of +/- 10 °C/min for analyzing the re-crystallization temperature. Later, in the second cycle, the specimens were again heated to 275 °C for recording the melt temperature, and subsequently cooled to -30 °C at a rate of +/- 10 °C/min. The peak melting temperature (T_m), enthalpy of melting (ΔH_m), peak crystallization temperature (T_C) and enthalpy of crystallization (ΔH_C) were analyzed.

The crystalline structure of the PEPA blend and nanocomposites was analyzed using wide angle X-ray diffraction (WAXRD). Panalytical X'Pert Pro diffractometer was employed for obtaining the WAXRD spectra of the injection molded specimens, with the step size of 0.020° s^{-1} and a scanning angle range of 2θ = 5-40°. The power setting of the XRD apparatus were 40 mA and 40 kV in reflection mode.

Attenuated total reflection-Fourier transform infrared spectroscopy (ATR-FTIR) analysis of the pristine CNTs, pure PA1010, HDPE/PA1010 blend and blend nanocomposites was carried out using the injection molded samples by employing a Bruker VERTEX 70 spectrometer, equipped with a DLaTGS detector with KBr window, diamond ATR accessory and KBr beam-splitter. The ATR accessory had a ZnSe crystal with a refractive index of about 2.4@1000 cm^{-1} and a depth penetration of almost 2 µm at 45° angle of incidence. The IR spectra in the 4000-600 cm^{-1} range were plotted by employing the OPUS software at 8 cm^{-1} resolution with an average scans of 64.

The elongation at break, tensile stress at yield and tensile modulus of the PEPA blend and nanocomposites were analyzed using Instron 5567 universal testing machine at the strain rate of 10 mm/min. The analysis was carried out at room temperature in accordance with ASTM D 638. The dumbbell-shaped samples of 4 mm width, 2 mm thickness and 53 mm length were employed. For the evaluation of the mechanical properties, Win Test Analysis software was employed. An average of five values was reported.

The Raman spectroscopic analysis was carried out utilizing the confocal microRaman spectrometer Horiba LabRAM HR, equipped with a 514 nm laser source.

The thermal conductivity analysis of the blend and nanocomposites was carried out utilizing the laser flash equipment LFA 447 from Netzsch, Germany (operated with the assistance of Nanoflash software). This technique is appropriate for conducting the non-contact measurements of thermal diffusivity, heat capacity and thermal conductivity of different materials. Thermal conductivity (λ) of the samples was obtained using the relation $\lambda = \alpha.\rho.Cp$ where α, ρ and Cp are the thermal diffusivity, density and specific heat of the samples, respectively.

4.3 Results and Discussion

Calorimetric properties of pristine PA1010, PEPA blend and blend nanocomposites are outlined in Table 4.2. Figures 4.1 and 4.2 also illustrate the melting and cooling curves of these samples. The pristine PA1010 showed two melting peaks, one at 191 °C and the other at 199 °C. The observed multiple melting behaviors might be due to the re-crystallization during the DSC analysis. PA1010 is comprised of crystallite dispersion of different levels of perfection, which gets

Table 4.2 Calorimetric behavior of PEPA blend and PEPA blend nanocomposites

Composition	Heating			Cooling		
	ΔH_m (cal/g)	ΔH_m (J/g)	T_m (°C)	ΔH_c (cal/g)	ΔH_c (J/g)	T_c (°C)
Pure PA1010	6.3	26.4	191,199	10.6	44.2	177
PEPA blend	1.4	5.8	199	4.2	17.5	178
PEPACNT1	5.3	22.3	198	4.6	19.4	183
PEPACNT2	4.8	20.0	198	4.8	20.0	183
PEPACNT3	5.2	21.6	198	5.2	21.6	183
PEPACNT5	1.2	5.0	198	3.7	15.5	182

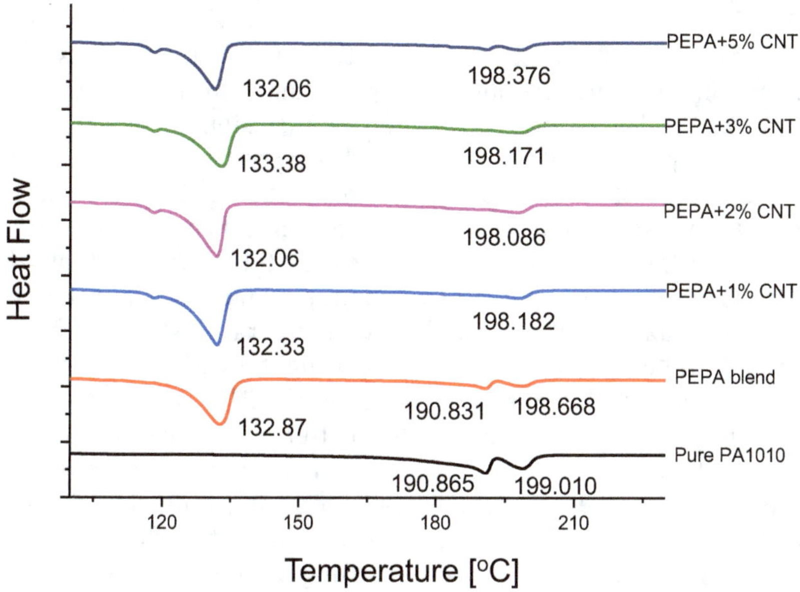

Figure 4.1 DSC melting curves of PEPA blend and nanocomposites.

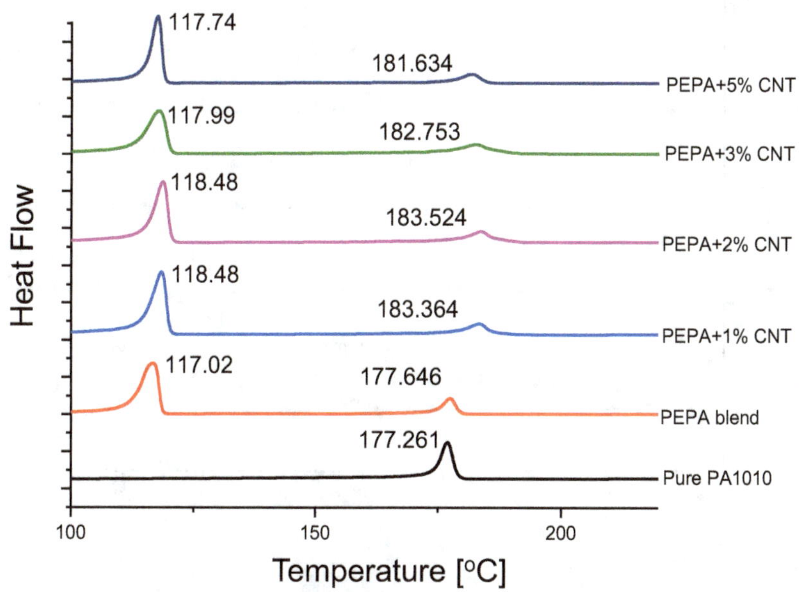

Figure 4.2 DSC cooling curves of PEPA blend and nanocomposites.

actively influenced by its previous thermal history. The peak located at 199 °C is the melting of highly stable crystals, whereas the peak at 191 °C might be attributed to the melting of unstable crystals [26]. The melting temperature and enthalpy obtained for HDPE were 133 °C and 21.7 cal/g, respectively. With the inclusion of the CNTs, the melting temperature of the PEPA blend composites exhibited an unchanged value of 198 °C for the PA phase and 132 °C for the PE phase. Conversely, an increase of ΔH_m (PA phase) could be observed for the CNTs filled PEPA nanocomposites till 3 wt%, relative to PEPA. This confirmed that the CNTs in the blend promoted the polymer crystallization. The crystallization temperatures of PEPA blend (both PA and PE phases) also increased slightly with the incorporation of CNTs, with the maximum value for 2% CNTs content.

Figure 4.3 presents the WAXRD patterns of pristine PA1010, pure HDPE, pure CNTs, PEPA blend and nanocomposites. HDPE exhibited two intense diffraction peaks, a strong reflection at 2θ = 24.3° and a weaker peak at 21.8°, corresponding to the (200) and (110) crystal planes, respectively. A strong Bragg peak concentrated around 0.34 nm, associated with the inter-shell spacing within the

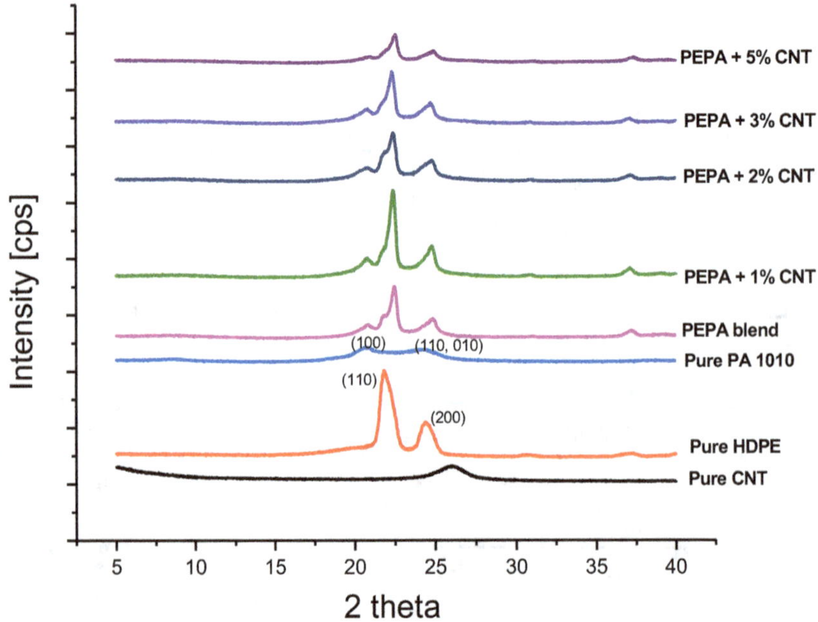

Figure 4.3 XRD patterns of pure PA1010, pure HDPE, pure CNTs, PEPA blend and PEPA blend nanocomposites.

nanotubes, was observed in the diffraction pattern of the MWCNTs. The WAXRD spectrum of the pristine PA1010 exhibited two distinct diffraction signals, at $2\theta = 24.4°$ and $20.5°$, associated with the (110/010) and 100 planes, respectively, of the α-phase crystal [27,28]. In the case of PEPA blend, the diffraction peak of (110) crystal plane of HDPE overlapped with the PA1010 diffraction peak of (100) at $2\theta = 22.3°$, and the diffraction peak of (200) crystalline plane of HDPE overlapped with the PA1010 diffraction peak of (110/010) at $2\theta = 24.7°$. XRD spectra nanocomposites exhibited a slight enhancement in the peak positions, though there was no specific effect of CNT concentration.

As the infrared (IR) bands in polymers are sensitive to molecular conformations, chain packing framework and hydrogen bonding, the IR analysis was carried out for examining the structural changes in the HDPE/PA1010 blend by the incorporation of CNTs. Pristine PA1010 exhibited the N-H stretching modes (amide A) of hydrogen bonded N-H groups at 3305 cm^{-1} and the carbonyl stretching vibrations (amide I) of hydrogen bonded carbonyl groups at 1636 cm^{-1} [29,30]. The coupling of C-N stretching and N-H in-plane bending modes (amide II) was observed at 1538 cm^{-1}. From Figure 4.4, it was

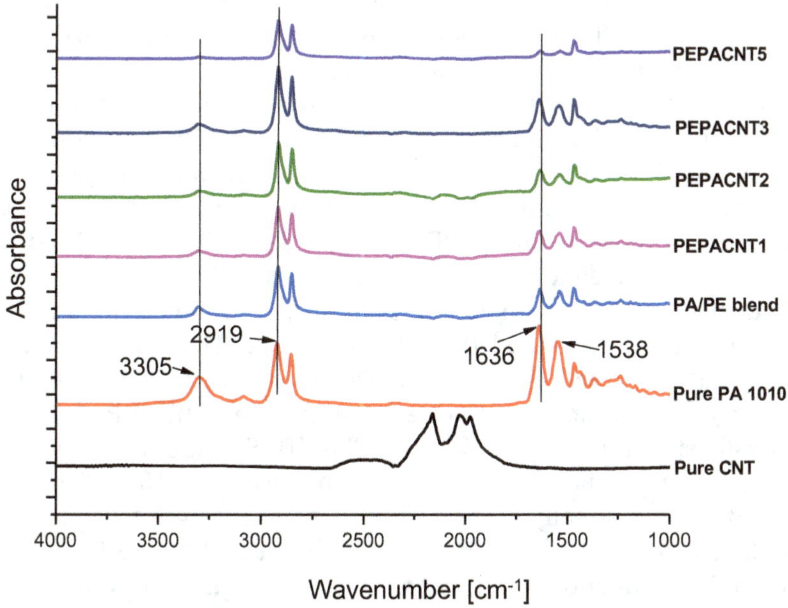

Figure 4.4 FTIR spectra of pure PA1010, pure CNTs, PEPA blend and PEPA blend nanocomposites.

noted that the carbonyl stretching vibrations (amide I) of hydrogen bonded carbonyl groups at 1636 cm^{-1} in the nanocomposites showed a decrease in intensity, relative to the HDPE/PA1010 blend. It was also quantitatively analyzed by determining the area ratio of this band (A_{1636}) to the C-H stretching vibration band seen at 2919 cm^{-1}. As presented in Table 4.3, the A_{1636}/A_{2919} ratio for HDPE/PA1010 blend was higher than the HDPE/PA1010 blend nanocomposites, clearly revealing that the particular amide bonds in the disordered amorphous area of the HDPE/PA1010 blend were subjected to chemical interactions in the presence of CNTs.

Table 4.3 A_{1636}/A_{2925} for pure PA1010, PEPA and nanocomposites

Material	A_{1636}	A_{2919}	A_{1636}/A_{2925}
Pure PA1010	0.476	0.368	1.29
PEPA blend	0.202	0.331	0.60
PEPACNT1	0.188	0.324	0.58
PEPACNT2	0.225	0.392	0.57
PEPACNT3	0.209	0.394	0.53
PEPACNT5	0.065	0.241	0.27

The mechanical properties of the blend nanocomposites are associated with different factors inclusive of the blend properties, impact of the nanofiller on the host polymer, blend morphology, positioning of the nanofiller inside the blend and influence of the filler on the state of adhesion at the blend's interface. The tensile properties of the CNTs filled HDPE/PA1010 nanocomposites are presented in Table 4.4. It was noted that the CNTs significantly enhanced the Young's modulus of the HDPE/PA1010 blend (Figure 4.5). For instance, the tensile modulus was increased to 1743 MPa for 5% CNTs filled HDPE/PA1010 nanocomposite, thus, showing an increment of 39% as compared to the HDPE/PA1010 blend. The molecular architecture as well as phase behavior perform a critical role in defining the chain orientation and tensile strength. The unfilled PA1010 had a tensile strength of about 50 MPa. The HDPE/PA1010 blend exhibited a lower tensile strength relative to pristine PA1010. The addition of CNTs gradually decreased the tensile strength to 26 MPa for 1% CNTs incorporation and 20 MPa in the case of 5% CNTs incorporation. The reduction in the strength can be attributed to the incompatible blend morphology. In addition, a gradual decrease in the elongation at break was noticed on incorporating the filler, thus, indicating enhanced brittleness.

Table 4.4 Tensile properties of PA1010, HDPE/PA1010 blend and nanocomposites

Material	Elongation at break (%)	Tensile stress at yield (MPa)	Tensile modulus (MPa)
Pure PA1010	19.8	50	1084
PEPA blend	7.2	28	1254
PEPACNT1	3.9	26	1425
PEPACNT2	2.9	23	1460
PEPACNT3	2.7	22	1479
PEPACNT5	1.9	20	1743

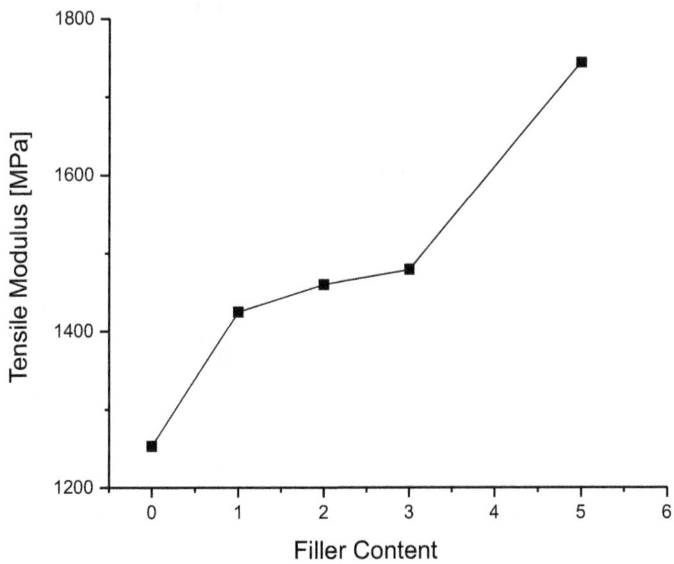

Figure 4.5 Tensile modulus of HDPE/PA1010 blend and different CNT based HDPE/PA1010 nanocomposites.

In Figure 4.6, the Raman spectrum of pure CNTs exhibited two typical graphite bands at 1574 cm^{-1} (G band due to in-plane vibrations of sp^2 bonded carbon atoms) and 1320 cm^{-1} (D band related to sp^3 bonds). An additional band, known as the G" band, was also present in the Raman range at 2642 cm^{-1} due to the D band overtone [31]. The spectrum of HDPE (Figure 4.7) exhibited conventional Raman modes of PE. The Raman bands observed at ~1069 cm^{-1} and ~1137 cm^{-1} could be attributed to the C-C stretching modes in the

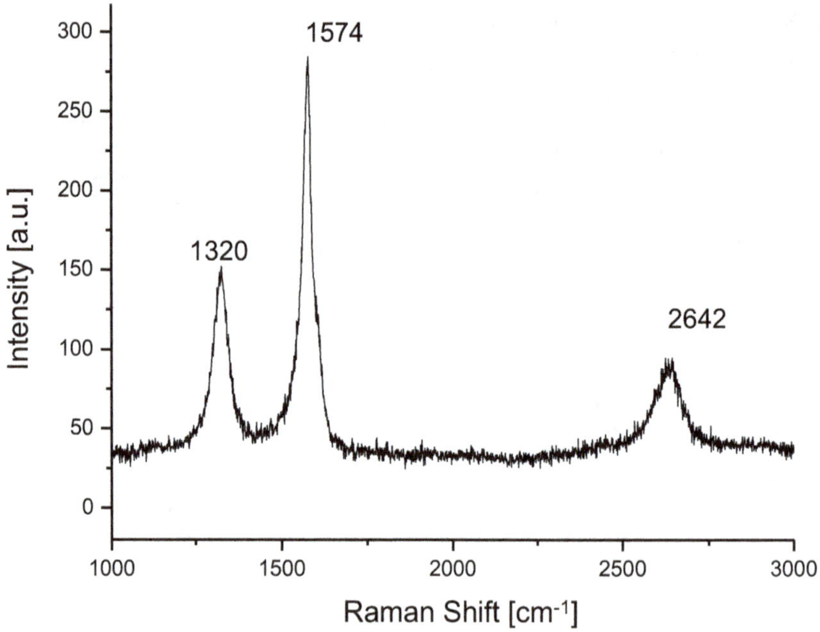

Figure 4.6 Raman spectrum of pristine CNTs.

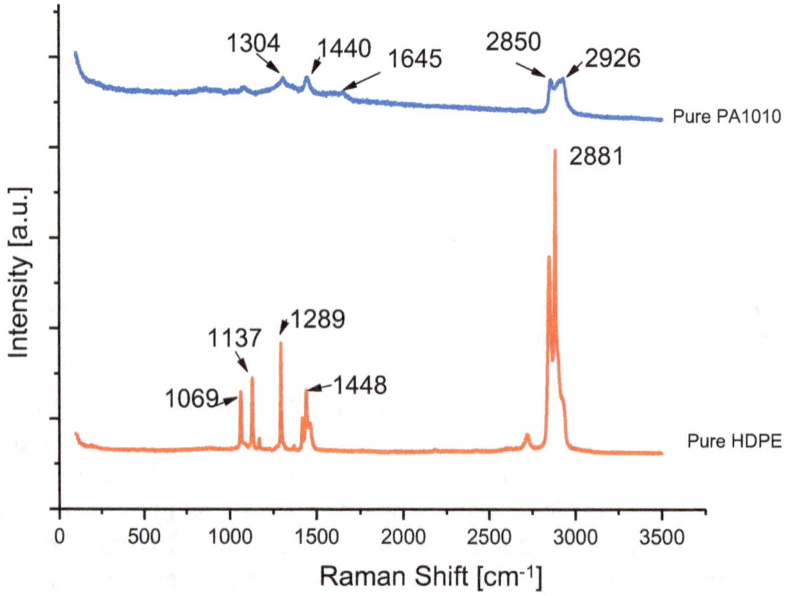

Figure 4.7 Raman spectra of pure HDPE and pure PA1010.

crystalline and anisotropic phases of PE. PA1010 spectrum consisted of the characteristic bands of PA, where the bands seen at 1645 cm^{-1} (amide I) and 1304 cm^{-1} (amide III) were associated with the presence of the amide groups in the polymer. The major bands observed at 2850 cm^{-1} and 2926 cm^{-1} were associated with the symmetric and asymmetric CH_2 stretching vibrations, respectively. Also, from Figure 4.8, it was observed that the Raman spectrum of 5 wt% CNTs filled HDPE/PA1010 nanocomposite was moderately influenced by the CNT bands. The analysis of Raman I_D/I_G ratio (I_D – Raman intensity of D peak; I_G – Raman intensity of G peak) of pure CNTs and of 5 wt% CNTs filled HDPE/PA1010 nanocomposite was performed to determine the interactions between the nanofiller and blend, as presented in Table 4.5. For the nanocomposite with 5% CNTs content, the value of I_D/I_G was greater, relative to MWCNTs. The observed higher I_D/I_G ratio of the PEPACNT5 nanocomposite indicated the possible generation of covalent interaction between the HDPE/PA1010 blend and MWCNTs, which moderately breaks the sp^2 carbon framework and subsequently generates a stronger interaction between them.

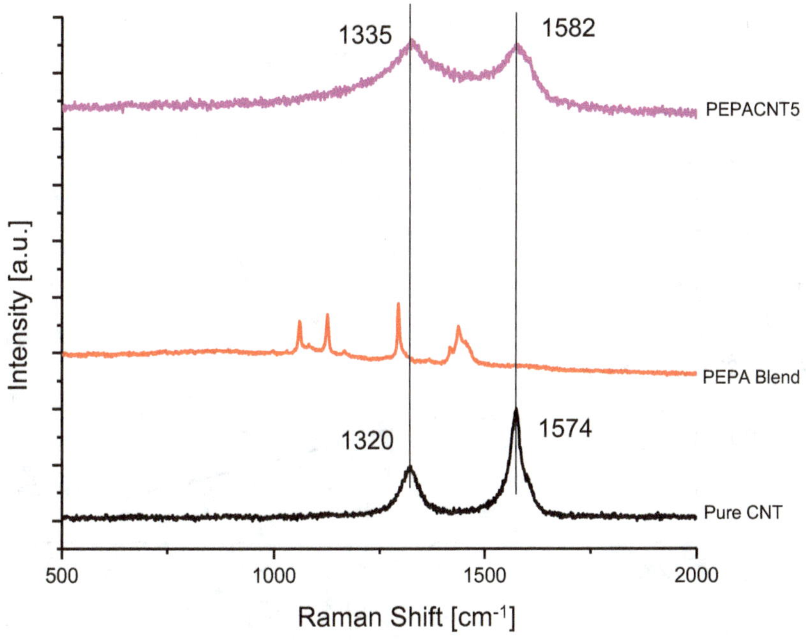

Figure 4.8 Raman spectra of pure CNTs, PEPA blend and 5 wt% CNTs filled HDPE/PA1010 nanocomposite.

Table 4.5 Raman intensity ratio (I_D/I_G) of pristine CNT and 5 wt% CNTs filled nanocomposite

Material	D band (cm^{-1})	G band (cm^{-1})	I_D	I_G	Raman Intensity ratio (I_D/I_G)
Pure CNTs	1320	1574	0.480	0.993	0.483
PEPACNT5	1335	1582	0.986	0.951	1.036

The thermal conductivity of suspended graphene is approximately 5000 W/m.K [32], whereas pristine HDPE and PA1010 have low conductivity. Thus, the outcome of the incorporation of CNTs on the thermal conductivity of HDPE/PA1010 blend was analyzed, as presented in Figure 4.9 and Table 4.6. The presence of CNTs enhanced the thermal conductivity of the HDPE/PA1010 blend significantly, as a function of filler content. Also, the conductivity was observed o decrease with temperature. At a temperature 60 °C, the composites with 3 and 5 wt% filler content attained the values 0.21 W/m.K and 0.236 W/m.K respectively, indicating 30% and 47% enhancement relative to the HDPE/PA1010 blend.

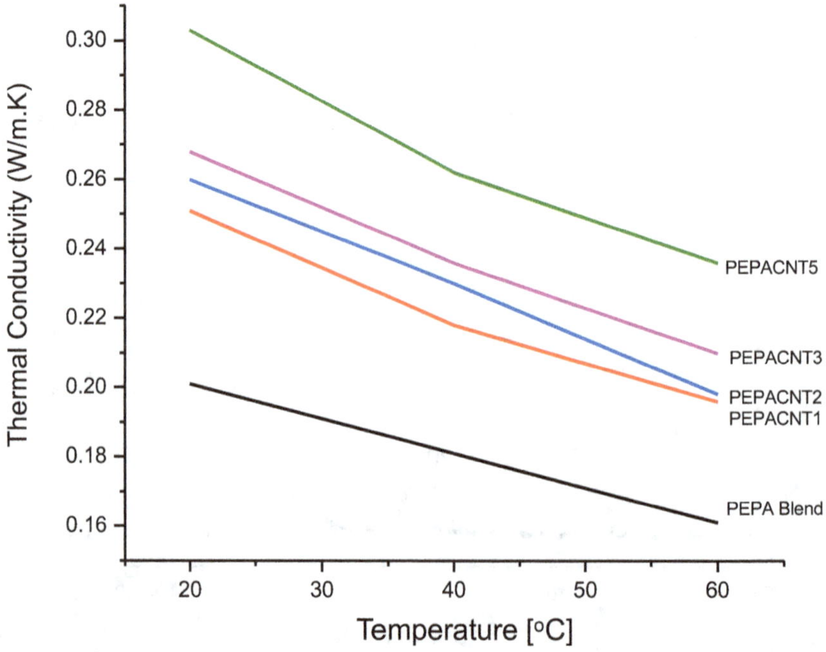

Figure 4.9 Thermal conductivity analysis of PEPA blend and PEPA blend nanocomposites.

Table 4.6 Thermal conductivity of PEPA blend and PEPA blend nanocomposites

Temp.	PEPA Blend	PEPACNT1	PEPACNT2	PEPACNT3	PEPACNT5
20 °C	0.201	0.251	0.260	0.268	0.303
40 °C	0.181	0.218	0.230	0.236	0.262
60 °C	0.161	0.196	0.198	0.210	0.236

4.4 Conclusions

In this study, the thermo-mechanical properties and structural characteristics of polyethylene (HDPE)/polyamide1010 (PA1010) nanocomposites containing the multi-walled carbon nanotubes (MWCNTs) have been analyzed. The inclusion of the CNTs did not affect the peak melting temperatures, though a slight increase in ΔH_m could be observed. The diffraction peaks of the blend also exhibited a slight increase in the 2θ values in the nanocomposites. The A_{1636}/A_{2919} ratio from the FTIR spectra was greater for HDPE/PA1010 blend, relative to the nanocomposites, indicating that the particular amide bonds in the disordered amorphous area of HDPE/PA1010 blend were subjected to chemical interaction in the presence of CNTs. CNTs significantly enhanced the Young's modulus of the HDPE/PA1010 blend, with the composite with 5% CNTs showing an increment of 39%. CNTs also enhanced the thermal conductivity of the HDPE/PA1010 blend significantly, as a function of filler content.

References

1. *Micro-and Nanostructured Multiphase Polymer Blend Systems: Phase Morphology and Interfaces*, Harrats, C., Thomas, S., and Groeninckx, G. (eds.), CRC Press, USA (2005).
2. Jordan, J., Jacob, K. I., Tannenbaum, R., Sharaf, M. A., and Jasiuk, I. (2005) Experimental trends in polymer nanocomposites - A review. *Materials Science and Engineering A*, **393**(1), 1-11.
3. Bordes, P., Pollet, E., and Avérous, L. (2009) Nano-biocomposites: biodegradable polyester/nanoclay systems. *Progress in Polymer Science*, **34**(2), 125-155.
4. Marchildon, K. (2011) Polyamides–still strong after seventy years. *Macromolecular Reaction Engineering*, **5**(1), 22-54.

5. Carraher, Jr., C. E., Morrison, A., Roner, M. R., Moric-Johnson, A., Al-Huniti, M. H., and Miller, L. (2015) Metallocene-containing polyesters from reaction of 3, 5-pyridinedicarboxylic acid and metallocene dihalides and their preliminary ability to inhibit cancer cell growth. *Journal of the Chinese Advanced Materials Society*, **3**(4), 310-327.
6. Zhishen, M., Qingbo, M., Jinhua, F., Hongfang, Z., and Donglin, C. (1993) Crystal structure and thermodynamic parameters of nylon-1010. *Polymer International*, **32**(1), 53-60.
7. Zhang, S. L., Wang, G. B., Jiang, Z. H., Wang, D., Ma, R. T., and Wu, Z. W. (2005) Impact properties, phase structure, compatibility, and fracture morphology of polyamide-1010/thermoplastic poly (ester urethane) elastomer blends. *Journal of Polymer Science, Part B: Polymer Physics*, **43**(10), 1177-1185.
8. Lu, Z., Wu, A., Ou, X., Zhang, S., Niu, J., Ji, S., and Ling, Y. (2017) Enhanced anti-aging and mechanical properties of polyamide 1010 by sol-hydrothermal synthetic titanium dioxide-coated kaolinite addition. *Journal of Alloys and Compounds*, **693**, 381-388.
9. Gonzalez-Montiel, A., Keskkula, H., and Paul, D. R. (1995) Impact-modified nylon 6/polypropylene blends: 1. Morphology-property relationships. *Polymer*, **36**(24), 4587-4603.
10. Mittal, V. (2008) Mechanical and gas permeation properties of compatibilized polypropylene–layered silicate nanocomposites. *Journal of Applied Polymer Science*, **107**(2), 1350-1361.
11. Carretero-González, J., Retsos, H., Verdejo, R., Toki, S., Hsiao, B. S., Giannelis, E. P., and López-Manchado, M. A. (2008) Effect of nanoclay on natural rubber microstructure. *Macromolecules*, **41**(18), 6763-6772.
12. Wang, K., Liang, S., Deng, J., Yang, H., Zhang, Q., Fu, Q., Dong, X., Wang, D., and Han, C. C. (2006) The role of clay network on macromolecular chain mobility and relaxation in isotactic polypropylene/organo-clay nanocomposites. *Polymer*, **47**(20), 7131-7144.
13. Bréchet, Y., Cavaille, J. Y., Chabert, E., Chazeau, L., Dendievel, R., Flandin, L., and Gauthier, C. (2001) Polymer based nanocomposites: Effect of filler-filler and filler-matrix interactions. *Advanced Engineering Materials*, **3**(8), 571-577.
14. Favier, V., Cavaille, J. Y., Canova, G. R., and Shrivastava, S. C. (1997) Mechanical percolation in cellulose whisker nanocomposites. *Polymer Engineering and Science*, **37**(10), 1732-1739.
15. Ma, L., Niu, H., Cai, J., Zhao, P., Wang, C., Lian, Y., Bai, X., and Wang, W. (2014) Optical, electrochemical, photoelectrochemical and electrochromic properties of polyamide/graphene oxide with various feed ratios of polyamide to graphite oxide. *Journal of Materials Chemistry C*, **2**(12), 2272-2282.
16. Zheng, D., Tang, G., Zhang, H. B., Yu, Z. Z., Yavari, F., Koratkar, N.,

Lim, S. H., and Lee, M. W. (2012) In situ thermal reduction of graphene oxide for high electrical conductivity and low percolation threshold in polyamide 6 nanocomposites. *Composites Science and Technology*, **72**(2), 284-289.

17. Ganesan, Y., Peng, C., Lu, Y., Loya, P. E., Moloney, P., Barrera, E., Yakobson, B. I., Tour, J. M., Ballarini, R., and Lou, J. (2011) Interface toughness of carbon nanotube reinforced epoxy composites. *ACS Applied Materials and Interfaces*, **3**(2), 129-134.
18. Utracki, L. A., Dumoulin, M. M., and Toma, P. (1986) Melt rheology of high density polyethylene/polyamide-6 blends. *Polymer Engineering and Science*, **26**(1), 34-44.
19. Chen, C. C., Fontan, E., Min, K., and White, J. L. (1988) An investigation of instability of phase morphology of blends of nylons with polyethylenes and polystyrenes and effects of compatibilizing agents. *Polymer Engineering and Science*, **28**(2), 69-80.
20. Hobbs, S. Y., Bopp, R. C., and Watkins, V. H. (1983) Toughened nylon resins. *Polymer Engineering and Science*, **23**(7), 380-389.
21. Willis, J. M., and Favis, B. D. (1988) Processing-morphology relationships of compatibilized polyolefin/polyamide blends. Part I: The effect of an Ionomer compatibilizer on blend morphology. *Polymer Engineering and Science*, **28**(21), 1416-1426.
22. Koulouri, E. G., Georgaki, A. X., and Kallitsis, J. K. (1997) Reactive compatibilization of aliphatic polyamides with functionalized polyethylenes. *Polymer*, **38**(16), 4185-4192.
23. Holsti-Miettinen, R. M., Perttilä, K. P., Seppälä, J. V., and Heino, M. T. (1995) Oxygen barrier properties of polypropylene/polyamide 6 blends. *Journal of Applied Polymer Science*, **58**(9), 1551-1560.
24. Al-Saleh, M. H. (2016) Carbon nanotube-filled polypropylene/polyethylene blends: compatibilization and electrical properties. *Polymer Bulletin*, **73**(4), 975-987.
25. Meincke, O., Kaempfer, D., Weickmann, H., Friedrich, C., Vathauer, M., and Warth, H. (2004) Mechanical properties and electrical conductivity of carbon-nanotube filled polyamide-6 and its blends with acrylonitrile/butadiene/styrene. *Polymer*, **45**(3), 739-748.
26. Mittal, V., Chaudhry, A. U., and Luckachan, G. E. (2014) Biopolymer–thermally reduced graphene nanocomposites: structural characterization and properties. *Materials Chemistry and Physics*, **147**(1), 319-332.
27. Xu, Z., and Gao, C. (2010) In situ polymerization approach to graphene-reinforced nylon-6 composites. *Macromolecules*, **43**(16), 6716-6723.
28. Li, Y., and Yan, D. (2001) Preparation, characterization and crystalline transition behaviors of polyamide 4 14. *Polymer*, **42**(11), 5055-5058.
29. Yoshioka, Y., Tashiro, K., and Ramesh, C. (2003) Structural change

in the Brill transition of Nylon m/n (2) conformational disordering as viewed from the temperature-dependent infrared spectral measurements. *Polymer*, **44**(20), 6407-6417.

30. Yoshioka, Y., and Tashiro, K. (2003) Structural change in the Brill transition of Nylon m/n (1) Nylon 10/10 and its model compounds. *Polymer*, **44**(22), 7007-7019.
31. Bokobza, L., and Zhang, J. (2012) Raman spectroscopic characterization of multiwall carbon nanotubes and of composites. *Express Polymer Letters*, **6**(7), 601-608.
32. Lehman, J. H., Terrones, M., Mansfield, E., Hurst, K. E., and Meunier, V. (2011) Evaluating the characteristics of multiwall carbon nanotubes. *Carbon*, **49**(8), 2581-2602.

5

Graphene Reinforced Polyamide/Polyethylene Blend Nanocomposites: Structural Characterization and Properties

5.1 Introduction

The morphology of the blends relies upon various parameters like viscosity ratio, composition and interfacial tension of the component phases as well as processing conditions at the time of blending [1-4]. Apart from the blending of distinct polymers, an alternative technique for improving the polymer performance and applications is the inclusion of the solid particles by melt mixing the components [4].

In the recent past, there has been increasing interest in the blends of polyamides (PAs) and polyolefins. Most of the commercial PAs are based on the petroleum-based monomers. However, an advancing interest in the development as well as commercialization of PAs based on bio-based monomers has been noted during the past few decades [5-11]. In this respect, PA1010 is a commercially developed engineering PA matrix possessing superior properties like effective strength, toughness, abrasion resistance and elasticity [12-16].

The development of polymer composites with nanofiller systems provides enhanced thermal stability, modulus, electrical conductivity and tensile strength, along with low gas penetrability and flammability [17-20]. Graphene nanoplatelets (GNPs) are stacks of multi-layered graphene sheets [21,22] and have emerged as an effective nanofiller for polymer matrices. GNPs have a superior balance of cost and physical properties as compared to other fillers [23-26]. The GNP functionalization has been utilized as a successful technique to enhance the dispersion of particles and material properties [27,28]. Due to the nanometer sized laminated framework of GNP nanoplatelets, these can prevent the flow of heat as well as combustible gases and possess the ability for developing the flame retardancy and thermal properties of polymers. Nevertheless, the power-

Haleema Saleem and Vikas Mittal
Khalifa University of Science and Technology, Abu Dhabi, UAE

ful van der Waals interactions among the graphene sheets mean that the GNPs have the tendency to agglomerate in the polymer matrices. The uniform GNP dispersion in the polymers is very critical to achieve an optimum advancement in properties [29].

The PA/polyolefin blends possess superior impact strength [30-32], along with excellent thermal, mechanical as well as oil resistance properties [33-35]. In the recent past, Pötschke *et al.* [36], Cao *et al.* [37] and Mohamadi *et al.* [38] examined the influence of graphene on the polymer blend properties. Cao *et al.* [37] observed that with the addition of 0.5 wt% graphene oxide sheets in immiscible polyamide/polyphenylene oxide (PA/PPO, 90/10) blend, the droplet diameter of the dispersed minor phase (PPO) was significantly decreased by more than 1 order of magnitude, showing enhanced compatibility. Also, it was confirmed that the graphene oxide sheets could also function as reinforcing fillers in the polymer blends, thereby, causing a significant enhancement in the thermal stability and mechanical performance.

In this study, the mechanical, thermal and structural characteristics of GNP filled PA1010/HDPE nanocomposites have been analyzed, so as to develop this important class of blend nanocomposites further.

5.2 Experimental

5.2.1 Materials

HDPE grade PE BB2581 was supplied by Abu Dhabi Polymers Company Limited. It possessed a density of 0.958 g/cm^3 and melt flow of 0.40 g/10 min at a temperature 190 °C. Renewably sourced biodegradable polyamide PA1010 with the trade name Zytel® RS LC1200 BK385 (70 wt% renewably sourced ingredient) was provided by DuPont. It possessed a density of 1.03 g/cm^3 and melt viscosity of 500 Pa.s at 220 °C. The graphene nanoplatelets with the product name N002-PDR were purchased from Angstron Materials, USA. The material possessed a density of ≤2.20 g/cm^3 and specific surface area in the range 400-800 m^2/g.

5.2.2 Preparation of PA1010/HDPE Blend

PA1010/HDPE blend (1:1 ratio) was prepared by the melt mixing method. For this purpose, PA1010 (2 g) and HDPE (2 g) were added to Thermo Scientific Haake Minilab II twin conical screw extruder. Mixing temperature of 220 °C was employed, and the mixing was performed for 5 min in co-rotating mode at 100 rpm, with 4 g batch size.

5.2.3 Preparation of PA1010/HDPE/GNP Nanocomposites

PA1010/HDPE/GNP nanocomposites were generated by the melt mixing method by utilizing appropriate quantity PA1010, HDPE and GNPs, using Thermo Scientific Haake Minilab II twin conical screw extruder. Mixing temperature of 220 °C was employed, and the mixing was performed for 5 min in co-rotating mode at 100 rpm, with 4 g batch size. A series of PA1010/HDPE/GNP nanocomposites were generated by varying the amount of GNPs as 1, 2, 3 and 5 wt%, as detailed in Table 5.1.

Table 5.1 Composition of the PA1010/HDPE blend nanocomposites

Blend/Nanocomposites	PA1010 (%)	HDPE (%)	GNP (%)
Pure PA1010 + HDPE blend	50.0	50.0	0.0
PA1010/HDPE + 1% GNPs	49.5	49.5	1.0
PA1010/HDPE + 2% GNPs	49.0	49.0	2.0
PA1010/HDPE + 3% GNPs	48.5	48.5	3.0
PA1010/HDPE + 5% GNPs	47.5	47.5	5.0

Subsequent to extrusion, the injection molding of the samples was performed using Thermo Scientific Haake Minijet Pro injection molding machine. The mold and cylinder temperatures were fixed as 125 °C and 235 °C, respectively. Further, the injection and post pressure were set as 570 and 470 bars, respectively, for 10 s. Different specimen shapes, like dumbbell and circular discs, were generated for each composition. The nanocomposites were designated as PAPE (PA1010/HDPE blend), PAPEGNP1 (PA1010/HDPE + 1 wt% GNPs), PAPEGNP2 (PA1010/HDPE + 2 wt% GNPs), PAPEGNP3 (PA1010/HDPE + 3 wt% GNPs) and PAPEGNP5 (PA1010/HDPE + 5 wt% GNPs).

5.2.4 Characterization

The tensile properties of the PA1010/HDPE blend and PA1010/HDPE blend nanocomposites were analyzed using Instron

5567 Universal Testing Machine (UTM) at 10 mm/min strain rate. The tests were performed in accordance with ASTM D 638 at room temperature. The dumbbell-shaped specimens of 4 mm width, 53 mm length and 2 mm thickness were utilized. Win Test Analysis software was used to examine the magnitude of the mechanical properties, and an average of five specimens was recorded.

Calorimetric behavior of PAPE blend and PA1010/HDPE/GNP nanocomposites was examined utilizing differential scanning calorimetry (DSC) under N_2 atmosphere. Utilizing the TRIOS software, the DSC thermograms were attained using TA Discovery DSC. The extruded samples of about 3-8 mg weight were used for the characterization. DSC analysis was carried out under the non-isothermal conditions. Two heating and cooling cycles were performed. During the first cycle, the specimens were heated from 25 °C to 275 °C and subsequently cooled to -30 °C at a rate of +/- 10 °C/min in order to analyze the re-crystallization temperature. Subsequently, during the second cycle, the samples were re-heated to 275 °C in order to record the melt temperature, and later cooled to -30 °C at a rate of +/- 10 °C/min. Overall, peak melting temperature (T_m), peak crystallization temperature (T_C), enthalpy of melting (ΔH_m) and enthalpy of crystallization (ΔH_C) were examined.

Attenuated total reflection-Fourier transform infrared spectroscopy (ATR-FTIR) analysis of the pure GNPs, pristine PA1010, PA1010/HDPE blend and PA1010/HDPE/GNP nanocomposites was performed on the injection molded specimens utilizing Bruker VERTEX 70 spectrometer, equipped with a diamond ATR accessory, DLaTGS detector with KBr window and KBr beam-splitter. The ATR accessory possessed a ZnSe crystal having a refractive index of about 2.4@1000 cm^{-1} and a depth penetration of almost 2 μm at 45° angle of incidence. IR spectra in the 4000-600 cm^{-1} range were plotted employing the OPUS software at 8 cm^{-1} resolution with 64 average scans.

Panalytical X'Pert Pro diffractometer was employed for performing the wide angle X-ray diffraction (WAXRD) of pristine PA1010, HDPE, pure GNPs, PA1010/HDPE blend and PA1010/HDPE/GNP nanocomposites, using the scanning angle range of 2θ = 5-40° and step size of 0.020° s^{-1}. The analysis was carried out in reflection mode, and the power settings of the instrument were 40 kV and 40 mA.

The thermal conductivity of the specimens was analyzed utilizing laser flash analysis (LFA) (LFA 447 from Netzsch, operated with the help of Nanoflash software). Thermal conductivity (λ) was calculated employing the relation λ= α.ρ.Cp where α, ρ and Cp are thermal diffusivity, density and specific heat of the specimens, respectively.

The Raman spectroscopic analysis was performed employing confocal microRaman spectrometer Horiba LabRAM HR, equipped with 514 nm laser source.

5.3 Results and Discussion

The tensile properties of the materials are presented in Table 5.2. The incorporation of GNPs was observed to significantly enhance the tensile modulus of the PA1010/HDPE blend (Figure 5.1). For

Table 5.2 Tensile properties of PA1010/HDPE blend and PA1010/HDPE/GNP nanocomposites

Material	Elongation at break (%)	Tensile stress at yield (MPa)	Tensile Modulus (MPa)
Pure PA1010	19.8	50	1084
PAPE blend	7.2	28	1254
PAPEGNP1	4.3	25	1309
PAPEGNP2	2.1	19	1450
PAPEGNP3	1.2	12	1612
PAPEGNP5	1.1	11	1639

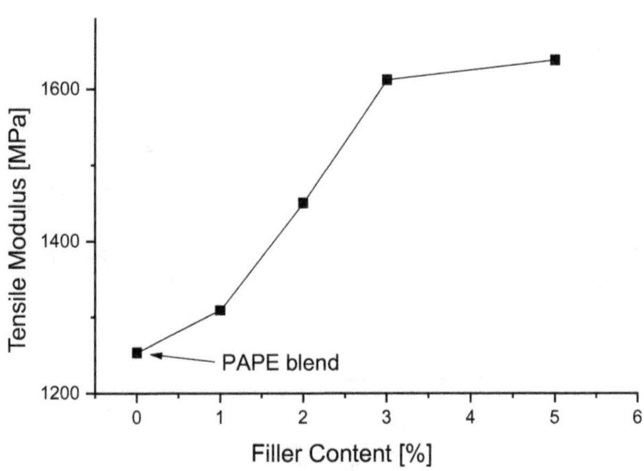

Figure 5.1 Tensile modulus of PA1010/HDPE blend and PA1010/HDPE/GNP nanocomposites.

instance, the PAPE blend exhibited the tensile modulus of 1254 MPa, which was enhanced to 1638 MPa for 5% GNPs reinforced PA1010/HDPE, thereby, displaying an increase of 31%. The pristine PA1010 exhibited a tensile strength of 50 MPa. The strength was observed to significantly reduce in the PA1010/HDPE blend, as compared to pristine PA1010. Incorporation of GNPs gradually reduced the tensile strength further to 24.48 MPa for 1% GNPs addition and 10.61 MPa for 5% GNPs inclusion. Further, it was confirmed that the PA1010/HDPE blend exhibited a gradual reduction in the elongation at break, as a function of GNP concentration.

Calorimetric properties of the materials are outlined in Table 5.3. Figures 5.2 and 5.3 also demonstrate the melting and cooling curves

Table 5.3 Calorimetric behavior of PA1010/HDPE blend and PA1010/HDPE/GNP nanocomposites

Composition	Heating			Cooling		
	ΔH_m (cal/g)	ΔH_m (J/g)	T_m (°C)	ΔH_c (cal/g)	ΔH_c (J/g)	T_c (°C)
Pure PA1010	6.3	26.4	191	10.6	44.2	177
PAPE blend	1.4	5.8	199	4.2	17.5	178
PAPEGNP1	1.4	5.7	198	4.0	16.8	182
PAPEGNP2	3.8	15.9	199	4.7	19.7	182
PAPEGNP3	2.9	12.3	199	4.5	18.8	183
PAPEGNP5	3.9	16.3	198	5.3	22.2	183

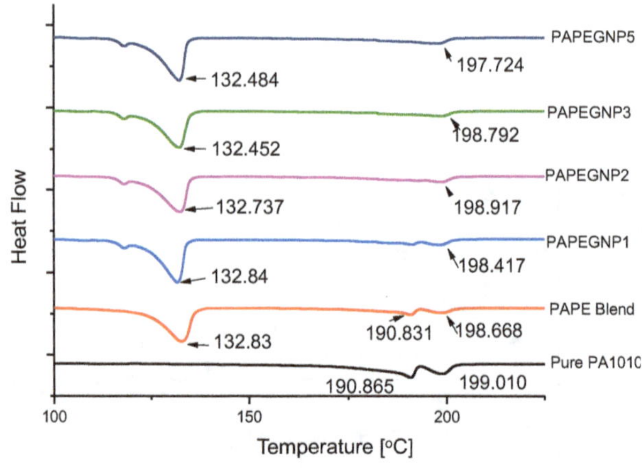

Figure 5.2 DSC melting curves of PA1010/HDPE blend and PA1010/HDPE/GNP nanocomposites.

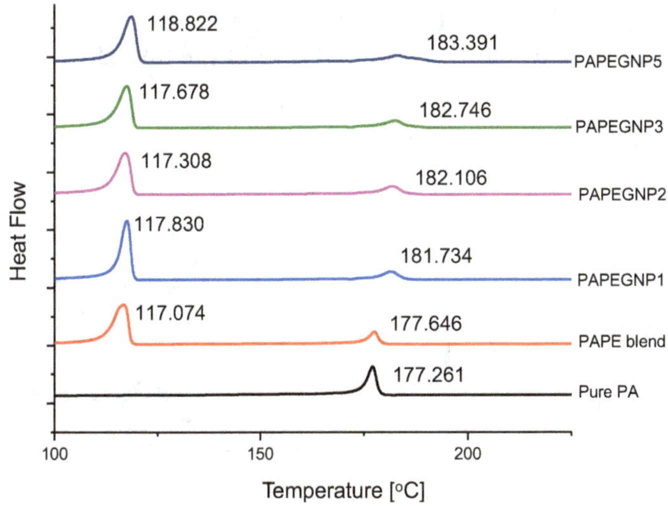

Figure 5.3 DSC cooling curves of PA1010/HDPE blend and PA1010/HDPE/GNP nanocomposites.

of the samples. PA1010 displayed two melting peaks at 191 °C and 199 °C. The observed multiple melting behavior may be due to the re-crystallization at the time of the DSC analysis. The peak situated at 191 °C might have resulted due to the melting of unstable crystals, whereas the peak at 199 °C can be attributed to the melting of highly stable crystals [39]. On incorporation of GNP nanoplatelets, the peak melting temperature of the PA phase in the composites displayed an unaltered value of 198 °C. A slight increment in the ΔH_m values (PA phase) was noted for the composites, relative to the PA1010/HDPE blend. It indicated that the presence of GNP nanofiller in the blend moderately promoted the polymer crystallization. The crystallization temperature of the PA1010/HDPE blend also improved marginally on incorporating GNPs, with the maximum value for the nanocomposite with 5% GNPs.

Pure PA1010 displayed the carbonyl stretching vibrations (amide I) of hydrogen bonded carbonyl groups at 1636 cm^{-1} and the N-H stretching modes (amide A) of hydrogen bonded N-H groups at 3305 cm^{-1} (Figure 5.4) [40,41]. Coupling of C-N stretching as well as N-H in-plane bending modes (amide II) was noted at 1538 cm^{-1}. It was also found that the carbonyl stretching vibrations (amide I) of hydrogen bonded carbonyl groups, the peak seen at 1636 cm^{-1}, in the PA1010/HDPE/GNP nanocomposites displayed a reduction in the peak intensity in the case of PAPEGNP1 and PAPEGNP2, as compared to the PA1010/HDPE blend. However, for the PAPEGNP3 and PAPEGNP5 nanocomposites, an increase in the intensity of the peak

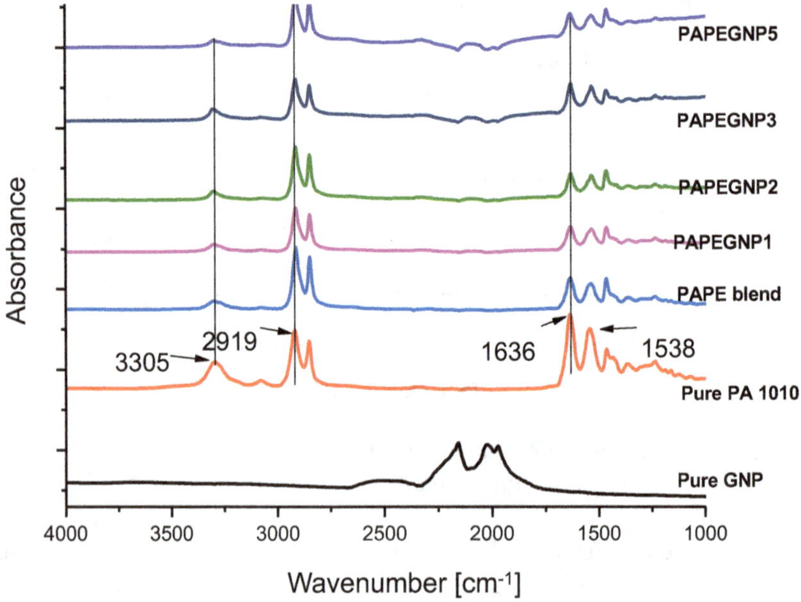

Figure 5.4 FTIR spectra of pristine GNPs, pure PA1010, PA1010/HDPE blend and PA1010/HDPE/GNP nanocomposites.

at 1636 cm^{-1} was observed, relative to the PA1010/HDPE blend. This was quantitatively examined by calculating the area ratio of this band (A_{1636}) relative to the C-H stretching vibrations observed at 2919 cm^{-1} (Table 5.4). A_{1636}/A_{2919} for PA1010/HDPE blend was observed to be lower as compared to the nanocomposites. It indicated that the particular amide bonds in the disordered amorphous phase of the PA1010/HDPE blend underwent chemical interaction

Table 5.4 A_{1636}/A_{2925} for pure PA1010, PA1010/HDPE blend and PA1010/HDPE/GNP nanocomposites

Material	A_{1636}	A_{2919}	A_{1636}/A_{2925}
Pure PA1010	0.476	0.368	1.29
PAPE blend	0.211	0.402	0.52
PAPEGNP1	0.181	0.295	0.61
PAPEGNP2	0.194	0.359	0.54
PAPEGNP3	0.318	0.356	0.89
PAPEGNP5	0.323	0.463	0.70

in the presence of GNP nanoplatelets.

Figure 5.5 demonstrates the WAXRD patterns of the materials. Pristine PA1010 exhibited two distinct diffractions peaks at 2θ = 24.5° and 20.7°, corresponding to the (110/010) and 100 planes, respectively of the α-phase crystal [42,43]. Two major diffraction peaks were seen in the XRD pattern of HDPE: a strong reflection at 21.9° and a less intensive peak at 24.5° 2θ, corresponding to the orthorhombic unit cell framework of (110) and (200) reflection

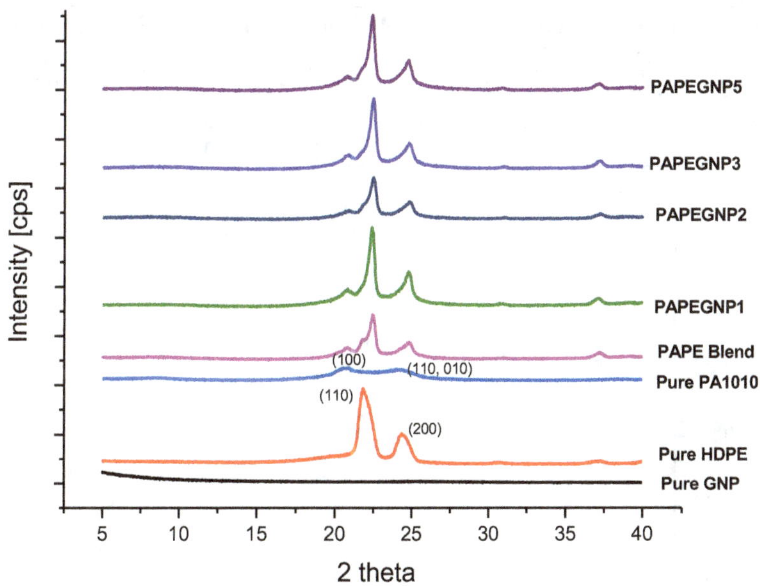

Figure 5.5 XRD patterns of pure PA1010, pure GNPs, PA1010/HDPE blend and PA1010/HDPE/GNP nanocomposites.

planes, respectively. For the PA1010/HDPE blend, the HDPE diffraction peak of (200) crystalline plane overlapped with the PA1010 diffraction peak of (110/010) at 2θ = 24.7°, and the HDPE diffraction peak of (110) crystal plane overlapped with the PA1010 diffraction peak of (100) at 2θ = 22.3°. For the PA1010/HDPE/GNP nanocomposites, the two diffraction peaks displayed higher values of 2θ, though there was no effect of the filler concentration on the peak positions. Further, the intensity of the diffraction peaks also showed a minor increment in the case of PA1010/HDPE/GNP nanocomposites, relative to the PA1010/HDPE blend. This indicated that the inclusion of GNP nanofiller to the PA1010/HDPE blend led to enhanced crystallinity.

The thermal conductivity of suspended graphene is almost 5000 W/m.K [44], thus, indicating the potential of conductivity improvement in polymers on the incorporating GNPs. The thermal conductivity of the blend and blend nanocomposites is depicted in Table 5.5 and Figure 5.6. The inclusion of GNPs was observed to enhance

Table 5.5 Thermal conductivity of PAPE blend and different PAPE nanocomposites

Temp.	PAPE Blend	PAPEGNP1	PAPEGNP2	PAPEGNP3	PAPEGNP5
20 °C	0.201	0.248	0.263	0.275	0.292
40 °C	0.181	0.219	0.230	0.244	0.260
60 °C	0.161	0.188	0.205	0.217	0.232

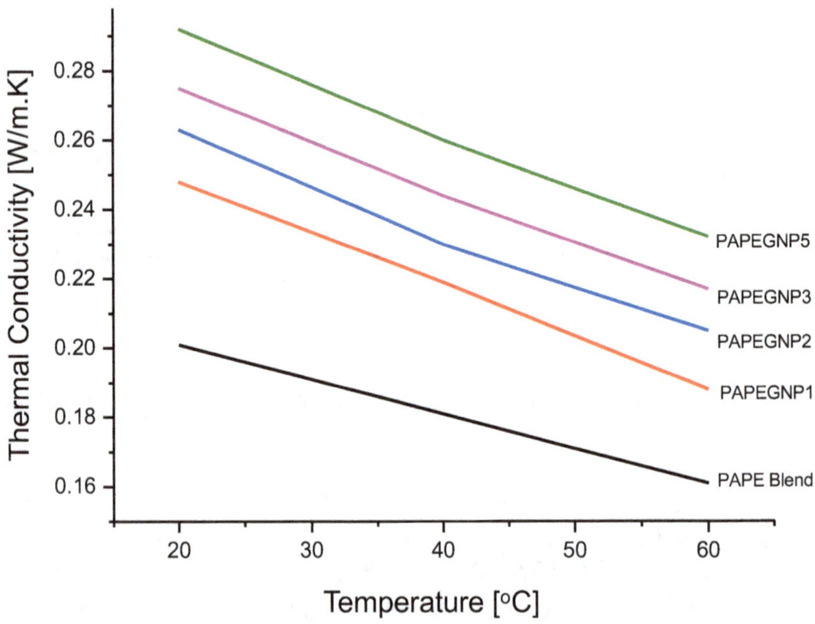

Figure 5.6 Thermal conductivity of PAPE blend and different PAPE nanocomposites.

the thermal conductivity of the PA1010/HDPE blend remarkably, as a function of filler concentration. At 60 °C, the nanocomposites with 3 wt% and 5 wt% GNP concentration exhibited the thermal conductivity values of 0.217 W/m.K and 0.232 W/m.K, respectively, indicating 35% and 44% increment, as compared to the HDPE/PA1010 blend.

Raman spectrum of pristine GNPs was collected in the range 1000-3000 cm^{-1}, as shown in Figure 5.7. The spectrum was observed to be similar to single crystal graphite [45]. The peak present at 1588 cm^{-1} was associated with the G band, whereas the peak at 1322 cm^{-1} corresponded to the D band. Also, the second-order Raman band (2D) was observed at 2656 cm^{-1}. HDPE spectrum presented typical Raman modes of PE (Figure 5.8). The bands noticed at approximately 1137 cm^{-1} and 1069 cm^{-1} were associated with the C-C stretching modes in the anisotropic and crystalline phases of the pure PE matrix. PA1010 spectrum comprised of the distinguishing

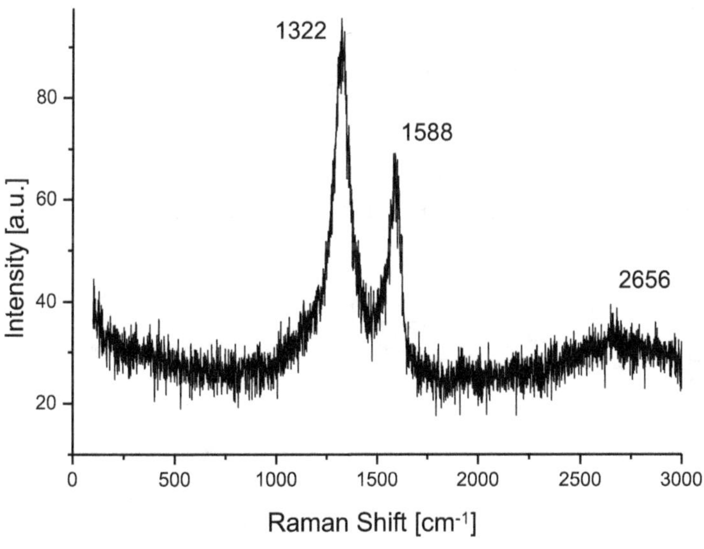

Figure 5.7 Raman spectrum of pure GNP.

bands of PA, in which the bands observed at 1304 cm^{-1} (amide III) and 1645 cm^{-1} (amide I) were related to the presence of the amide groups in the polymer matrix. The intensive bands observed at 2926 cm^{-1} and 2850 cm^{-1} corresponded to the asymmetric and symmetric CH_2 stretching vibrations, respectively. In the Raman spectra of the PA1010/HDPE/GNP nanocomposites (Figure 5.9), the bands were observed to be shifted to higher frequency, relative to the pristine GNPs. The aforementioned shift in the D and G bands confirmed the significant interaction between the GNPs and PA1010/HDPE blend. The analysis of the Raman I_D/I_G ratio (I_D – D-peak Raman intensity,

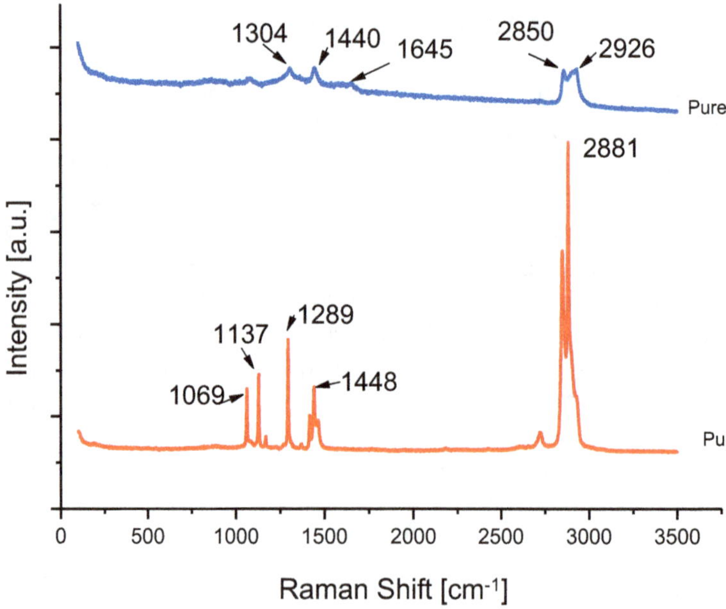

Figure 5.8 Raman spectra of pure HDPE and pure PA1010.

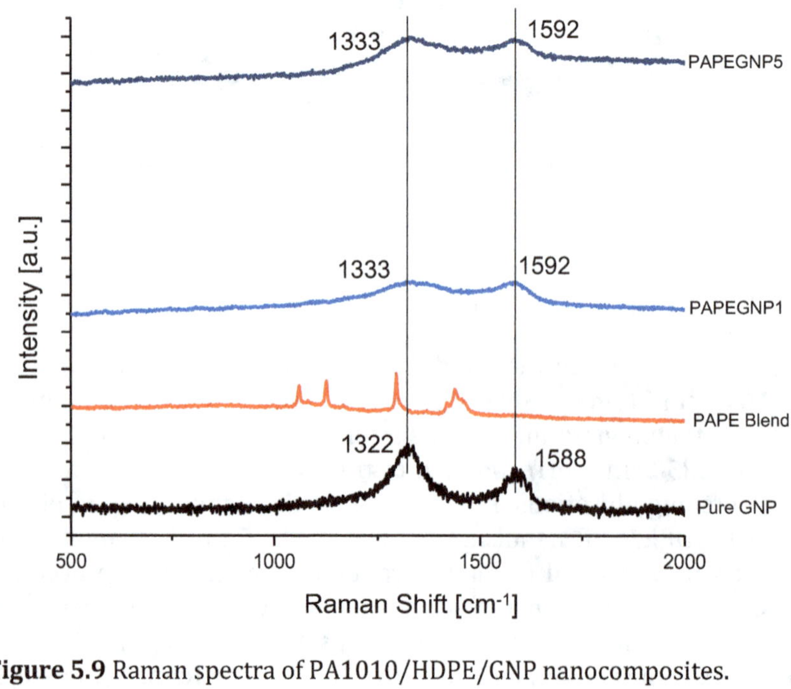

Figure 5.9 Raman spectra of PA1010/HDPE/GNP nanocomposites.

I_G – G-peak Raman intensity) of pure GNPs, PA1010/HDPE/1%GNP and PA1010/HDPE/5%GNP nanocomposites is detailed in Table 5.6. The value of I_D/I_G was greater in the case of PAPEGNP5 nanocomposite relative to PAPEGNP1. Higher I_D/I_G ratio indicated possible covalent interactions between the PA1010/HDPE blend and GNPs.

Table 5.6 Raman intensity ratio (I_D/I_G) of pristine GNPs and PA1010/HDPE nanocomposites

Material	D band (cm^{-1})	G band (cm^{-1})	I_D	I_G	Raman Intensity ratio (I_D/I_G)
Pure GNP	1322	1588	94	69	1.362
PAPEGNP1	1333	1592	3063	3055	1.002
PAPEGNP5	1333	1592	1591	1559	1.020

5.4 Conclusion

In this study, the mechanical, thermal properties and structural properties of polyamide1010 (PA1010)/high density polyethylene (HDPE) blend nanocomposites with graphene nanoplatelets (GNPs) have been evaluated. The incorporation of GNPs significantly enhanced the tensile modulus of the PA1010/HDPE blend, as a function of filler content. An increase of 31% in the modulus was observed for the composite with 5% filler content. The melting temperatures of the PA1010/HDPE blend and PA1010/HDPE/GNP nanocomposites did not indicate any significant change. A slight increment in the melt enthalpy was noted on GNPs addition, indicating enhanced crystallization in the composites. The A_{1636}/A_{2919} for the PA1010/HDPE blend was observed to be lesser as compared to the PA1010/HDPE/GNP composites. It indicated that the amide bonds in the disordered amorphous phase of the PA1010/HDPE blend underwent chemical interaction in the presence of GNP nanofiller. The XRD results also confirmed qualitatively the increased crystallinity on incorporating GNPs. Additionally, the presence of GNPs improved the thermal conductivity of the PA1010/HDPE blend, as a function of filler content. The Raman spectra of the PA1010/HDPE/GNP composites exhibited the shifting of the bands to higher frequency, relative to the pristine GNPs. The developed blend composites represent high value nanomaterials with a strong potential of use in a wide range of engineering applications.

References

1. Holsti-Miettinen, R., Seppälä, J., and Ikkala, O. T. (1992) Effects of compatibilizers on the properties of polyamide/polypropylene blends. *Polymer Engineering and Science*, **32**(13), 868-877.
2. La Mantia, F. P. (1993) Blends of polypropylene and nylon 6: Influence of the compatibilizer, molecular weight, and processing conditions. *Advances in Polymer Technology*, **12**(1), 47-59.
3. *Micro and Nanostructured Multiphase Polymer Blend Systems: Phase Morphology and Interfaces*, Harrats, C., Thomas, S., and Groeninckx, G. (eds.), CRC Press, USA (2005).
4. Jordan, J., Jacob, K. I., Tannenbaum, R., Sharaf, M. A., and Jasiuk, I. (2005) Experimental trends in polymer nanocomposites - A review. *Materials Science and Engineering A*, **393**(1), 1-11.
5. Bordes, P., Pollet, E., and Avérous, L. (2009) Nano-biocomposites: biodegradable polyester/nanoclay systems. *Progress in Polymer Science*, **34**(2), 125-155.
6. Carraher, Jr., C. E., Morrison, A., Roner, M. R., Moric-Johnson, A., Al-Huniti, M. H., and Miller, L. (2015) Metallocene-containing polyesters from reaction of 3, 5-pyridinedicarboxylic acid and metallocene dihalides and their preliminary ability to inhibit cancer cell growth. *Journal of the Chinese Advanced Materials Society*, **3**(4), 310-327.
7. Jia, Y., He, H., Peng, X., Meng, S., Chen, J., and Geng, Y. (2017) Preparation of a new filament based on polyamide-6 for three-dimensional printing. *Polymer Engineering and Science*, **57**(12), 1322–1328
8. Kausar, A. (2016) Polycarbonate/polypropylene-graft-maleic anhydride and nano-zeolite-based nanocomposite membrane: Mechanical and gas separation performance. *Advances in Materials Science*, **16**(4), 17-28.
9. Marchildon, K. (2011) Polyamides–still strong after seventy years. *Macromolecular Reaction Engineering*, **5**(1), 22-54.
10. Ma, L., Niu, H., Cai, J., Zhao, P., Wang, C., Lian, Y., Bai, X., and Wang, W. (2014) Optical, electrochemical, photoelectrochemical and electrochromic properties of polyamide/graphene oxide with various feed ratios of polyamide to graphite oxide. *Journal of Materials Chemistry C*, **2**(12), 2272-2282.
11. Zheng, D., Tang, G., Zhang, H. B., Yu, Z. Z., Yavari, F., Koratkar, N., Lim, S. H., and Lee, M. W. (2012) In situ thermal reduction of graphene oxide for high electrical conductivity and low percolation threshold in polyamide 6 nanocomposites. *Composites Science and Technology*, **72**(2), 284-289.
12. Zhishen, M., Qingbo, M., Jinhua, F., Hongfang, Z., and Donglin, C. (1993) Crystal structure and thermodynamic parameters of nylon-

1010. *Polymer International*, **32**(1), 53-60.
13. Zhang, S. L., Wang, G. B., Jiang, Z. H., Wang, D., Ma, R. T., and Wu, Z. W. (2005) Impact properties, phase structure, compatibility, and fracture morphology of polyamide-1010/thermoplastic poly (ester urethane) elastomer blends. *Journal of Polymer Science, Part B: Polymer Physics*, **43**(10), 1177-1185.
14. Lu, Z., Wu, A., Ou, X., Zhang, S., Niu, J., Ji, S., and Ling, Y. (2017) Enhanced anti-aging and mechanical properties of polyamide 1010 by sol-hydrothermal synthetic titanium dioxide-coated kaolinite addition. *Journal of Alloys and Compounds*, **693**, 381-388.
15. Gonzalez-Montiel, A., Keskkula, H., and Paul, D. R. (1995) Impact-modified nylon 6/polypropylene blends: 1. Morphology-property relationships. *Polymer*, **36**(24), 4587-4603.
16. Mittal, V. (2008) Mechanical and gas permeation properties of compatibilized polypropylene–layered silicate nanocomposites. *Journal of Applied Polymer Science*, **107**(2), 1350-1361.
17. Bréchet, Y., Cavaille, J. Y., Chabert, E., Chazeau, L., Dendievel, R., Flandin, L., and Gauthier, C. (2001) Polymer based nanocomposites: Effect of filler-filler and filler-matrix interactions. *Advanced Engineering Materials*, **3**(8), 571-577.
18. Favier, V., Cavaille, J. Y., Canova, G. R., and Shrivastava, S. C. (1997) Mechanical percolation in cellulose whisker nanocomposites. *Polymer Engineering and Science*, **37**(10), 1732-1739.
19. Carretero-González, J., Retsos, H., Verdejo, R., Toki, S., Hsiao, B. S., Giannelis, E. P., and López-Manchado, M. A. (2008) Effect of nanoclay on natural rubber microstructure. *Macromolecules*, **41**(18), 6763-6772.
20. Wang, K., Liang, S., Deng, J., Yang, H., Zhang, Q., Fu, Q., Dong, X., Wang, D., and Han, C. C. (2006) The role of clay network on macromolecular chain mobility and relaxation in isotactic polypropylene/organo-clay nanocomposites. *Polymer*, **47**(20), 7131-7144.
21. Kausar, A., Anwar, Z., Khan, L. A., and Muhammad, B. (2017) Functional graphene nanoplatelet reinforced epoxy resin and polystyrene-based block copolymer nanocomposite. *Fullerenes, Nanotubes and Carbon Nanostructures*, **25**(1), 47-57.
22. Kulkarni, H., Tambe, P., and Joshi, G. (2017) High concentration exfoliation of graphene in ethyl alcohol using block copolymer surfactant and its influence on properties of epoxy Nanocomposites. *Fullerenes, Nanotubes and Carbon Nanostructures*, **25**(4), 241-249.
23. Wang, F., Drzal, L. T., Qin, Y., and Huang, Z. (2016) Enhancement of fracture toughness, mechanical and thermal properties of rubber/epoxy composites by incorporation of graphene nanoplatelets. *Composites, Part A: Applied Science and Manufacturing*, **87**, 10-22.
24. Wang, F., Drzal, L. T., Qin, Y., and Huang, Z. (2015) Mechanical properties and thermal conductivity of graphene nanoplate-

let/epoxy composites. *Journal of Materials Science*, **50**(3), 1082-1093.

25. Ma, P. C., Siddiqui, N. A., Marom, G., and Kim, J. K. (2010) Dispersion and functionalization of carbon nanotubes for polymer-based nanocomposites: a review. *Composites, Part A: Applied Science and Manufacturing*, **41**(10), 1345-1367.
26. Cui, L. J., Wang, Y. B., Xiu, W. J., Wang, W. Y., Xu, L. H., Xu, X. B., Meng, Y., Li, L.Y., Gao, J., Chen, L. T., and Geng, H. Z. (2013) Effect of functionalization of multi-walled carbon nanotube on the curing behavior and mechanical property of multi-walled carbon nanotube/epoxy composites. *Materials and Design*, **49**, 279-284.
27. Anwar, Z., Kausar, A., Khan, L. A., and Muhammad, B. (2016) Modified graphene nanoplatelet and epoxy/block copolymer-based nanocomposite: physical characteristic and EMI shielding studies. *Nanocomposites*, **2**(3), 141-151.
28. Paszkiewicz, S., Szymczyk, A., Livanov, K., Wagner, H. D., and Rosłaniec, Z. (2015) Enhanced thermal and mechanical properties of poly (trimethylene terephthalate-block-poly (tetramethylene oxide) segmented copolymer based hybrid nanocomposites prepared by in situ polymerization via synergy effect between SWCNTs and graphene nanoplatelets. *Express Polymer Letters*, **9**(6), 509–524.
29. Terrones, M., Martín, O., González, M., Pozuelo, J., Serrano, B., Cabanelas, J. C., Vega-Díaz, S. M., and Baselga, J. (2011) Interphases in graphene polymer-based nanocomposites: Achievements and challenges. *Advanced Materials*, **23**(44), 5302-5310.
30. Utracki, L. A., Dumoulin, M. M., and Toma, P. (1986) Melt rheology of high density polyethylene/polyamide-6 blends. *Polymer Engineering and Science*, **26**(1), 34-44.
31. Chen, C. C., Fontan, E., Min, K., and White, J. L. (1988) An investigation of instability of phase morphology of blends of nylons with polyethylenes and polystyrenes and effects of compatibilizing agents. *Polymer Engineering and Science*, **28**(2), 69-80.
32. Hobbs, S. Y., Bopp, R. C., and Watkins, V. H. (1983) Toughened nylon resins. *Polymer Engineering and Science*, **23**(7), 380-389.
33. Willis, J. M., and Favis, B. D. (1988) Processing-morphology relationships of compatibilized polyolefin/polyamide blends. Part I: The effect of an ionomer compatibilizer on blend morphology. *Polymer Engineering and Science*, **28**(21), 1416-1426.
34. Koulouri, E. G., Georgaki, A. X., and Kallitsis, J. K. (1997) Reactive compatibilization of aliphatic polyamides with functionalized polyethylenes. *Polymer*, **38**(16), 4185-4192.
35. Holsti-Miettinen, R. M., Perttilä, K. P., Seppälä, J. V., and Heino, M. T. (1995) Oxygen barrier properties of polypropylene/polyamide 6 blends. *Journal of Applied Polymer Science*, **58**(9), 1551-1560.

36. Pötschke, P., Göldel, A., Gültner, M., and Liebscher, M. (2014) Localization Behavior of Carbon nanotubes and Graphene Nanoplatelets in Melt-Mixed Immiscible Polymer Blends, *POLYCHAR 22 World Forum on Advanced Materials*, South Africa.
37. Cao, Y., Zhang, J., Feng, J., and Wu, P. (2011) Compatibilization of immiscible polymer blends using graphene oxide sheets. *ACS Nano*, **5**(7), 5920-5927.
38. Mohamadi, S., and Sharifi-Sanjani, N. (2011) Investigation of the crystalline structure of PVDF in PVDF/PMMA/graphene polymer blend nanocomposites. *Polymer Composites*, **32**(9), 1451-1460.
39. Mittal, V., Chaudhry, A. U., and Luckachan, G. E. (2014) Biopolymer–thermally reduced graphene nanocomposites: structural characterization and properties. *Materials Chemistry and Physics*, **147**(1), 319-332.
40. Yoshioka, Y., Tashiro, K., and Ramesh, C. (2003) Structural change in the Brill transition of Nylon m/n (2) conformational disordering as viewed from the temperature-dependent infrared spectral measurements. *Polymer*, **44**(20), 6407-6417
41. Yoshioka, Y., and Tashiro, K. (2003) Structural change in the Brill transition of Nylon m/n (1) Nylon 10/10 and its model compounds. *Polymer*, **44**(22), 7007-7019.
42. Xu, Z., and Gao, C. (2010) In situ polymerization approach to graphene-reinforced nylon-6 composites. *Macromolecules*, **43**(16), 6716-6723.
43. Li, Y., and Yan, D. (2001) Preparation, characterization and crystalline transition behaviors of polyamide 4 14. *Polymer*, **42**(11), 5055-5058.
44. Lehman, J. H., Terrones, M., Mansfield, E., Hurst, K. E., and Meunier, V. (2011) Evaluating the characteristics of multiwall carbon nanotubes. *Carbon*, **49**(8), 2581-2602.
45. Tuinstra, F., and Koenig, J. L. (1970) Raman spectrum of graphite. *The Journal of Chemical Physics*, **53**(3), 1126-1130.

6

Effect of Molecular Structure on Mechanical, Thermal, Rheological and Morphological Properties of High Molecular Weight Polyethylene

6.1 Introduction

Polyethylene (PE) is one of the most prominent polyolefins with large number of applications due to its low cost, good processability and ease of recycling. In addition, it imparts properties to the products, such as non-toxicity, biocompatibility, and resistance against chemicals [1]. For instance, Figure 6.1 shows the height, amplitude and phase images of a high molecular weight PE material at two different magnifications; top row represents higher magnification, whereas the lower row presents the images taken at lower magnification [2]. Homogenous ultrastructure, as demonstrated in height and amplitude images, is clearly evident, which also results in relatively uniform mechanical properties (e.g., hardness and viscoelasticity), as indicated in the phase images. Furthermore, as observed in the higher magnification phase image, the crystalline domains in the polymer are narrow in size distribution and are dispersed uniformly, which lead to the observed polymer behavior.

The overall set of PE properties drives its numerous applications, however, to meet the rising challenges related to product specifications and requirements, it is required to further improve/tune the performance of the polyethylene matrix. Among different methodologies useful for this purpose, tuning the molecular characteristics during polymer synthesis represents a fundamental and effective approach. Recently, Ruff *et al.* [3-5] reported extensively about controlled polymerization of high molecular weight PE, especially using in-reactor blending approaches. In another study, Mittal *et al.* [6] synthesized polyethylene samples with different molecular weights, using coordination polymerization, to study the effect of the variation in molecular weight on the properties of the polymers as well as the

Gisha E. Luckachan[a], Gigi George[a], Christian Paulik[b] and Vikas Mittal[a]*
[a]Khalifa University of Science and Technology, Abu Dhabi, UAE; [b]Johannes Kepler University, Linz, Austria
**Current address: CMS College, Mahatma Gandhi University, Kerala, India*

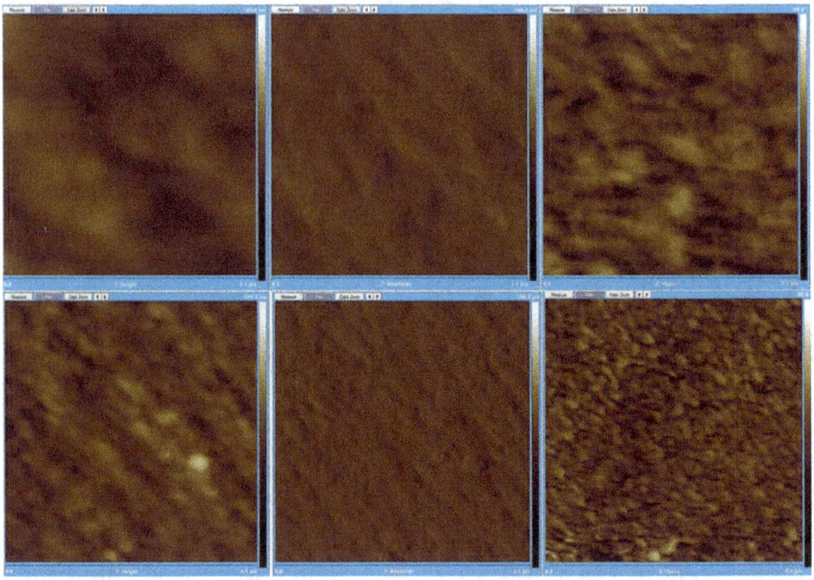

Figure 6.1 Atomic force microscopic height, amplitude, and phase images of pure polyethylene at higher (upper row) and lower (lower row) magnifications. Reproduced from Reference 2 with permission from Sage Publishers.

nanocomposites generated using these. The synthesized polymers exhibited similar crystal structure, however, the degree of crystallinity was observed to significantly decrease on increasing the polymer molecular weight, thereby indicating the hindrance to crystallization due to the extensive entanglements in the longer polymer chains. Due to a higher extent of crystallinity, the polymer with the lowest molecular weight was also observed to demonstrate a much higher tensile modulus.

A further comprehensive analysis of the structure-property correlations and the effect of polyethylene chains of different molecular weights is needed, which would be beneficial for the selection of polymer materials with specific molecular characteristics for corresponding applications. Thus, in the current study, high molecular weight polyethylene materials of different molecular weights and molecular weight distributions were analyzed for their structural properties using a multitude of characterization methods such as tensile testing, thermogravimetry, scanning calorimetry, diffraction, scanning electron microscopy, scattering, nuclear magmatic resonance, etc.

6.2 Experimental

6.2.1 Materials

PE samples with different molecular weights (PE1-PE4) were synthesized using the synthesis protocol reported earlier [3-5]. All samples were generated using Ziegler Natta catalyst systems with TEA as co-catalyst. Molecular weight of the samples was calculated using Mark-Houwink parameters from intrinsic viscosity measurements. PE1-PE3 were unimodal in molecular weight distributions, whereas PE4 had two molecular weight fractions, thus, bimodal in nature. The samples were obtained as fine white powders. The polymer samples were not further stabilized by the addition of stabilizers and were used as synthesized. Table 6.1 reports the peak molecular weights and nature of molecular weight distributions of the PE samples. The test specimens for mechanical, thermal and structural characterization were prepared using mini injection molding machine (HAAKE MiniJet PRO, Thermo Scientific) with a cylinder temperature of 180 °C, mold temperature of 125 °C, injection pressure of 430 bar for 10 s and post pressure of 500 bar for 6 s.

Table 6.1 Molecular weight distribution and average molecular weight of PE samples

Sample	molecular weight distribution	M_w (kg/mol)
PE1	unimodal	175
PE2	unimodal	252
PE3	unimodal	1000
PE4	bimodal	763

6.2.2 Characterization Methods

Differential scanning calorimetric (DSC) analysis of the PE samples was carried out under nitrogen atmosphere using Discovery Series DSC of TA Instruments. 3-8 mg amount of the samples was scanned at a rate of 10 °C/min from -60 to 200 °C and then 200 to 25 °C. The second heating and cooling represented the melting and crystallization thermograms respectively.

The thermal stability and decomposition characteristics of the PE samples were analyzed using thermogravimetric analyzer (TGA)

from Discovery Series of TA Instruments under nitrogen atmosphere. 3-8 mg amount of the polymer samples was heated from 25 to 700 °C at a rate of 10 °C/min and the mass loss was recorded as a function of temperature.

Tensile properties of the polymer samples were performed at room temperature using universal testing machine Instron 3345 with a crosshead speed of 10 mm/min according to ASTM D638. The results represented the average of 5 dumbbell shaped specimens. Melt rheological properties such as viscosity (η'), elasticity (η''), storage modulus (G') and loss modulus (G'') of PE samples were measured on AR 2000ex Rheometer from TA Instruments in a shear mode, according to ASTM D4440. Dynamic frequency sweep curves were obtained from 0.1 to 100 rad/s at 2% strain using disc shaped samples of 25 mm diameter and 1.5 mm thickness. The analysis was carried out at 180 °C in parallel plate geometry of 25 mm diameter with a geometric gap of 1.4 mm.

Small-angle X-ray scattering (SAXS) was performed on Bruker Nanostar from Bruker Instruments GmbH. SAXS analysis was performed at smaller diffraction angles, from $2\theta = 0.01°$ to 3°. Specimens were prepared by hot pressing and had <5 mm thickness.

X-ray diffraction measurements of the PE powders were carried out using an analytical powder diffractometer (X'Pert PRO) using Cu Kα radiation (λ= 1.5406 Å, 45 kV and 40 mA). Diffractograms of the samples were recorded in the range of 5-65° 2θ scale, with a step size of $0.02°.s^{-1}$. Integrated WAXS line width was used to calculate the crystallite size using the Scherrer equation. Degree of crystallinity were calculated by the following equation:

$$\%\ \text{crystallinity} = A_{cr} / (A_{cr} + A_{am})$$

where A_{cr} indicates area under the peak due to crystalline phase and A_{am} represents the area under the peak due to amorphous phase [7].

Solid state 13C NMR measurements on PE samples were carried out on a Bruker 400 MHz spectrometer at the resonance frequency of 100.628 MHz for 13C and 400.133 MHz for 1H. A 4 mm double resonance cross polarization magic angle spinning probe was used for the CP MAS experiments. Around 100 mg of sample was tightly packed into a 4 mm diameter cylindrical zirconium oxide rotor with a Kel-F end cap. 13C CP/MAS NMR spectrum was acquired with a cross-polarization contact time of 2 ms and a recycle delay of 5 s at a MAS speed of 10 kHz. Spin-lattice relaxation time (T1C) measurements were conducted by inversion recovery pulse sequences developed by Torchia and T1H was determined using saturation recovery pulse sequences [8,9]. All experiments were carried out at ambient probe temperature (25 °C) with high-power proton decoupling during acquisition. All 13C NMR spectra were assigned using adamantane as an external reference. LabRAM HR spectrometer (Horiba Jobin Yvon) was used to record the Raman spectra of the PE samples. For this purpose, laser light of 633 nm wavelength using He/Ne laser was used for excitation. A long working distance objective with magnification 50× was used both to collect the scattered light and to focus the laser beam on the sample surface. CCD (charge-coupled device) detector exposure time was 5 s and an average of 2 cycles was used to increase S/N ratio. Raman shifts were calibrated using Si wafer signal at 520.7 cm^{-1}.

Scanning electron microscope (SEM) (FEI Quanta, FEG250, USA) at accelerating voltages of 5-10 kV was used to study the microstructure of the PE samples. Prior to analysis, the sample surface was sputter-coated with a thin carbon layer.

6.3 Results and Discussion

Wide angle X-ray diffraction (WAXRD) analysis was conducted on the polyethylene samples to probe the crystal phase structures. The X-ray diffractograms shown in Figure 6.2a exhibit two prominent peaks at 2 theta 21° and 23° corresponding to orthorhombic (110) and (200) reflections respectively [10,11]. It suggested that the predominant crystalline form of all PE samples was orthorhombic form. The contribution from monoclinic (001) reflection could also observed at 2 theta 19° as small hump. A deconvoluted X-ray pattern shown in Figure 6.2b clearly demonstrated the contributions from non-crystalline phase along with orthorhombic and monoclinic phases. The different polymerization conditions used to generate the PE samples, thus, didn't change the peak positions of orthorhombic reflections. Table 6.2 shows the degree of crystallinity and crystallite sizes of PE samples calculated from the X-ray diffractograms. PE1 was observed to exhibit crystallinity of 66.9%, which enhanced to 76.4% for the PE2 sample. The PE3 material with the highest molecular weight exhibited the lowest degree of crystallinity, due to expected hindrance to chain packing in the case of longer polymer chains. Similar phenomena have also been reported earlier in literature [6]. The bimodal PE4 sample exhibited the percent crystallinity of 72.4, which was an aggregate of the two different molecular weight fractions in the polymer sample.

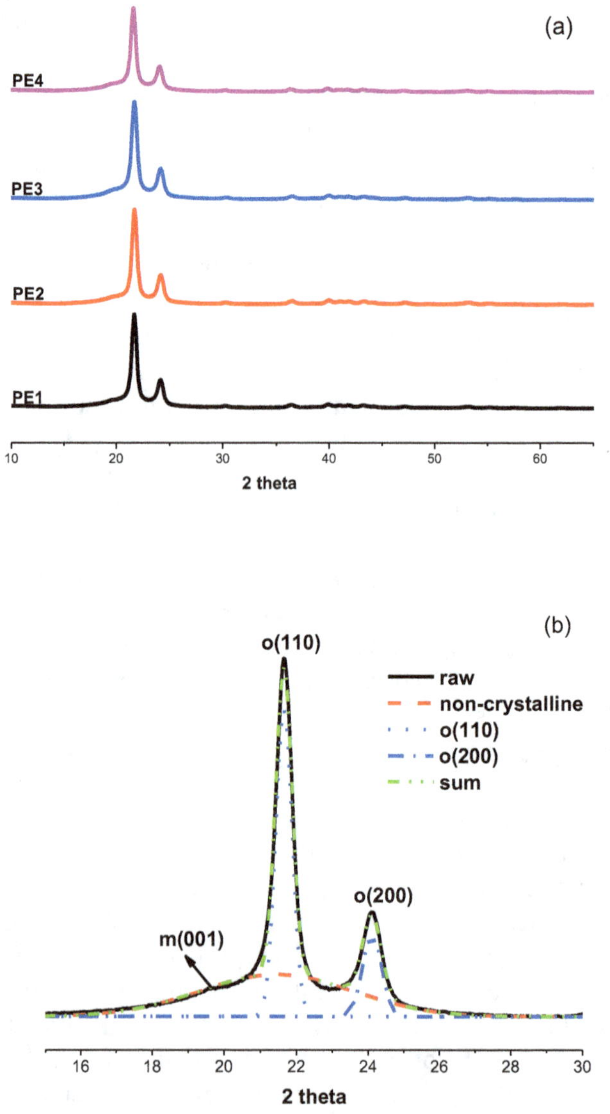

Figure 6.2 (a) WAXRD patterns of PE samples, (b) deconvolution of the pattern of PE1 sample into three different contributions from non-crystalline, orthorhombic crystalline and monoclinic phases.

Table 6.2 Degree of crystallinity and crystallite size obtained from WAXD patterns

Polymer	XRD crystallinity (%)	crystallite size (nm)
PE1	66.9	14.87
PE2	76.4	14.59
PE3	70.8	13.20
PE4	72.4	14.76

Table 6.3 also presents the calorimetric properties of PE samples, as measured using DSC. In addition, Figure 6.3 shows the melting and crystallization thermograms of the PE materials. The peak melting point of PE1-PE3 materials was observed to marginally enhance with molecular weight, whereas the peak crystallization temperatures exhibited a slight reduction on increasing molecular weight. For calculating the degree of crystallinity, the melt enthalpy of 293 J/g for pure crystalline high density polyethylene was taken [12]. The degree of crystallinity of the PE samples exhibited the same trend as observed from the WAXRD analysis, though the two techniques indicated some differences in the magnitude of the crystallinity values. Owing to the higher molecular weight, the degree of crystallinity of PE3 was measured to be 70%, which was the lowest among the PE samples. In addition, the lamella thickness was calculated using the Gibbs-Thomson equation [13]:

Table 6.3 Calorimetric properties of PE samples

Polymer	peak melting point, T_m (°C)	melt enthalpy, ΔH_m (J/g)	crystallinity (%)	peak crystallization temperature, T_c (°C)	crystallization enthalpy, ΔH_c (J/g)	lamella thickness (nm)
PE1	135	215	73	119	224	21.4
PE2	136	222	76	118	229	23.6
PE3	137	205	70	117	207	26.3
PE4	136	217	74	117	219	23.6

$$lamella\ thickness, l = cf * \left(\frac{2\sigma}{\Delta h_m}\right) * \frac{T_m^o}{T_m^o - T_m}$$

According to [14], cf is the correction factor and is equal to 0.87, σ is the surface free energy of the basal plane and equals 9.00 x 10^{-2} J/m^2, T_m^o is the equilibrium melting point of an infinite crystal and is equal to 418.7 K and $\triangle h_m$ is the enthalpy of fusion per unit volume and equals 2.86 x 10^8 J/m^3. The lamella thickness of the materials varied slightly with the melting point. Overall, it was confirmed that the molecular characteristics had an impact on the polymer crystallinity as well as lamella dimensions, which, in turn, would affect the physical properties of the polymers.

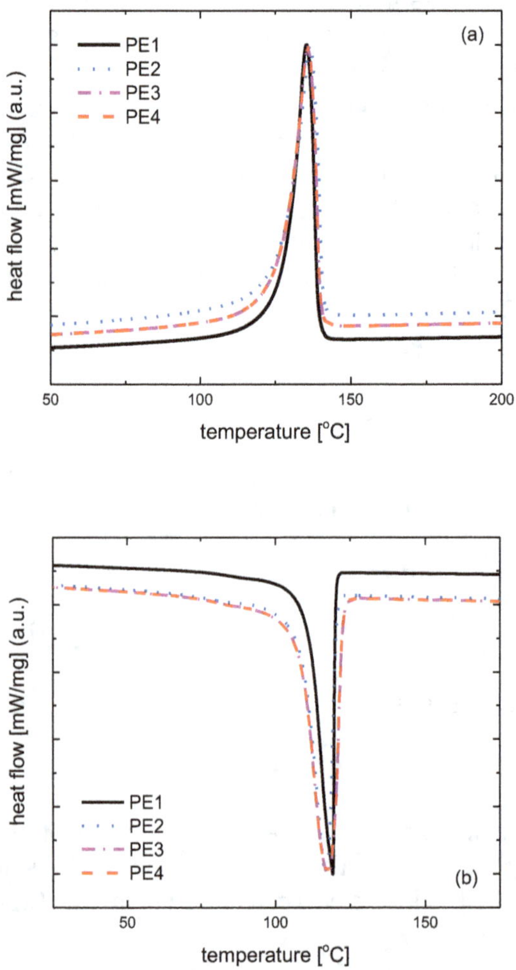

Figure 6.3 DSC (a) melting and (b) crystallization thermograms of PE materials.

Figure 6.4 also illustrates the SAXS runs for all the samples. As evident from Figure 6.4, all samples exhibited similar trend. In addition, the SAXS spectra exhibited lack of any visible peaks (i.e. q_{peak}). In a recent study, a weak signal indicating q_{peak} of 0.027027 Å^{-1} was reported for the similar PE samples [15]. Using similar q_{peak} value for calculating long period and lamella thickness, similar trend of lamella thickness as DSC was observed, though the values were lower in magnitude.

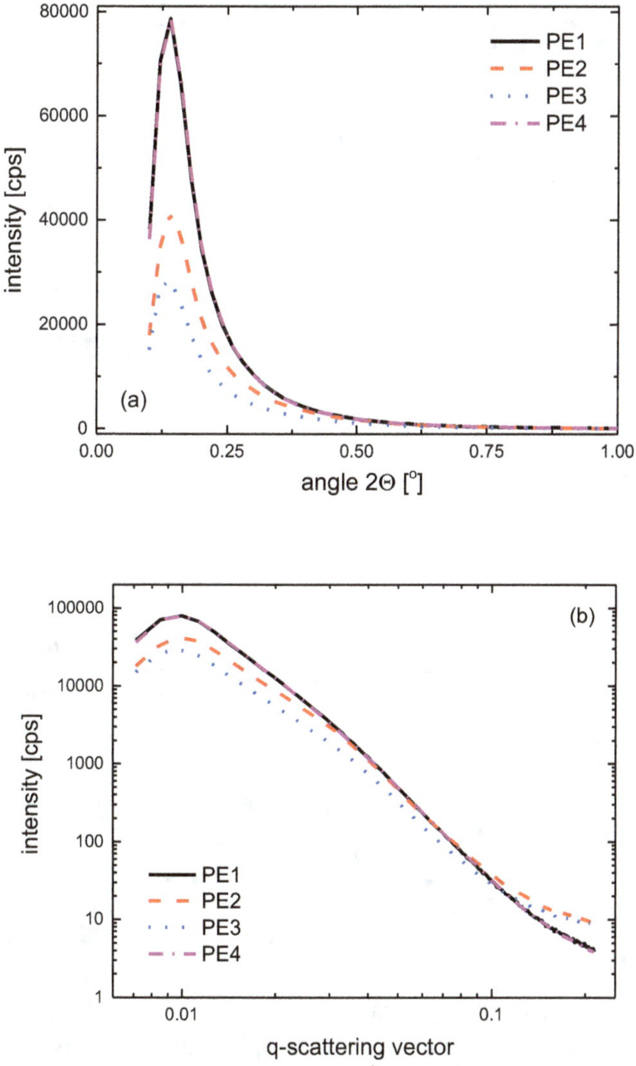

Figure 6.4 Small-angle X-ray scattering runs of the PE materials.

Figure 6.5a shows the CP MAS NMR spectra of unimodal and bimodal PEs. All spectra exhibited a single prominent peak at 33.0 ppm with two shoulder peaks centered at 34.4 ppm and 31.3 ppm. In order to clearly identify the shoulder peaks, the PE1 spectrum was deconvoluted in terms of three Lorentzians centered at 34.4 ppm, 33.0 ppm

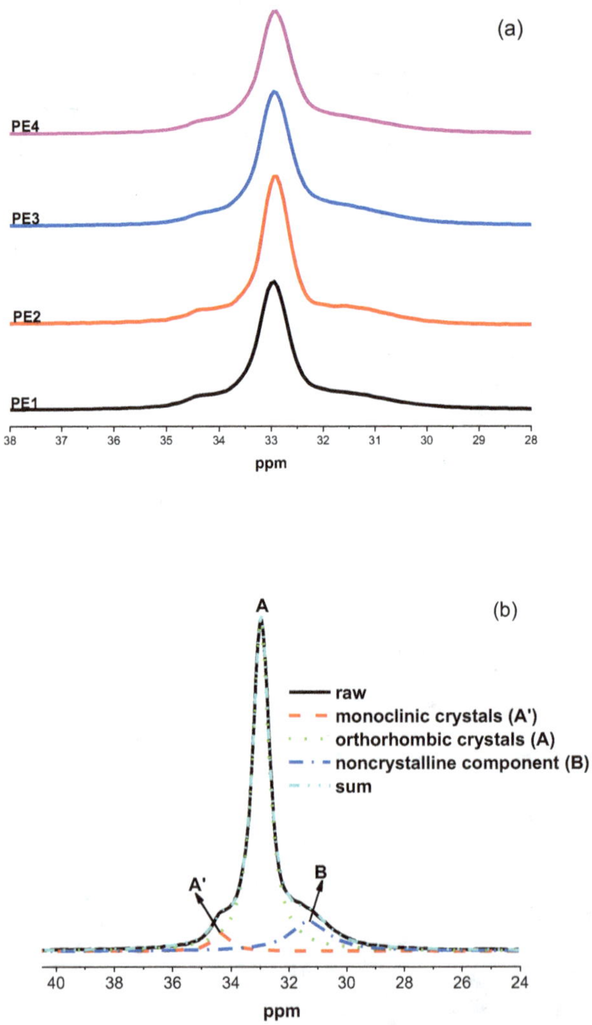

Figure 6.5 (a) 13C CP-MAS NMR spectra of PE samples, (b) component analysis of the total spectrum of PE1 sample. (A) crystalline component at 33.0 ppm (orthorhombic crystals); (A') crystalline component at 34.4 ppm (monoclinic crystals); (B) non-crystalline component at 31.3 ppm.

and 31.3 ppm by least-squares fitting [9]. The corresponding curves of the three components were observed to have good agreement with the polymer spectrum (Figure 6.5b). This, thus, confirmed the assignments of the three different components in the samples. Earl and VanderHart [16] assigned these peaks by spin-lattice relaxation time measurements; peak A (33.0 ppm) results from the *trans* zigzag architecture in the crystalline component, peak B (31.3 ppm) results from the presence of methylene carbons in the non-crystalline component, which are observed to undergo rapid transitions between different conformers. Further, the high frequency peak A' at 34.4 ppm is attributed to the presence of methylene carbons in the monoclinic crystal region [16-18]. The less resolved non-crystalline line at 31.3 ppm in comparison with the crystalline line at 33.0 ppm indicates the high degree of crystallinity of PE samples irrespective of the molecular characteristics. Table 6.4 also shows the mass fractions of each component in the unimodal and bimodal PEs measured from the integrated intensity fractions of each component in the spectra in Figure 6.5. The total content of the crystalline component calculated from the CP MAS NMR spectra agreed well with the degree of crystallinity determined using XRD analysis. Total fraction of crystalline

Table 6.4 The integrated intensity fractions of respective components calculated from CP MAS NMR spectra

Polymer	peak A (33.0 ppm) (%)	peak A' (34.4 ppm) (%)	peak B (31.3 ppm) (%)	total fraction of crystalline component (A & A') (%)
PE1	57.5	11.0	31.3	68.6
PE2	67.8	11.0	21.2	78.8
PE3	64.0	9.2	26.8	73.2
PE4	63.7	10.4	25.9	74.1

components of PE2 was higher than PE1. The fraction for PE3 and PE4 was observed to be in between PE1 and PE2, as observed earlier from WAXRD and DSC studies. It was observed that PE1 and PE2 contained the same fraction of monoclinic crystalline components. On the other hand, a higher fraction of orthorhombic phase was observed in PE2 than PE1 sample. Both amorphous and monoclinic fractions transferred to orthorhombic phase in the bimodal PE. The monoclinic crystal phase decreased and non-crystalline phase increased for PE3 sample as compared with PE4 sample. Furthermore, analysis of T1C data was performed on the PE samples to analyze the different phases in the orthorhombic crystal form by using inversion recovery pulse sequences [8,19]. The T1C values obtained by the least-squares fitting of decay curve are listed in Table 6.5. It could be observed that the crystalline component at 33.0 ppm exhibited three different T1C's. The unimodal PE1 exhibited a longest T1C of orthorhombic crystals, although it was less as compared with the previous literature reports on UHMWPE films (1000-3000 s) [9,20]. It might be associated with the small crystallite size of nearly 15 nm. However, the T1C of PE2 sample, which exhibited highest crystallinity among all PEs, exhibited a lower value. It can be explained based on the crystallite size which was lower than PE1 sample due to the probable fragmentation of PE2 during the polymerization process [3,4]. The T1C of PE3 and PE4 were higher than the unimodal PE2 sample, though the crystallite size of PE3 was lesser than PE2 which indicated that the crystallites of these PEs resided in stable positions which restricted the free mobility of molecular chains. Average proton T1 relaxation times of all PE samples measured by saturation recovery pulse sequences are also presented in Table 6.5. The PE1 sample containing more amorphous content exhibited the highest T1H value compared with PE2 sample, which had the lowest amorphous content. Such a contradictory behavior indicated the possibility of proton spin

Table 6.5 T1C and T1H spin lattice relaxation time (in seconds) data of all PE samples

	$T1_{C1(orthorhombic)}$	$T1_{C2\ (orthorhombic)}$	$T1_{C3\ (orthorhombic)}$	$T1_H$
PE1	2.02±0.4	101.83±10	758.15±54	1.69
PE2	1.40±0.2	58.31±6	612.94±28	1.58
PE3	1.40±0.2	85.53±8	651.32±49	1.46
PE4	1.62±0.3	93.15±5	686.43±35	1.59

diffusion between the non-crystalline and crystalline components. It could be correlated with the crystallize sizes of each polymer. As the crystallite size increased, an increase in T1H values occurred. Therefore, the PE3 sample having the lowest crystallite size exhibited a lower T1H and PE1 sample having highest crystallite size had a higher T1H value.

The "fingerprint" region from 950 cm^{-1} -1550 cm^{-1} in the Raman spectra of the PE samples is presented in Figure 6.6. The bands at 1060 cm^{-1} as well as 1130 cm^{-1} could be attributed to the asymmetric

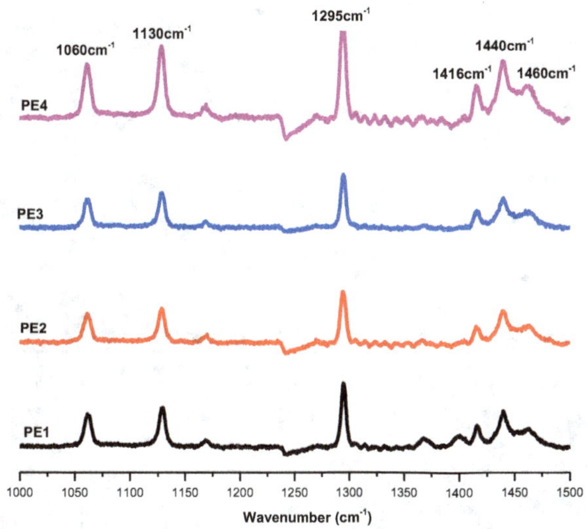

Figure 6.6 "Finger print" region of Raman spectra of PE samples.

and symmetric C-C stretching vibrations. It included all trans C-C bonds located in both crystalline and amorphous phases. The main spectral feature at 1295 cm^{-1} was represented by the twisting vibrations of the unit -CH_2-, this band mainly arising from the crystalline part of the polymeric structure. Pezzottie [21] and Hahn *et al.* [22] reported a contribution of C-C vibrational bands from the amorphous part of the polymeric structure at 1080 cm^{-1} [15,16]. Such a band in the polyethylene samples was absent indicating their high crystallinity. Also, the content of amorphous part was too low for the accurate morphological analyses. The characteristic vibrations in the region between 1350 cm^{-1} -1500 cm^{-1} were attributed to the -CH_2- bending mode. Here, the first band at 1416 cm^{-1} was attributed to crystalline phase and the other two bands at 1440 cm^{-1} and 1460 cm^{-1} were assigned to the amorphous phases [23]. Both unimodal and bimodal polyethylene exhibited all characteristic spectral features without any changes in the peak positions.

The PE samples were also analyzed using SEM to gain further insights into their chain morphology. SEM images of unimodal and bimodal PE samples are presented in Figures 6.7 and 6.8. At lower magnification, a nodular-fibrilar structure was identifiable for all PE samples. At high magnification, it was observed that the macroscopic particles and nodules consisted of spherical sub-particles connected by

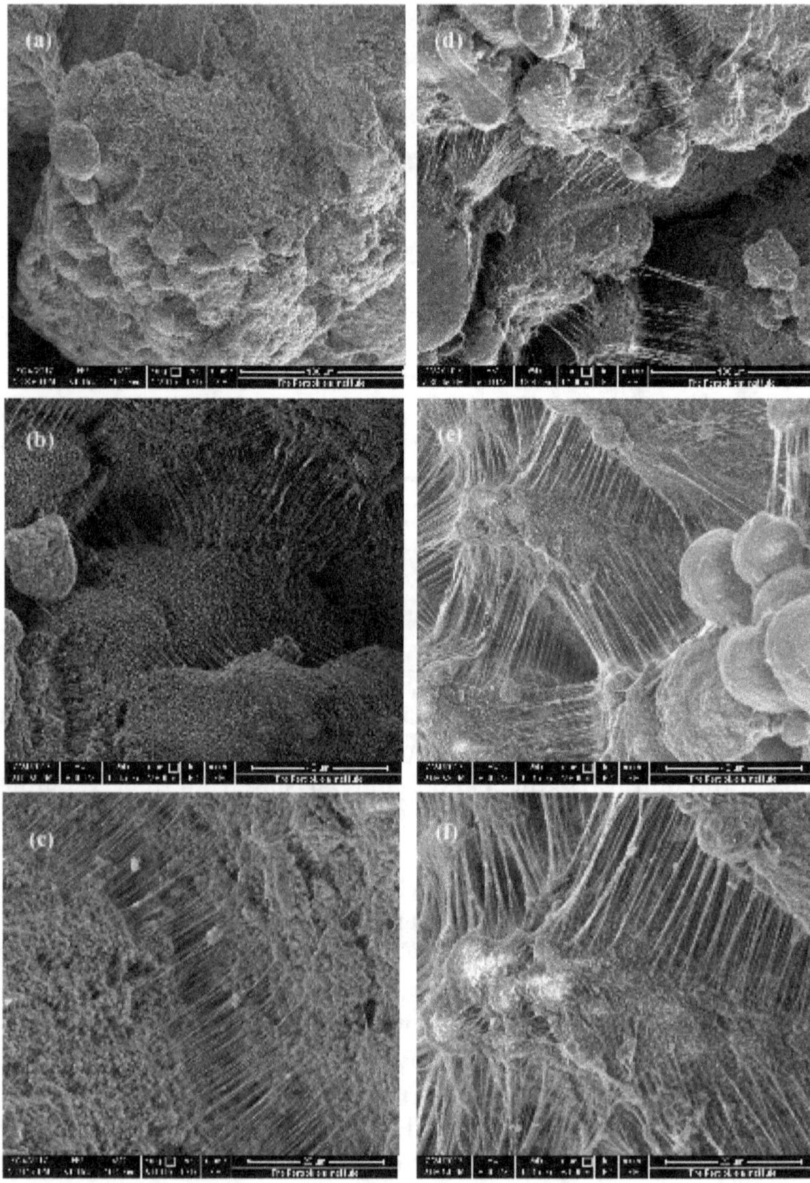

Figure 6.7 SEM images of PE1 (a,b,c) and PE2 (d,e,f) samples.

the fibrils. The sub-particles in the nodules were arranged in more orderly fashion in PE1 with short fibrils running out from the sub-particles (Figures 6.7 b & c). However, for PE2, the fibrils were more elongated, which might be due to the high growth stress which would

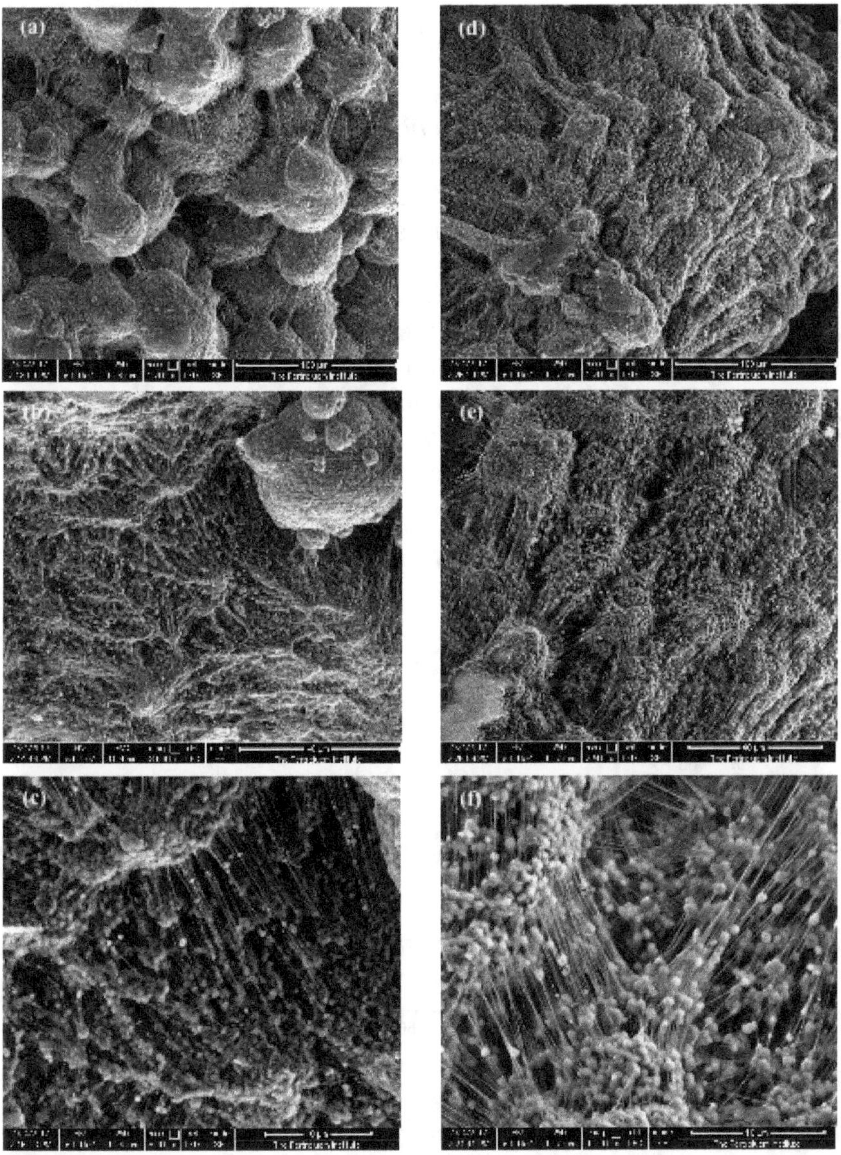

Figure 6.8 SEM images of PE3 (a,b,c) and PE4 (d,e,f) samples.

stretch out the fibrils from the sub-particles during the polymerization, as visible in Figures 6.7 d, e & f [4]. SEM images of PE3 also exhibited a nodular-fibrilar structure at low magnification, whereas PE4 showed ant-hill morphology (Figures 6.8 a & d). Both structures at high magnification exhibited a cobweb structure fully packed with sub-particles and fibrils. Here, the fibrils were stretched out from the sub-particles in a more disordered manner compared with PE1 sample images.

The thermal degradation behavior of the polyethylene samples is demonstrated in Figure 6.9. All samples exhibited one step degradation pattern, however, some differences were evident in the onset and peak degradation temperatures. PE1 had the lowest onset and peak degradation temperatures, whereas these were the highest for bimodal PE4. PE2 and PE3 were observed to have the thermal degradation performance in between PE1 and PE4 matrices. Comparing PE1-PE3, the thermal degradation performance was observed to have weak relation with the molecular weight. However, superior thermal performance for the bimodal PE4 sample with lower molecular weight than PE3 also indicated that other molecular characteristics affected the performance too.

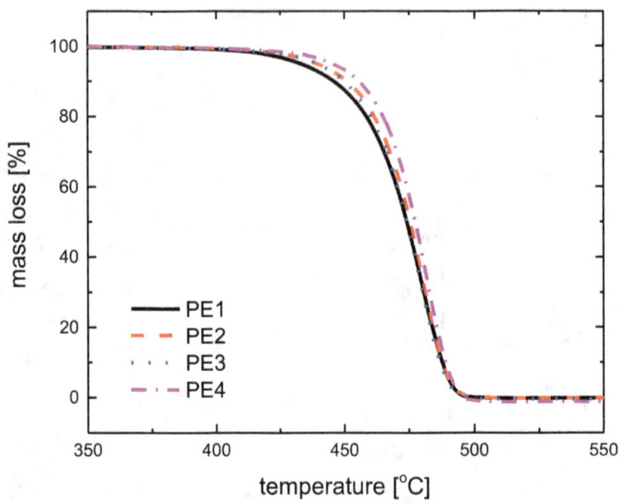

Figure 6.9 TGA thermograms of PE materials.

Table 6.6 reports the tensile properties of the PE materials. In general, the tensile modulus was proportional to the extent of crystallinity in the polymers. PE1 exhibited a modulus of 854 MPa, which was enhanced to 1224 MPa for PE2 due to enhancement in the extent of crystallinity. PE3, due to the least crystallinity, had a lowest modulus of 602 MPa. Due to the presence of bimodality in PE4, the tensile modulus had an average value of 688 MPa. The tensile strength of the PE polymers also exhibited similar trend as the modulus, with PE2 having the strength of 44.4 MPa, in comparison to 37.5 and 25.3 MPa for PE1 and PE3 respectively. Due to the lowest extent of crystallinity, PE3 had the highest elongation of 45.1%, whereas PE2 exhibited the lowest elongation of 16.7%. Thus, the tensile properties of the materials were strongly impacted by the molecular characteristics and corresponding crystallinity.

Table 6.6 Tensile modulus, strength and elongation at break of PE materials

Polymer	tensile modulus (MPa)	elongation at break (%)	tensile strength (MPa)
PE1	854±43	25.4±2.4	37.5±1.5
PE2	1224±99	16.7±1.7	44.4±4.7
PE3	602±31	45.1±1.9	25.3±2.4
PE4	688±27	27.2±3.4	29.2±2.6

Further insights into the effect of molecular characteristics on the flow behavior of the polymers were gained through shear rheology. Figure 6.10 shows the strain sweep measurements for the samples at 1 rad/s and 180 °C. The samples were stable at least till 5% strain, thus, a value of 2% strain was chosen for the frequency sweep analysis. Furthermore, the shear modulus was observed to increase with molecular weight, which was different in behavior as compared to tensile modulus. For instance, PE4 with higher molecular exhibited nearly two orders of magnitude higher storage modulus than PE1. As the role of crystallinity was absent due to melt formation, the observed effect of storage modulus would, thus, be due to higher degree of entanglements of higher molecular weight chains [6]. The stability of the melt was also observed to reduce with molecular weight. Figure 6.11 shows the storage and loss moduli of PE samples as a function of angular frequency. As expected, PE1 exhibited the lowest storage modulus among the polymer matrices over the whole angular frequency range. For instance, at an angular frequency of 0.1 rad/s, PEI exhibited a storage modulus of 10828 Pa, which was enhanced to 29579 and 405182 Pa for PE2 and PE3 respectively. Accordingly, the bimodal PE4 sample had the storage modulus of 131468 Pa at 0.1 rad/s. Thus, the extent of molecular entanglements owing to different molecular weight chains impacted the melt behavior significantly. In

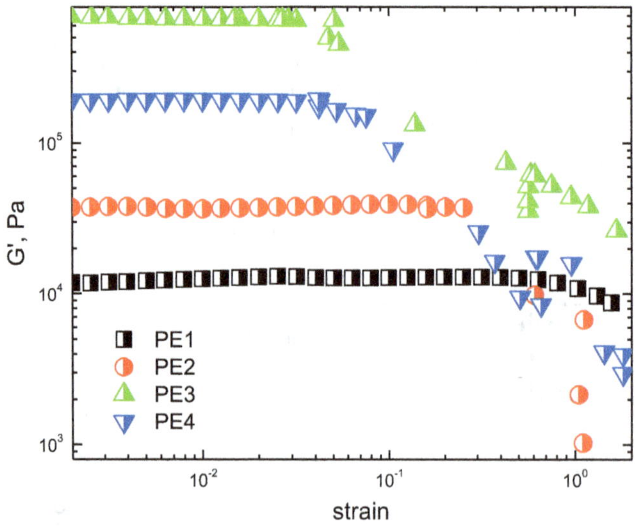

Figure 6.10 Storage modulus of the PE materials as a function of strain at 180 °C and 1 rad/s.

addition, the storage modulus of the materials increased with angular frequency, however, the rate of increase was observed to diminish at higher angular frequency values. The loss modulus of the materials also exhibited similar pattern as storage modulus. However, the loss modulus curves of the materials were observed to merge into a single curve at higher angular frequency indicating absence of molecular weight effect. In addition, PE3 and PE4 curves had lower slope as compared to PE1 and PE2 indicating approach towards frequency independent behavior. Furthermore, the loss modulus of the materials was lower in magnitude as compared to storage modulus, with the difference enhancing with increasing molecular weight. Figure 6.12 also presents the viscosity and elasticity of the PE materials as a function of angular frequency. Similar to the modulus, the viscosity and elasticity of the materials were also a function of molecular weight. A significant reduction in these properties was also observed at higher angular frequency values leading to merging of the different curves. Though the viscosity of the materials was significantly enhanced with increase in molecular weight, the materials were still processable using the same processing protocols. Also, no specific effect of the bimodality of the PE4 sample on the processing requirements could be

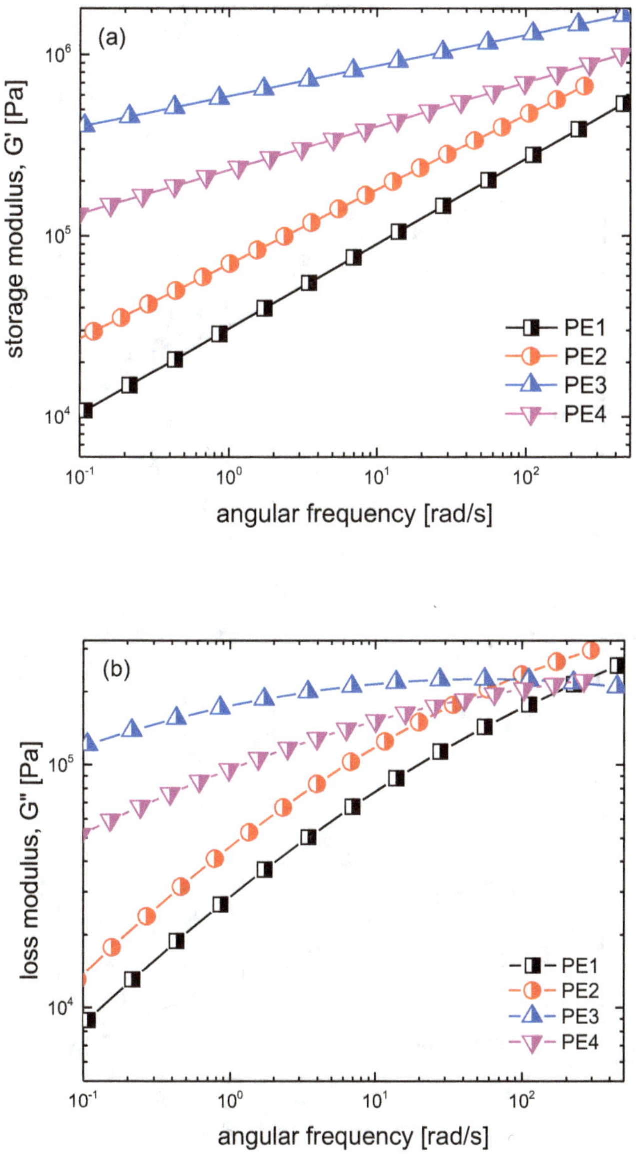

Figure 6.11 (a) Storage and (b) loss moduli of the PE materials as a function of angular frequency (at 2% strain and 180 °C).

observed. Finally, the relaxation time of the polymer melt was also assessed from the ratio of elasticity and product of angular frequency

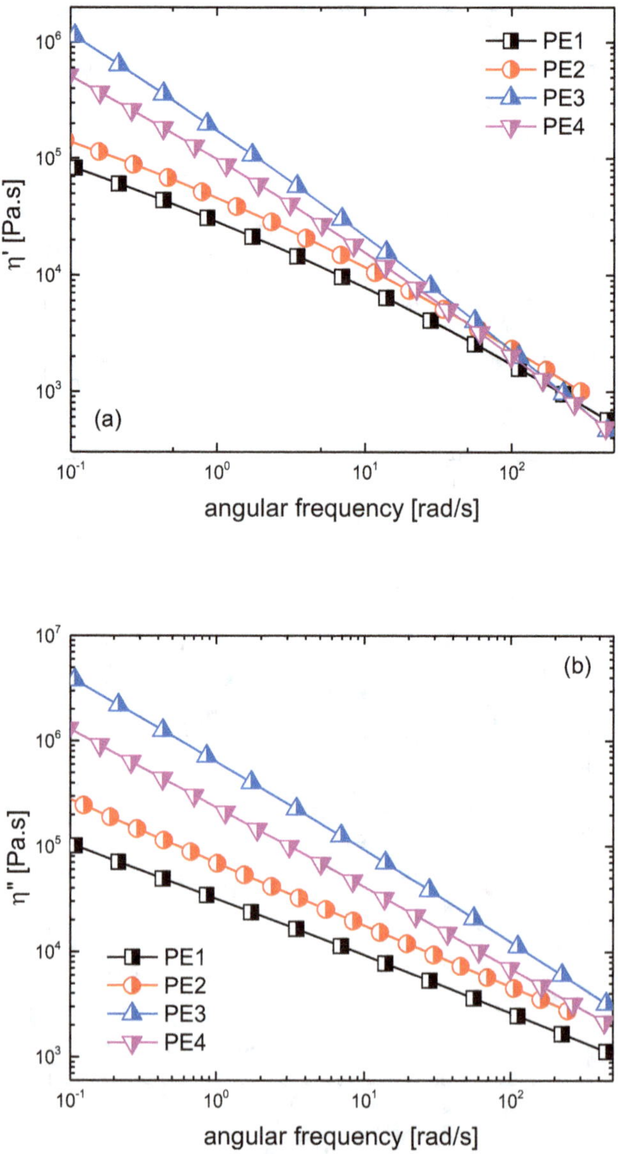

Figure 6.12 (a) Viscosity and (b) elasticity of the PE materials as a function of angular frequency (at 2% strain and 180 °C).

and viscosity, as shown in Figure 6.13 as a function of angular frequency. As the increase in molecular weight is generally associated with the increase in the relaxation time, PE1, thus, exhibited a lower relaxation time, followed by PE2, PE4 and PE3 respectively. The estimated relaxation time from the polymer melts can be envisaged to have different physical importance as compared to the relaxation time observed for the solid samples through solid state NMR. In the case of solid samples, the crystallinity and its extent is expected to have a significant effect on the polymer behavior, whereas entanglement of polymer chains might be the dominating effect in the case of polymer melts.

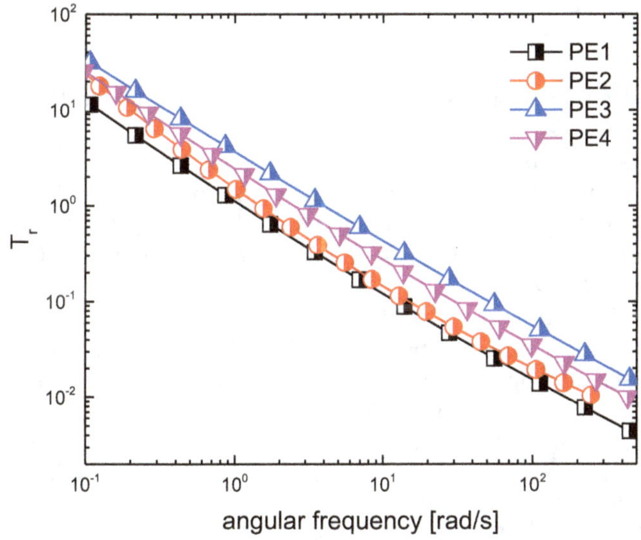

Figure 6.13 Relaxation time of PE polymers as a function of angular frequency.

6.4 Conclusions

In the current study, the effect of molecular characteristics on the mechanical, thermal, rheological and morphological properties of high molecular weight polyethylene materials was studied. The molecular weight and distribution though didn't change the orthorhombic crystal form in the polymers, however, crystallinity, crystallite size and lamellae thickness of the polymers were affected. In addition, the molecular characteristics also had an effect on the relaxation times, thermal stability as well as morphology of the polymeric materials. Owing to the different degrees of crystallinity in the materials, the tensile properties (i.e., modulus and strength) of the materials exhibited significant differences. Likewise, the chain entanglements owing to different molecular weights also correspondingly affected the storage and loss moduli, as well as viscosity and chain relaxation properties. In conclusion, the differences in molecular characteristics had a significant impact on the overall polymer behavior and its quantification can help to achieve an accurate control on the polymer properties for specific applications.

References

1. Brydson, J. A. (1999) *Plastics Materials,* Butterworth-Heinemann, UK.
2. Mittal, V., and Matsko, N. (2014) Characterization of polyethylene-based multiphase systems containing zero- and two-dimensional nanoparticulate reinforcement materials by analytical electron and atomic force microscopy. *Journal of Thermoplastic Composite Materials,* **27**(6), 845-864.
3. Ruff, M., and Paulik, C. (2012) Controlling polyelefin properties by in-reactor blending, 1-polymerization process, precise kinetics, and molecular properties of UHMWPE polymers. *Macromolecular Reaction Engineering,* **6**, 302-317.
4. Ruff, M., and Paulik, C. (2013) Controlling polyolefin properties by in-reactor blending: 2. Particle design. *Macromolecular Reaction Engineering,* **7**(2), 71-83.
5. Ruff, M., Lang, W. R., and Paulik, C. (2013) Controlling polyolefin properties by in-reactor blending: 3. Mechanical properties. *Macromolecular Reaction Engineering,* **7**, 328-343.
6. Mittal, V., Kim, S., Neuhofer, S., and Paulik, C. (2016) Polyethylene/graphene nanocomposites: effect of molecular weight on mechanical, thermal, rheological and morphological properties. *Colloid and Polymer Science,* **294**, 691-704.
7. Moly, K. A., Radusch,H. J., Androsh, R., Bhagawan, S. S., and Thomas, S. (2005) Nonisothermal crystallisation, melting behavior and wide angle X-ray scattering investigations on linear low density polyethylene (LLDPE)/ethylene vinyl acetate (EVA) blends: effects of compatibilisation and dynamic crosslinking. *European Polymer Journal,* **41**(6), 1410-1419.
8. Bin, Y., Ma, L., Adachi, R., Kurosu, H., and Matsuo, M. (2001) Ultradrawing of low molecular weight polyethylene-ultra-high molecular weight polyethylene blend films prepared by gelation/crystallization from semi-dilute solutions. *Polymer,* **42**, 8125-8135.

9. Joo, Y. L., Han, O. H., Lee, H. K., and Song, J. K. (2000) Characterization of ultra-high molecular weight polyethylene nascent reactor powders by X-ray diffraction and solid state NMR. *Polymer*, **41**, 1355-1368.
10. Russell, K. E., Hunter, B. K., and Heyding, R. D. (1997) Monoclinic polyethylene revisited. *Polymer*, **38**(6), 1409-1414.
11. Russell, K. E., Hunter, B. K., and Heyding, R. D. (1993) Polyethylenes revisited: WAXS and the phase structure. *European Polymer Journal*, **29** (2/3), 211-217.
12. Joshi, M., Butola, B. S., Simon, G., and Kukaleva, N. (2006) Rheological and viscoelastic behavior of HDPE/octamethyl-POSS nanocomposites. *Macromolecules*, **39**(5), 1839-1849.
13. Cheng, J. J. (2008) Mechanical and Chemical Properties of High Density Polyethylene: Effects of Microstructure on Creep Characteristics, Ph.D. Thesis, University of Waterloo, Canada.
14. Crist, B., and Mirabella, F. M. (1999) Crystal thickness distribution from melting homopolymers or random copolymers. *Journal of Polymer Science, Part B: Polymer Physics*, **37**, 3131-3140.
15. Fawaz, J., Deveci, S., and Mittal, V. (2016) Molecular and morphological studies to understand slow crack growth (SCG) of polyethylene. *Colloid and Polymer Science*, **294**, 1269-1280.
16. Earl, W. L, and VanderHart, W. L. (1979) Observations in solid polyethylenes by carbon-13 nuclear magnetic resonance with magic angle sample spinning. *Macromolecules*, **12**, 762-767.
17. Akiyama, S., Komoto, T., and Ando, I. (1990) Structure of ultrahigh-molecular weight polyethylene in the solid state as studied by variable-temperature 13C CP/MAS NMR spectroscopy. *Journal of Polymer Science, Part B: Polymer Physics*, **28**, 587-594.
18. Ando, I., Sorita, T., Yamanobe, T., Komoto, T., Sato, H., Deguchi, K., and Imanari, M. (1985) Chain conformation of polyethylene in the solid state studied by 13C cross-polarization/magic angle spinning nuclear magnetic resonance spectroscopy. *Polymer*, **26**, 1864-1868.
19. Cook, J. T. E., Klein, P. G., Ward, I. M., Brain, A. A., Farrar, D. F., and Rose, J. (2000) The morphology of nascent and moulded ultra-high molecular weight polyethylene. Insights from solid-state NMR, nitric acid etching, GPC and DSC. *Polymer*, **41**, 8615-8623.
20. Shimizu, Y., Harashina, Y., Sugiura, Y., and Matsuo, M. (1995) Morphology and molecular mobility of several kinds of polyethylene specimens by 13C solid-state NMR and x-ray diffraction techniques. *Macromolecules*, **28**, 6889-6901.
21. Pezzotti, G. (2017) Raman spectroscopy of biomedical polyethylenes. *Acta Biomaterialia*, **55**, 28-99.
22. Hahn, D. W., Wolfarth, D. L., and Parks, N. L. (1997) Analysis of polyethylene wear debris using micro-Raman. *Journal of Biomedical Materials Research*, **35**, 31-37.

23. Affatato, S., Modena, E., Carmignato, S., and Taddei, P. (2013) The use of Raman spectroscopy in the analysis of UHMWPE uni-condylar bearing systems after run on a force and displacement control knee simulators. *Wear*, **297**, 781-790.

7

Polyethylene Nanocomposites with Surface Treated Sepiolite

7.1 Introduction

Polymer nanocomposites have been extensively studied over the last two decades owing to the significant enhancement of a wide spectrum of material properties at low filler volume fractions (e.g., mechanical, thermal, gas barrier and flame retardation properties) [1,2]. Use of polar compatibilizers is a most commonly employed technique to achieve polarity in the nonpolar polymer matrices which helps to achieve improvements in filler exfoliation, however, the use of compatibilizers also leads to occasional lowering of mechanical performance [3]. Ideally, if the filler surface is optimally designed and modified, it would avoid the need to alter the polymer polarity and, thus, would help to retain the properties of the homopolymer. Various techniques have been developed in recent years to achieve optimum surface modification of the filler, including grafting of polymer chains 'from' or 'to' the surface [4-10], chemical reactions of the long chain alcohols or acids with the carboxyl or hydroxyl groups ionically bound to the filler surface [11-16], physical adsorption of the organic molecules or polymer chains on the filler surface [17-20], etc. The techniques result in modified fillers which have much better chances of complete exfoliation at nanoscale in the polymer matrices. These advanced fillers do not require the use of any additional low molecular weight compatibilizer or surfactant for nanocomposite synthesis [21].

Mittal [9] reported filler surface modifications carrying methacrylic functional groups and free radical initiator, which were used to polymerize lauryl methacrylate in order to further improve the organic coverage of the filler platelets by grafting polymer chains either 'to' or 'from' the clay surface. Significance of the amount of monomer or initiator attached on the surface, time, as well as temperature used for polymerization was successfully quantified. For instance, Figure

M. R. Vengatesan[a], Sevim Isci[b] and Vikas Mittal[a,]*
[a]Khalifa University of Science and Technology, Abu Dhabi, UAE; [b]Istanbul Technical University, Istanbul, Turkey

7.1 presents the diffraction analysis of organically modified filler with initiator present in an amount corresponding to 70% surface coverage. The filler had the 001 basal plane spacing of 1.9 nm, which remained unchanged when reacted with lauryl methacrylate at 60 °C for 72 h under non-living conditions, as is evident from a minor increase in organic weight loss. However, when reacted at 120 °C under the living conditions, the *d*-spacing was increased to 3.8 nm. It indicated that the effective control over the termination reactions by using the living agent was beneficial in enhancing the basal plane spacing of the filler platelets. Thus, it confirmed the successful generation of the polymethacrylate brushes effectively grafted from the filler surface.

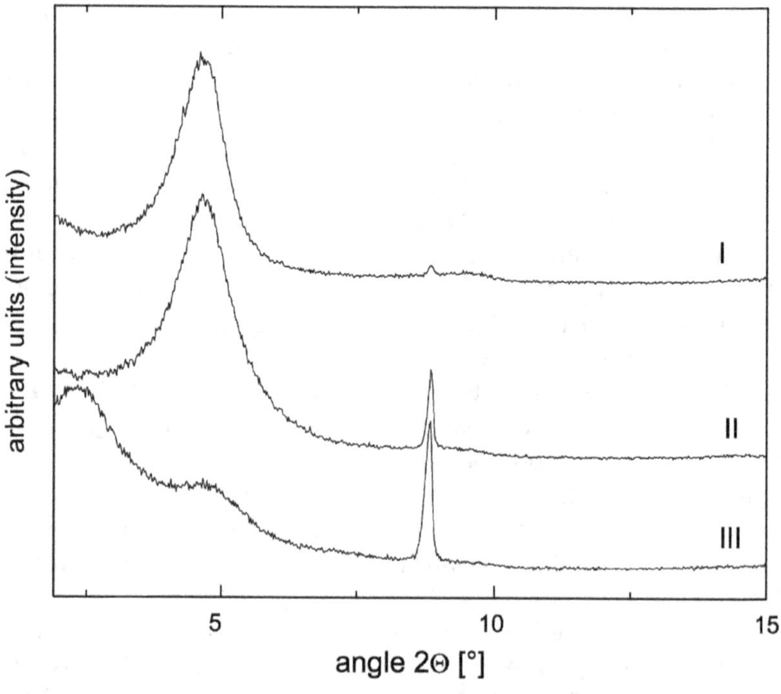

Figure 7.1 X-ray diffractograms of the (I) filler modified by using bicationic initiator in an amount corresponding to 70% surface coverage, (II) filler of I when reacted under non-living conditions at 60 °C, and (III) filler of I when polymerized under living conditions at 120 °C. The peak at 8.84° 2θ corresponds to the 001 basal plane spacing of mica platelets used as internal standard. Reproduced from Reference 9 with permission from Elsevier.

In another study, Mittal [16] further explored the surface reactions approach for the achievement of higher filler basal spacing in order to decrease the electrostatic interactions and, thus, to ease the filler exfoliation in the polymer matrix. Filler platelets were modified with modifiers of different chemical architectures in order to study the effect on the extent of surface esterification reactions with long chain fatty acids, such as dotriacontanoic or lacceroic acid. Figure 7.2 shows the comparison of the modified filler before and after the surface reaction. As observed, the higher extent of organic matter on the clay surface led to a significant improvement in the basal plane spacing, which leads to higher susceptibility of these filler stacks to exfoliate in polymer when compounded together with the polymer under shear in the compounder.

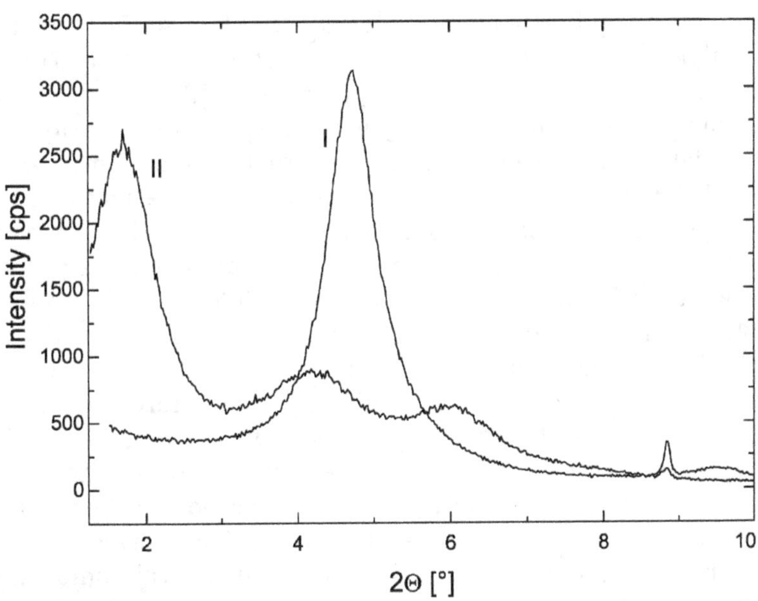

Figure 7.2 X-ray diffractograms of I: modified filler and II: filler of I after surface esterification reaction with dotriacontanoic acid. The peak at 8.84 2θ corresponds to the 001 basal plane spacing of mica platelets used as internal standard. Reproduced from Reference 16 with permission from Elsevier.

Mittal and Herle [20] also reported the potential of physical adsorption of long chain solvent-soluble molecules and polymers on the surface of the filler. These molecules (e.g., long alkyl chain alcohols) could adsorb on the surface by forming H bonds with the OH groups present either in the inside structure of the filler crystals or on the edges of the filler platelets. Figure 7.3 shows the thermal degradation behavior of the pre-modified filler subsequently physically adsorbed with varying amount of poly(vinylpyrrolidone) (PVP). The physical adsorption led to thermal behavior enhancement as the peak degradation temperature of the PVP-modified clay increased to 332 °C (curve II in Figure 7.3a). This increased thermal resistance is particularly beneficial when compounding the filler at high compounding temperatures required by polymer matrices like polyethylene. The mechanism of adsorption was suggested as follows: the more electronegative oxygen atoms in the PVP chains can form hydrogen bonds with OH groups present either on edges or in the internal structure of filler.

High density polyethylene (HDPE) is a well-known thermoplastic material possessing good mechanical properties and has been extensively used in different industrial applications [22]. In order to improve the properties of HDPE polymer, incorporation of various nanomaterials has been reported. This leads to the reinforcement effect on the HDPE performance, thus, resulting in better thermal, mechanical, rheological and barrier properties [23-26]. Specifically, nanomaterials such as silica [27], alumina [28], mica [29], clays [30-32], natural fibers [33,34], carbon black [35], CNT [36,37], graphene [38,39], etc. have been used as filler materials for the polyethylene matrix. Among these, clay materials are more economical and commercially acceptable class of nano-fillers for the reinforcement of polymer matrices so as to expand their spectrum of applications. Clay based polymer nanocomposites have been reported to exhibit a good balance of thermal, mechanical as well as the rheological properties [40]. Different types of clay materials such as montmorillonite, bentonite, hectorite, sepiolite, etc., differing in structural composition and morphology, have been used as reinforcing fillers for thermoplastic materials [41-44].

Sepiolite is one of the most useful minerals in the category of clay materials with 1-D morphology. It is a non-swelling, natural, abundant and low-cost clay having theoretical unit cell formula $(Si_{12}Mg_8O_{30})(OH)_4(OH_2)_4{\cdot}8H_2O$ (crystal structure depicted in Figure 7.4). It belongs to the class of layered silicates comprising of hydrated magnesium silicate having micro-fibrous needle like morphology. As compared to phyllosilicates (BET surface area 82±1 m^2/g), it has high surface area (BET 374±7 m^2/g) and contains large amount of silanol groups present at the surface [45,46]. The silanol groups have been proven effective as sepiolite acts as reinforcing and compatibilizing agent for different polymer matrices owing to coupling reactions. In a related study, Chen *et al.* [47] prepared well dispersed polyurethane/sepiolite nanocomposite using *in-situ* polymerization technique. Sepiolite improved the tensile strength and modulus of polyurethane due to the strong interfacial interaction between the organic

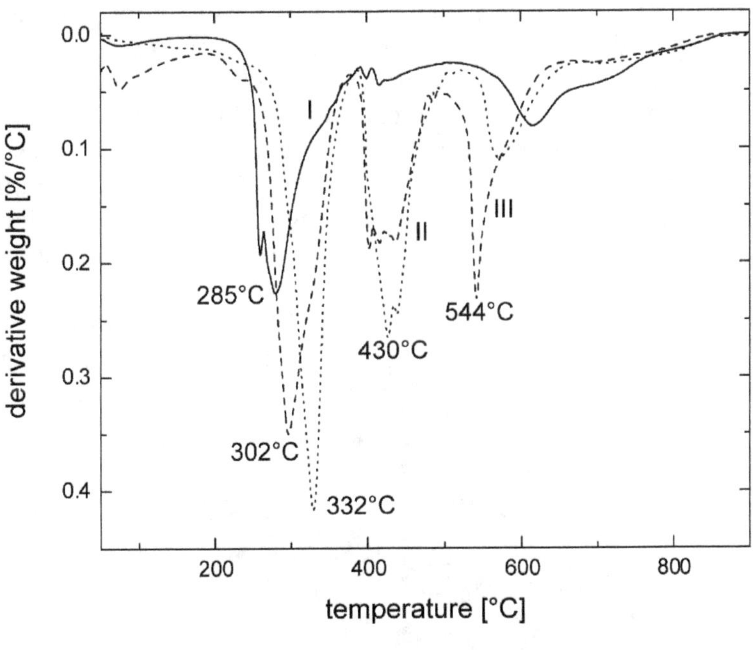

Figure 7.3 (a) Derivative weight thermograms of modified filler before adsorption (I), after adsorption with 1.5 g of initially added PVP (for 2 g of filler) (II) and after adsorption with 1 g of initially added PVP (for 2 g of filler) (III). Reproduced from Reference 20 with permission from Elsevier.

and inorganic phases [47]. Similarly, it has also been utilized as a nano-filler for other polymer matrices [48-51].

The incorporation of the unmodified sepiolite clay into the polymer matrices may lead to an aggregated morphology due to the mismatch of surface polarity between the polymer and inorganic phases, thus, resulting in deterioration of the composite properties. Therefore, the dispersion/defibrillation of the 1-D sepiolite crystals in the polymer matrix and the interfacial interactions between the polymer and sepiolite phases are the major factors required to be optimized to enhance thermal stability and mechanical properties of the nanocomposites [52]. In the case of HDPE based composites, no natural interfacial interaction between the organic and inorganic phases exists, thus, surface modification of the sepiolite particles is solely responsible for the minimization of the contact between sepiolite nanofibers and correspondingly enhanced interfacial interaction/contacts between inorganic and organic phases [53]. The surface of the sepiolite nano-fibrils is generally modified with different types of organic materials such as organosilanes, cationic, anionic and non-ionic surfactants and polymers through covalent, H-bonding and physical adsorption (van der Waals interaction) methods [54-58]. As mentioned earlier, the surface modification of the filler particles through the physical adsorption method is an acceptable strategy for use in polymer nanocomposites, even though the binding of the modified molecules to the filler surface is not covalent in nature. Similarly, physical adsorption method for the surface modification of sepiolite

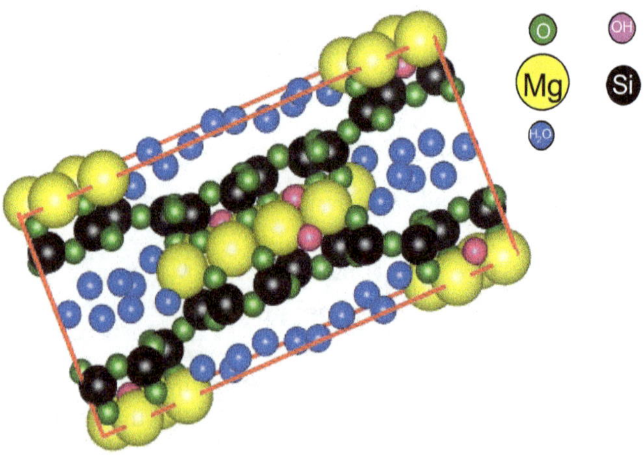

Figure 7.4 Schematic representation of sepiolite crystal structure.

crystals using different electrolytes and uncharged polymers can also be useful in its development as functional filler for nanocomposites [58,59]. The adsorption of the uncharged polymer on the sepiolite surface leads to an 'entropy-driven' process. Therefore, the adsorbed polymer is generally observed to be randomly coated on the clay surface. However, the higher molecular weight of the adsorbing polymer leads to overall significant extent of surface adsorption due to van der Waals interactions, thereby, resulting in organo-modified sepiolite.

By the use of physically adsorbed organic modification molecules of amphiphilic nature, the interfacial interaction of the filler particles with the polymer matrix can be achieved. Thus, the nature of the polymers adsorbed on the sepiolite surface significantly influences the dispersion/defibrillation process and interfacial adhesion with the polymer matrix which, thus, results in the variation in the mechanical, thermal as well the rheological properties of the resulting nanocomposites [60].

In the present work, the influence of a variety of polymeric surface modifications physically adsorbed on the surface of sepiolite was studied on the thermal, mechanical and rheological properties of the HDPE nanocomposites. The various modifications included poly(vinyl pyrrolidone) (PVP), poly(vinyl butyral) (PVB), polyethylene glycol (PEG), polyurethane (PU) and ethylene vinyl acetate-graft-maleic anhydride (EVA*g*MA). The amount of filler was fixed to 5 wt% and the composites were generated using industrially accepted melt mixing method without the use of any external compatibilizers. The compounding conditions were carefully chosen to ensure that good polymer-filler intermixing was obtained in the compounder under shear, however, the thermal degradation of the polymer and surface modification was avoided.

7.2 Experimental

7.2.1 Materials

High-density polyethylene (BB2581) was supplied by Abu Dhabi Polymers Company Limited (Borouge), UAE. It was in the form of white pellets with a specific gravity of 0.958. Sepiolite clay was received from Sigma Aldrich. It was in the form of fine white powder and was used as received for surface modification trials. Polyurethane liquid, polyvinylpyrolidone, polyvinylbutaral and polyethylene glycol with an average M_w of 8000 (PEG8000) were procured from Sigma Aldrich and were used without further purification. Ethylene vinyl acetate-graft-maleic anhydride copolymer (density 0.950 g/cm^3, melt flow index (190 °C/2.16 kg) 16 g/10 min and peak melting temperature 71 °C) with a trade name of Fusabond C190 was obtained from DuPont. The standard solvents such as toluene and acetone for surface modification experiments were also procured from Sigma Aldrich and were used as received. Deionized water with low conductivity was used for the surface modification trials performed in water.

7.2.2 Surface Modification of Sepiolite

The surface of pristine sepiolite (S-P) was modified with EVA*g*MA, PEG, PU, PVB and PVP polymers. For modification with PU, PU was dispersed in deionized water at a concentration of 1 mg/ml at 70 °C. Sepiolite (2% w/w) was added to the polymer dispersion and shaken for 2 h and ultra-sonicated for 5 min. The content was further stirred overnight. The sample was washed with deionized water followed by acetone. The sample was finally dried at 50 °C. The final product was termed as S-PU clay. The same protocol was used to prepare PEG and PVP modified sepiolite in the presence of deionized water and termed as S-PEG and S-PVP respectively. For the modification of sepiolite with EVA-g-MA and PVB, toluene and acetone were used as solvents respectively, while keeping the other parameters constant. The EVA*g*MA and PVB modified materials were denoted as S-EVA*g*MA and S-PVB respectively.

7.2.3 Preparation of HDPE/Sepiolite Nanocomposites

Pristine and surface modified sepiolite were melt mixed with HDPE using mini twin screw conical extruder (HAAKE MiniLab, Thermo Scientific) at a temperature of 180 °C using a rotation speed of 80 rpm for 10 minutes. Before compounding, the fillers and polymer were dried in a vacuum oven at 60 °C overnight so as to remove any traces of surface bound moisture, which can lead to thermal degradation during high temperature compounding. In addition, the materials were physically mixed prior to melt mixing in the compounder. The test specimens for the mechanical and rheological analysis were prepared using mini injection molding machine (HAAKE MiniJet PRO, Thermo Scientific) at 180 °C at 400 bar for 10 s. The mold temperature was maintained as 50 °C. Care was taken to cool all samples by following the same procedure so as not to affect the polymer crystallinity.

7.2.4 Materials Characterization

IR spectra of pristine and modified sepiolite were collected on Nicolet iS10 spectrometer equipped with SmartiTR diamond ATR accessory (angle of incidence of 45°), DTGS KBr detector and KBr beam splitter. It had a diamond ATR crystal (index of refraction 2.4 at 1000 cm^{-1}) and a depth of penetration of 2 μm at 1000 cm^{-1} for a sample with refractive index of 1.5. IR Spectra were recorded by the OMNIC software in the 4000-525 cm^{-1} region with a resolution of 4 cm^{-1} from 16 scans.

Differential scanning calorimetric (DSC) analysis of the nanocomposites was carried out using a Discovery DSC from TA instruments under nitrogen atmosphere. 3-7 mg of sample was heated to 200 °C using a heating rate of 10 °C/min. Subsequently, the sample was cooled to -60 °C using a cooling rate of 5 °C/min and kept under isothermal conditions at -60 °C for 1 min. The sample was re-heated to 200 °C using a heating rate of 10 °C/min and finally cooled to 25 °C using the same rate. The thermal degradation behavior of the samples was studied using a Discovery thermogravimetric analyzed (TGA) from TA instruments under nitrogen atmosphere. Samples weighing 3-7 mg were analyzed from 25 °C to 700 °C by using the heating rate of 10 °C/min.

Tensile properties such as tensile strength, modulus and extension at break of composite samples were evaluated at ambient temperature on universal testing machine (Instron 3345) according to ASTM D638 at a crosshead speed of 10 mm/min. The dumbbell shaped test specimen were used and the reported values represented the average of five test specimens. The melt rheological characteristics of HDPE and its nanocomposites were analyzed in shear mode on AR 2000ex Rheometer from TA Instruments according to ASTM D4440. A parallel plate configuration was used with plate diameter of 25 mm and a geometric gap of 1.4 mm. Injection molded disc shaped samples of 25 mm diameter and 1.5 mm thickness were used for the analysis. The samples were scanned in the dynamic oscillatory mode from 0.1 to 100 rad/s at 180 °C with a constant strain of 4%.

Analytical powder diffractometer (X'Pert PRO) (45 kV, 40 mA) with Cu-Kα radiation (wavelength = 1.5406 A°) was utilized for X-ray diffraction analysis of the polymer and nanocomposites samples. The diffraction of the samples was measured at a rate of 0.0170 deg/s and 2θ scale ranging from 5 to 60°.

For the bright field transmission electron microscopy (TEM) analysis of the nanocomposite samples, Philips CM 20 (Philips/FEI, Eindhoven) electron microscope at 120 kV and 200 kV accelerating voltages was used. Thin sections of 70-90 nm thickness were microtomed from the sample blocks using a Leica Ultracut E microtome (Leica, Austria) equipped with a diamond knife (Diatome, Switzerland) at a temperature of -120 °C. The so-obtained sections were supported on 100 mesh grids sputter coated with a 3 nm thick carbon layer. The modified fillers were analyzed in Hitachi field emission in-lens S-900 high resolution scanning electron microscope (SEM) at accelerating voltages of 10-20 kV. In addition, the block faces of the nanocomposite specimens were also investigated at ambient conditions using ZEISS Axioplan light microscope (Zeiss, Germany) equipped with Zeiss Axio Cam ICc 1 CCD camera in reflected polarized light. Zeta potential measurements of the modified sepiolite materials were performed using ZetaPALS device.

7.3 Results and Discussion

In the current study, 1-D nano-fibrils of sepiolite mineral were surface modified with a variety of polymeric modifiers using physical adsorption approach. The choice of the modifications was based on tuning the polarity of the sepiolite surface so as to achieve good degree of delamination in the polyethylene matrix. In addition to other polymers, copolymers like EVA*g*MA were also used for this purpose as the ethylene component can be expected to have good intermixing with the HDPE matrix, whereas the polar groups would interact with filler surface, thereby leading to phase compatibilization and filler delamination. As mentioned earlier, such an approach of filler modification eliminates the need of the use of conventional low molecular weight compatibilizers, which occasionally negatively affect the crystallinity and mechanical properties of the nanocomposites. Even though the surface of the filler in the case of physical adsorption approach would be partially covered by the polymeric modifier owing to the steric hindrance to the high molecular weight chains, however, it can still be expected to significantly change the surface behavior of the filler particles. In addition, the amount of the modification needed for the purpose is much lower than the amount of the compatibilizer required to attain good filler dispersion.

7.3.1 Properties of Modified Fillers

Table 7.1 demonstrates the zeta potential values of the pristine and modified sepiolite. Pristine sepiolite in water was observed to have the zeta potential value of -19.5 mV, which indicated that the sepiolite particles tend to flocculate in water. In the case of the surface modification with water soluble polymers like PEG and PVP, decrease in the zeta potential values for the modified sepiolite was observed. For instance, the zeta potential was reduced to -13.1 for S-PEG, whereas it was observed to be even lower (-10.2 mV) for S-PVP. It indicated that the polymer chains partially covered the surfaces of the sepiolite particles. The sepiolite particles modified with other polymers such as PU, PVB and EVA*g*MA in water dispersions were observed to flocculate and the zeta potential values of these materials increased.

Table 7.1 Zeta potential values of the pristine and modified sepiolite

Sample	Zeta Potential (mV)
S-P	-19.5±1.0
S-PEG	-13.1±0.5
S-PVB	-28.8±3.7
S-PU	-29.2±0.9
S-PVP	-10.2±1.5
S-EVA*g*MA	-20.6±3.4

The SEM analysis of the modified sepiolite also provided more insights into their morphology. Figure 7.5 shows the SEM images of the pristine sepiolite and sepiolite modified with PEG and PVP. The particles were observed to be uniformly dispersed as primary particles without forming any large sized aggregates. On the other hand, Figure 7.6 presents the SEM images of the sepiolite modified with PVB, PU and EVA*g*MA. Filler particles were observed to form large aggregates in these modified sepiolite materials, thus, indicating that the different surface modifications resulted in varying extent of bridging of primary filler particles. This behavior is also confirmed in the high magnification SEM images of the modified sepiolite, as shown in Figure 7.7. At high magnification too, the PEG and PVP modified sepiolite particles exhibited dispersed morphology, whereas the morphology was more aggregated in the case of PVB, PU and EVA*g*MA modified sepiolite. However, it should be noted the observed aggregates may be only loosely held aggregated structures, which may be easily broken when compounded with polymers under shear. Thus, all modified fillers exhibited potential of uniform dispersion in the filler matrix due to the surface modification making these more susceptible to exfoliation.

The interaction between the adsorbed polymer and clay surface was determined with FTIR. As mentioned earlier, the non-ionic polymers create mostly hydrogen bonding with the clay surface. Any changes in the characteristic filler peaks after modification may indicate the formation of hydrogen bonds between the polymer and sepiolite surface. The band at 3684 cm^{-1} in the IR spectra corresponded

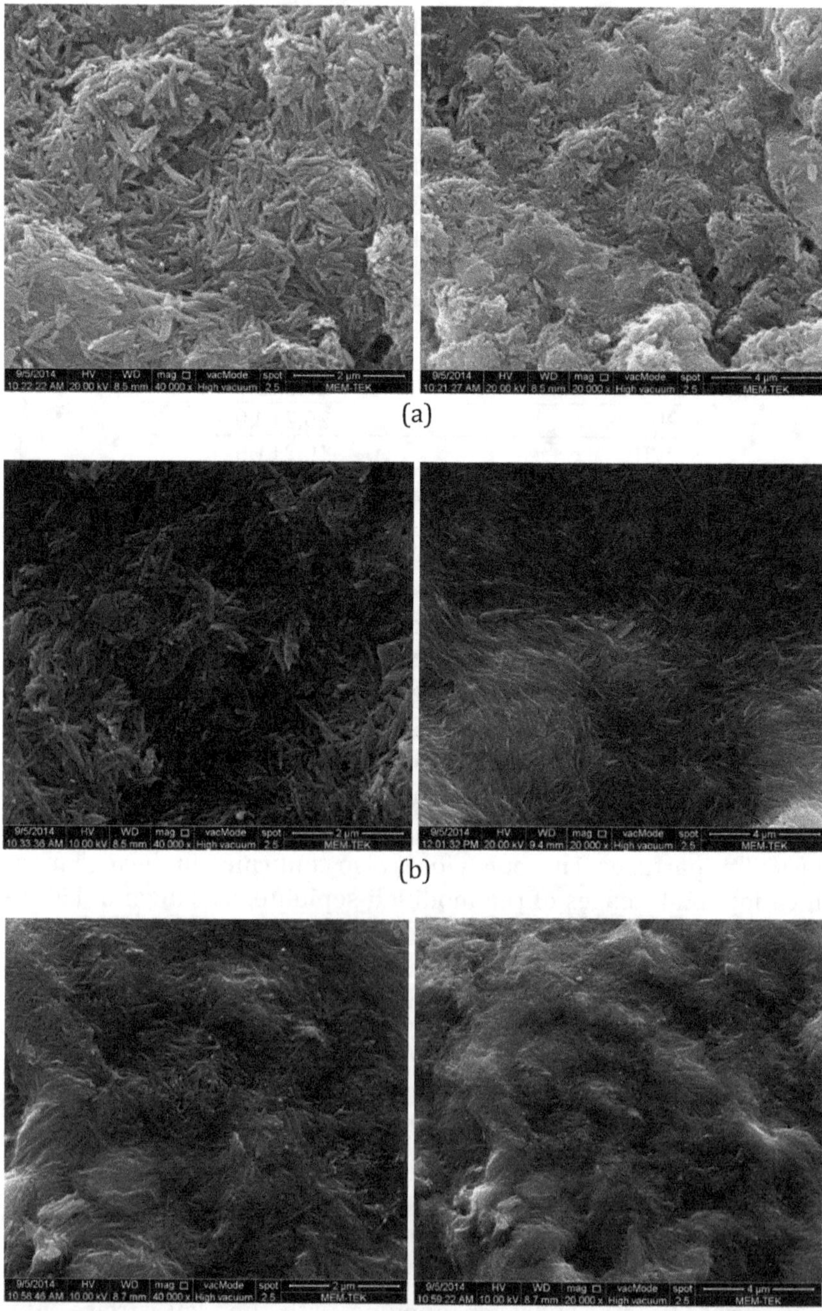

Figure 7.5 SEM images of (a) S-P, (b) S-PEG and (c) S-PVP at two different magnifications.

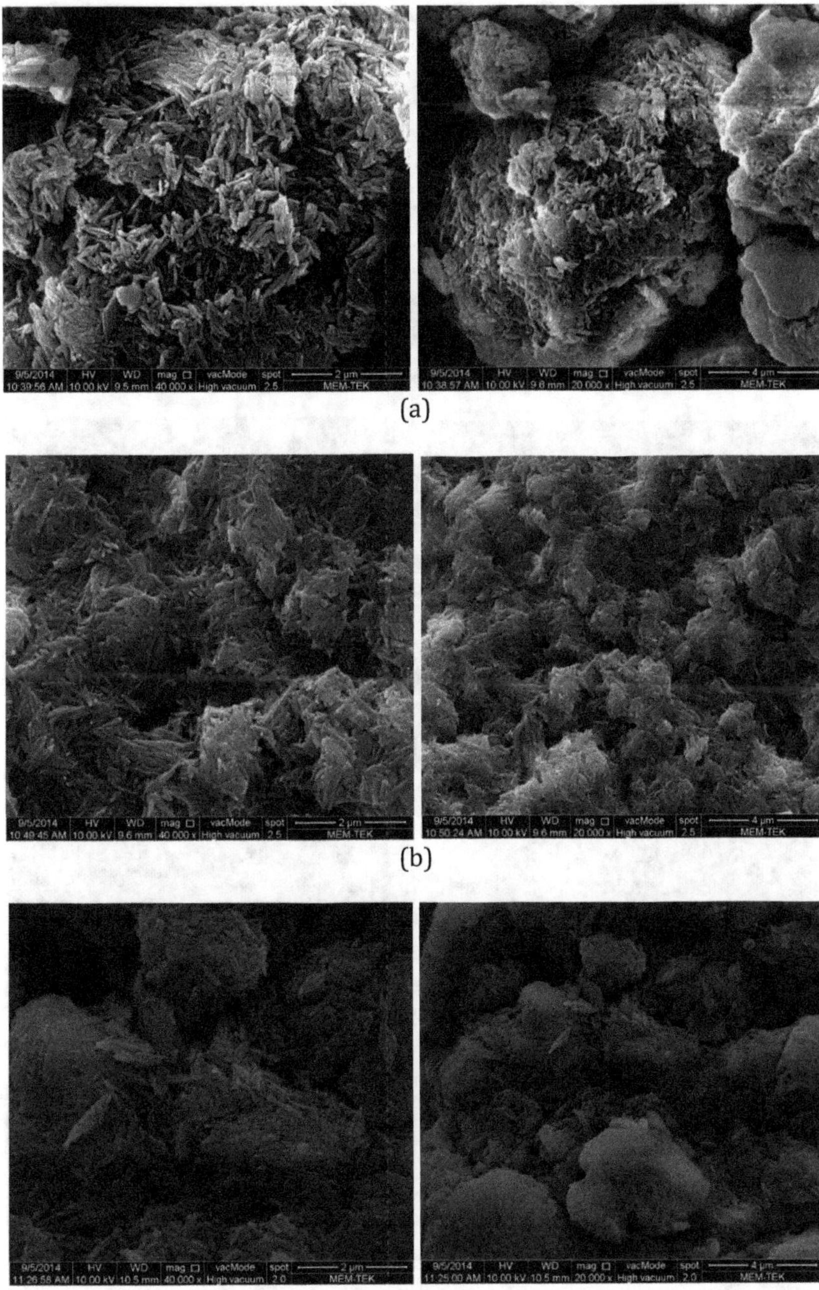

Figure 7.6 SEM images of (a) S-PVB, (b) S-PU and (c) S-EVA*g*MA at two different magnifications.

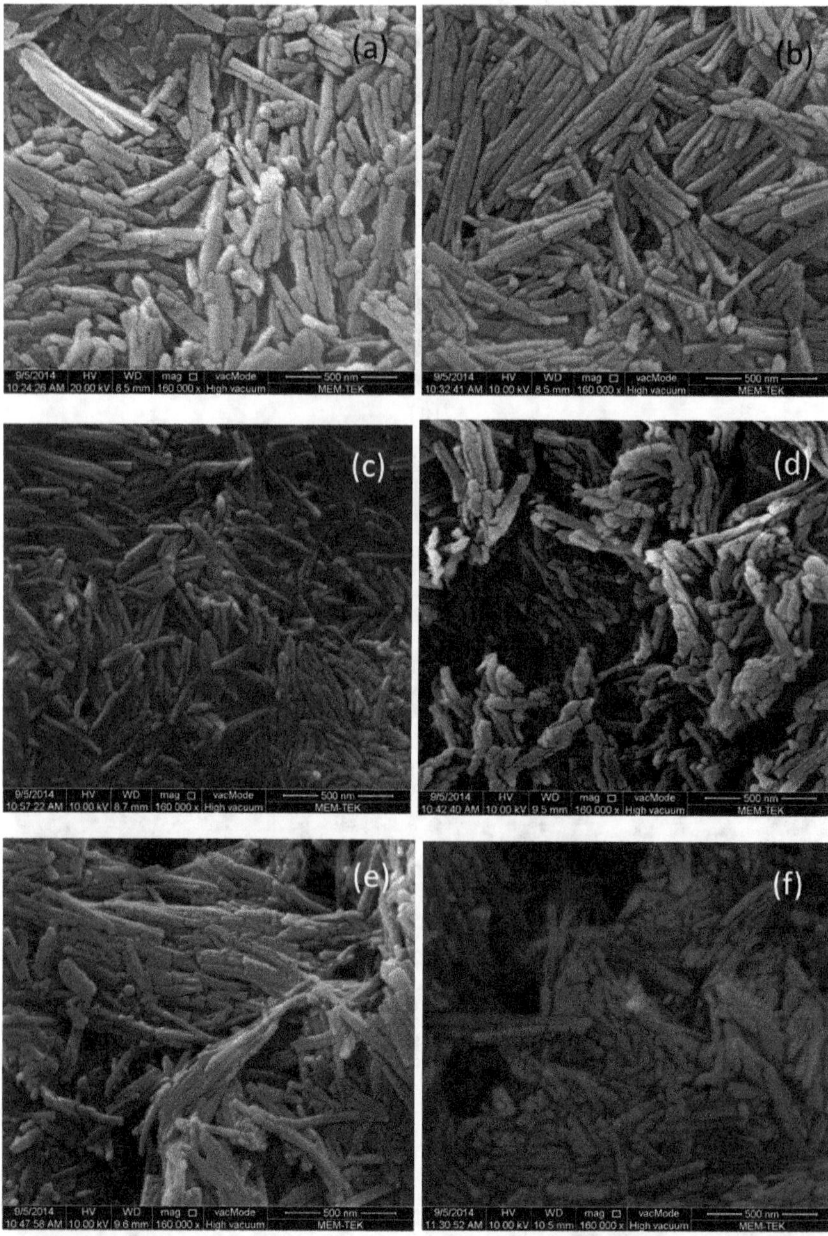

Figure 7.7 SEM micrographs of (a) S-P, (b) S-PEG, (c) S-PVP, (d) S-PVB, (e) S-PU and (f) S-EVA*g*MA minerals at high magnification.

to the stretching (vOH) vibrations of hydroxyl groups (belonging to Mg_3OH) attached to Mg ions located in the interior blocks of natural sepiolite. No change in its position was observed after modification of the surface with polymeric chains. However, the other OH deformation peaks at 3550 cm^{-1} changed significantly with polymer adsorption, as shown in Table 7.2. Also, the H-O-H deformation IR peaks were observed to change with polymer addition on the mineral surface. These observations indicated the formation of H-bonds between the polymer chains and sepiolite surface, which can benefit the composite properties effectively when the modified fillers are incorporated in the HDPE matrix.

Table 7.2 Variation in the IR bands after surface modification of sepiolite with different modifiers

Sample	OH deformation	H-O-H deformation		Si-O stretching
S-P	3550	1693	1658	1022
S-PEG	3562	1742	1657	1015
S-PVB	3565	--	1669	1019
S-PU	3565	1696	--	1018
S-PVP	3564	--	1657	1022
S-EVAgMA	3570	1671	--	1019

7.3.2 Nanocomposite Properties

Figure 7.8a presents the X-ray diffraction pattern of pristine sepiolite. Pristine sepiolite exhibited an intense reflection at $2\theta = 7.48°$ (110) with a *d*-spacing of 1.08 nm. It is well known that the *d*-spacing of the clay platelets is influenced by the organic modifier [61]. In the present work, as sepiolite was modified with different polymers through van der Waals interactions, thus, the basal plane spacing is expected to increase for the modified sepiolite samples (data not provided). However, as observed in the SEM micrographs, the surface coating of the filler particles by the polymer chains was not extensive, thus, a significant increment in the basal plane spacing cannot be expected. In addition, the polymer coating on the sepiolite surface might also slightly reduce its diffraction intensity. However, it was of larger interest to study the changes in the basal plane spacing of the modified sepiolite in the composites, the diffraction patterns of which are compared with pure polyethylene in Figure 7.8b. The diffraction pattern of polyethylene exhibited characteristic diffraction peaks at 2θ = 22.0° (110) and 24.3° (200). Both of these characteristic diffraction

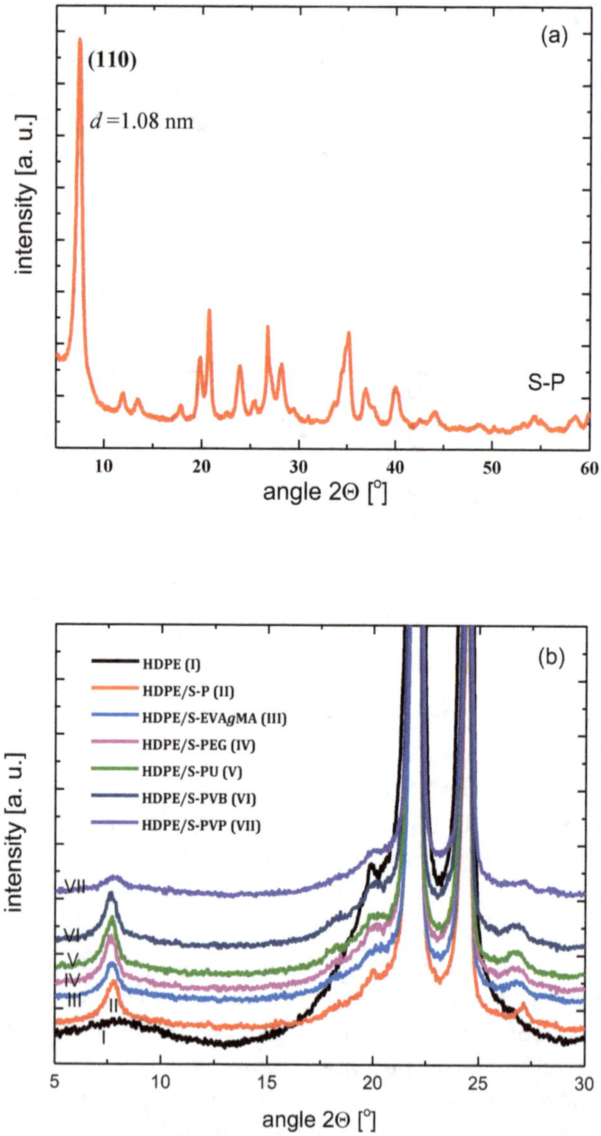

Figure 7.8 X-ray diffractograms of (a) S-P and (b) HDPE/sepiolite composites.

peaks of HDPE were also observed in the nanocomposites, which indicated that the incorporation of sepiolite did not affect the crystal structure of HDPE. The characteristic peak of sepiolite at 2θ = 7.48° was observed to shift by a small extent in the composite with HDPE, thus, indicating some polymer intercalation inside the filler interlayers, even in the absence of any filler surface modification. The peak represented a *d*-spacing of 1.13 nm in the composite, as compared to 1.08 nm in the pristine filler. The characteristic peak of sepiolite was observed to shift further, though only marginally, in the nanocomposites generated with modified sepiolite filler. For example, the sepiolite characteristic peak (110) in the case of S-PEG reinforced HDPE nanocomposite exhibited the *d*-spacing value of 1.16 nm. Similarly, other composites also exhibited slight increase in the basal plane spacing of the filler. In addition, the intensity of the diffraction peak in the S-PVP composite was also significantly reduced, thus, also qualitatively signifying the enhanced degree of filler dispersion. It indicated firstly that the surface modification of sepiolite led to enhanced degree of polymer intercalation. Secondly, the dispersion level of sepiolite was observed to be influenced by the surface coated polymer, though the differences in the resulting basal plane spacing in the composites with different modified fillers were small. As X-ray does not quantify the extent of filler intercalation and exfoliation, thus, it is also likely that varying degrees of filler exfoliation were also achieved as a result of surface modification, which is required to be analyzed through microscopy as well as by correlating with the property enhancements.

Figure 7.9 shows the TGA thermograms of pure HDPE, and nanocomposites with pristine and modified sepiolite. The thermal degradation temperatures for 10% and 40% weight loss and char yield at 700 °C are also presented in Table 7.3. It was observed that the addition of both unmodified and modified sepiolite in HDPE resulted in a significant improvement in the thermal stability. It indicated that effective heat transfer took place from the polymer chains to the sepiolite nano-fibrils, therefore, delaying the thermal degradation of the organic matter. Other studies have also reported the formation of a physical network by sepiolite with HDPE interphase, thus, resulting in enhanced thermal stability [61-63]. Among the nanocomposites with modified sepiolite, S-PVB containing composite exhibited inferior thermal stability, though it was still significantly improved as compared to pure polymer. Composites containing S-PEG and S-PVP exhibited highest temperature for 10% and 40% weight loss. The observed temperatures for these composites were 25 °C and >20 °C higher than the corresponding temperature for 10% and 40% weight

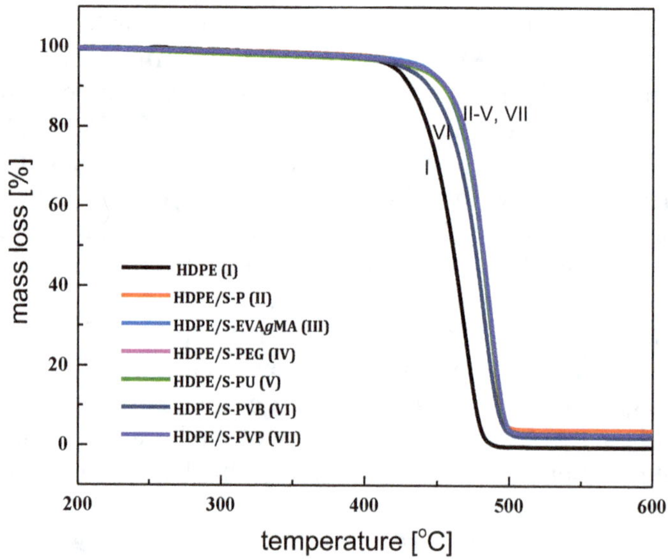

Figure 7.9 TGA thermograms of HDPE and HDPE/sepiolite nanocomposites.

loss of pure HDPE. It indicated that more uniform surface modification observed in SEM for S-PEG and S-PVP (no bridging of the primary particles) would have resulted in uniform filler delamination in the polymer matrix, thus, resulting in enhancement of the thermal performance of the polymer. However, higher thermal stability of the adsorbed polymer may have also resulted in the observed superior performance of the S-PEG and S-PVP containing composites. The char yield of the nanocomposites was also higher than that of pristine HDPE. Thus, sepiolite acted as the thermal degradation barrier at the interface with polymer, which restricted the thermal degradation of polymer at higher temperature, the effect which was further enhanced on the surface modification of sepiolite [64]. The dynamic TGA was also carried out at the compounding temperature of 180 °C as a function of time so as to ascertain the suitability of the modified sepiolite minerals for high temperature processing and applications. No significant weight loss was observed over the test duration (equivalent to the compounding time), thus, indicating that the adsorption of filler surface with the polymer did not result in early initiation of degradation.

Table 7.3 Thermal stability and char yield of HDPE/sepiolite nanocomposites

Sample	T^d_{10} (°C)	T^d_{40} (°C)	Char yield at 700 °C (%)
Pure HDPE	431	456	0.00
HDPE/S-P	454	476	3.9
HDPE/S-EVA*g*MA	454	476	2.9
HDPE/S-PEG	456	478	2.9
HDPE/S-PU	454	477	2.4
HDPE/S-PVB	444	471	2.1
HDPE/S-PVP	456	477	2.9

T^d_{10} (°C): 10% weight loss temperature
T^d_{40} (°C): 40% weight loss temperature

Figure 7.10 represents the melting and crystallization thermograms of pure HDPE, as well as composites with unmodified and modified sepiolite. The percentage of crystallinity of the polymer in the HDPE/sepiolite nanocomposites was calculated by using the following equation and the calorimetric values are also depicted in Table 7.4.

$$Xc\ (\%) = \frac{\Delta H_m}{(1-f)\Delta H_m^o} \times 100$$

where X_c is the percentage crystallinity, ΔH_m is the melting enthalpy of the polymer (second heating cycle) and ΔH_m^o is the melt enthalpy of 100% crystalline HDPE, taken as 293 J/g [65]. The addition of 5 wt% of unmodified sepiolite increased the extent of crystallinity of the HDPE (67.8% in the composite as comparison to 61% for pure polymer) which indicated that the sepiolite nano-fibrils acted as a nucleating agent for the crystallization of the HDPE matrix. It was also indicated by the increase in peak crystallization temperature in the composite to 120.6 °C from 119.3 °C for pure polymer. Variations were observed in the calorimetric behavior of nanocomposites with modified sepiolite. Composites with S-PEG, S-PVP and S-PU sepiolite exhibited similar increment in the melt enthalpy and degree of crystallinity as the composite with S-P. On the other hand, composites with S-EVA*g*MA and S-PVB exhibited lower melt enthalpy values (176 J/g) and correspondingly lower degree of crystallinity (63.2%). Same

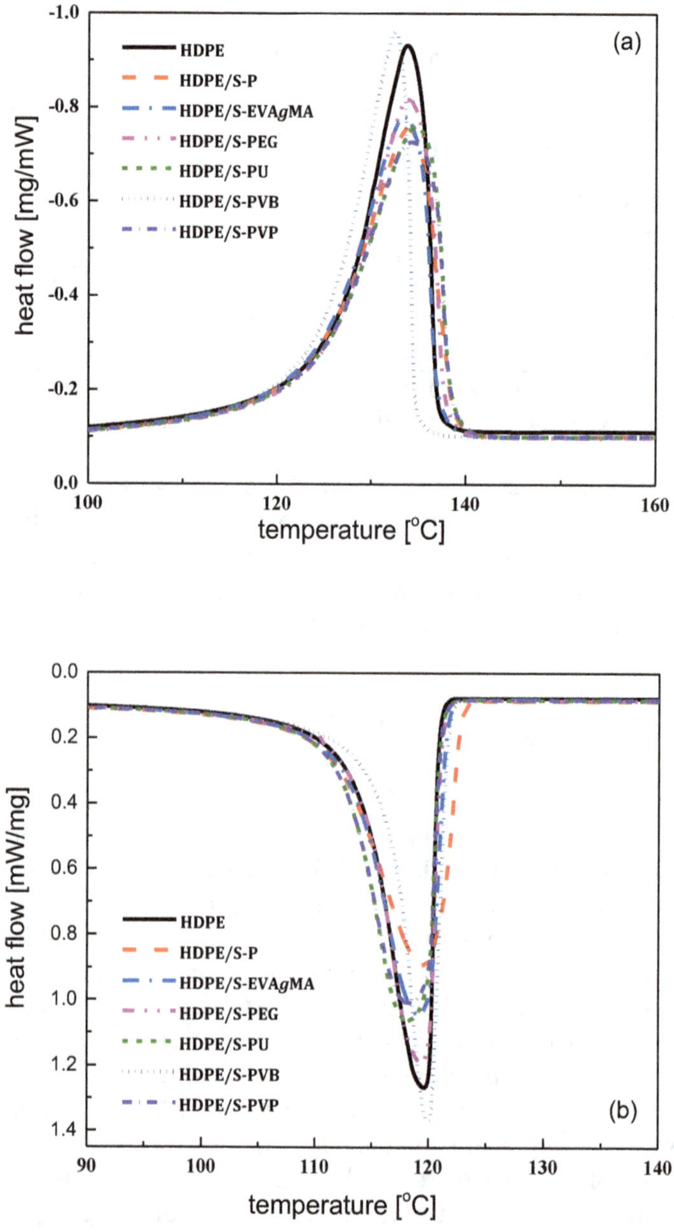

Figure 7.10 DSC (a) melting and (b) crystallization thermograms of HDPE and nanocomposites with pristine and modified sepiolite.

effect has also been observed in PA6 based polymer/organo-modified sepiolite nanocomposites [53]. However, it should be noted that the degree of crystallinity in the composites with S-EVA*g*MA and S-PVB was still higher than the pure polymer. Thus, the presence of EVA*g*MA and PVB chains adsorbed on the sepiolite surface had lower extent of nucleating efficiency of the filler during the polymer crystallization. The observed changes in the crystallinity are also expected to affect the mechanical performance, along with the effect of filler dispersion.

Table 7.4 Calorimetric behavior of HDPE, HDPE/sepiolite HDPE/modified sepiolite composites

Sample	T_m (°C)	ΔH_m (J/g)	T_c (°C)	ΔH_c (J/g)	X_c (%)
Pure HDPE	167.1	181.8	119.3	175.7	61.0
HDPE/S-P	168.2	188.9	120.6	175.2	67.8
HDPE/S-EVA*g*MA	167.1	176.0	119.2	170.0	63.2
HDPE/S-PEG	167.6	188.8	119.7	176.4	67.8
HDPE/S-PU	167.4	188.0	119.0	177.0	67.5
HDPE/S-PVB	166.4	176.0	120.1	164.0	63.2
HDPE/S-PVP	166.2	185.2	119.0	174.5	66.5

The tensile properties of the HDPE/sepiolite nanocomposites are demonstrated in Table 7.5. Pure HDPE had a tensile modulus and strength of 757 and 18.3 MPa. The addition of 5 wt% S-P significantly improved the modulus and strength of HDPE to 930 and 25.2 MPa respectively. Thus, the incorporation of even pristine sepiolite exhibited significant stress transfer and reduced the crack formation in HDPE at higher load which resulted in increased modulus. In addition, the high aspect ratio of the sepiolite nano-fibrils must have underwent some degree of defibrillation due to shear in compounder, thus, resulting in good dispersion in HDPE. The nanocomposites with modified sepiolite (except S-EVAgMA) exhibited even further enhancement in the modulus, thus, indicating that the surface modification of sepiolite surface led to enhanced polymer-filler interactions, thereby, resulting in effective filler dispersion. The modified filler may have also allowed effective delamination of the filler stacks under shear in the compounder. Thus, as also mentioned earlier, even though the diffraction peak of the filler did not change significantly in the polymer composites with modified sepiolite, however, a good degree of filler dispersion could still be achieved owing to combination of polymer-filler interactions as well as shearing in the compounder, thus, resulting in overall superior mechanical performance. Comparing the properties of composites with sepiolite modified with different polymers, it was observed that the mechanical properties of the nanocomposite were affected by the nature of modifier molecules. Maximum enhancement in the modulus was observed for the S-PEG containing composite which exhibited the modulus of 1117 MPa. It was followed by the composite with S-PVP which also exhibited significantly higher modulus of 1039 MPa. S-PVB and S-PU containing composites also enhanced the tensile modulus, though the values

Table 7.5 Mechanical properties of HDPE/sepiolite nanocomposites

Sample	Tensile modulus (MPa)	Tensile stress at yield (MPa)	Elongation at break (%)
Pure HDPE	757 (±10)	18.3 (±3)	10.3 (±2)
HDPE/S-P	930 (±10)	25.2 (±3)	8.4 (±2)
HDPE/S-EVA*g*MA	864 (±10)	25.4 (±2)	8.7 (±2)
HDPE/S-PEG	1117 (±10)	25.3 (±2)	9.4 (±2)
HDPE/S-PU	966 (±10)	24.7 (±2)	10.0 (±2)
HDPE/S-PVB	977 (±10)	24.1 (±2)	9.5 (±2)
HDPE/S-PVP	1039 (±10)	25.0 (±2)	10.3 (±2)

were only slightly higher than the corresponding composite with S-P. Incorporation of S-EVA*g*MA, on the other hand, exhibited the least increase in the modulus of PE. Thus, induced differences in crystallinity as well as polymer-filler interactions resulted in overall good degree of modulus enhancement in most of the nanocomposites. In addition, the observed enhancements were also attained without the addition of any external compatibilizer for filler compatibilization, which generally leads to matrix plasticization, thus, leading to deterioration of the properties. It should also be noted that the polymers were only physically adsorbed on the filler surface and were uncharged and non-reactive in nature. Thus, such non-reactive polymer chains on the filler surface are expected to improve the interfacial adhesion with the HDPE matrix through the van der Waals forces. In addition, the overall structural architecture and polarity of these adsorbed chains would drive such interactions with the non-polar polymer matrix. In cases like S-EVA*g*MA, probably there was a significant polarity mismatch with the polymer matrix, which resulted in lower extent of property enhancement. On the other hand, PEG chains on the filler surface may have resulted in good degree of intermixing with the polymer at the interface. Also, the PEG chains on the sepiolite surface have been reported to increase the free volume of the nanocomposite and superficially endorse the stress transfer at the interface, thus, resulting in a higher modulus value [66]. Overall, it should be noted that the interactions suggested in this study are weak in nature due to non-polar nature of the matrix and the surface modification of filler also contributed to effective kinetic mixing of the phases during shearing. The tensile strength of composites with surface modified sepiolite was observed to be similar to the composite with S-P. The elongation at break in the composites with S-PEG, S-PVP, S-PU and S-PVB minerals was almost similar to the pure polymer, which indicated that the incorporation of the filler did not lead to reduction in the toughness of the polymer. It also signified a good degree of filler dispersion in the matrix which did not result in strain hardening of the polymer chains. On the other hand, the composites with S-P and S-EVA*g*MA exhibited lower elongation due to less optimum filler dispersion and presence of large filler aggregates.

Melt rheological properties of the nanocomposites were studied using shear rheometer at 180 °C with 4% strain rate. Figure 7.11 shows the frequency sweep curves of complex viscosity, storage modulus and loss modulus the HDPE/sepiolite composites, in comparison with pure polymer. The complex viscosity of the HDPE and composites decreased with increasing ω, indicating the existence of shear thinning behavior which arises due to the pseudo-plastic nature [67]. This suggested that the nanofiller was finely dispersed in the HDPE matrix. The HDPE nanocomposites with modified sepiolite exhibited lower η^* values as compared to the composite with unmodified sepiolite. This is attributed to the low surface friction between the HDPE matrix and surface modified sepiolite, which resulted in interlayer slipperiness. Among the composites with surface modified fillers, S-EVA*g*MA containing composite exhibited the lowest viscosity. This might be due to the orientation of the copolymer chains on the sepiolite surface with the flow direction of HDPE matrix and probably disentangling from each other, thus, resulting in the reduced viscosity [68]. As the modification is only physically bound to the filler surface, a fraction of these chains can also be expected to dissociate from the filler surface at melt shearing conditions, thus, disturbing the polymer-matrix interface, which can also result in lower viscosity (and modulus) due to reduced stress transfer to the filler particles. In addition, the materials were observed to exhibit lower values of η^* at

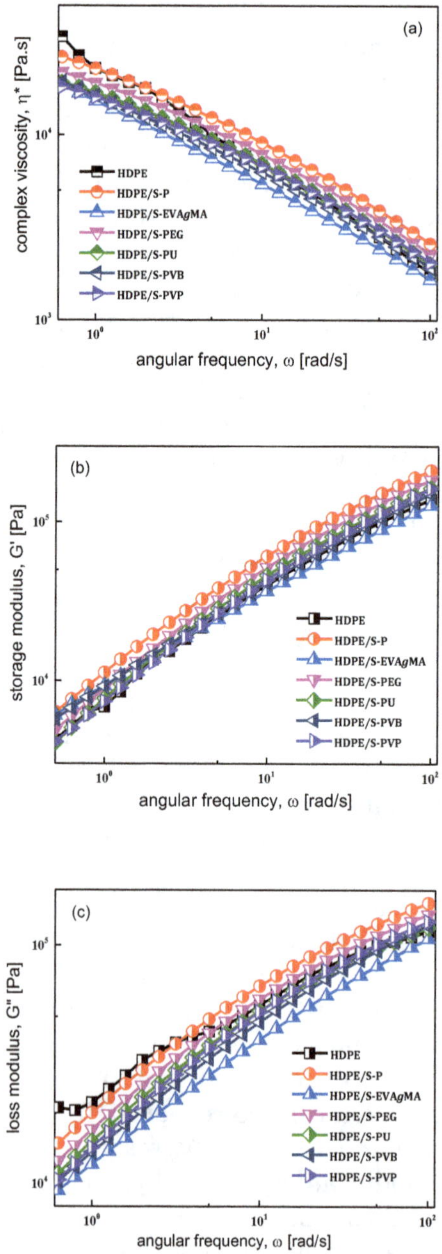

Figure 7.11 (a) Complex viscosity, (b) storage modulus and (c) loss modulus of the HDPE/sepiolite composites as a function of angular frequency.

higher angular frequencies. This is due to the fact that the molecules become completely oriented to the flow direction, thus, resulting in the generation of stable entanglements in shorter time period, which leads to reduction in the resistance to flow [69]. As observed in Figures 7.11 b and c, the G′ and G″ values of the materials increased with increase in angular frequency which indicated that the system behavior was from liquid like to solid like. The composite with unmodified sepiolite exhibited higher values of G' and G" than the composites with modified sepiolite due to the formation of physical network by the unmodified sepiolite inside the HDPE matrix phase. Similar results have also been reported for the composites with pristine and modified carbon nanotubes [70]. On the other hand, in the case of nanocomposites with surface modified sepiolite, the adsorbed polymer chains on the surface of sepiolite may result in lower G' and G" after entanglement with the polymer phase. Such an effect is not observed during the tensile analysis due to the solid state of the sample, where the entanglement of the modifier chains with the polymer matrix results in enhanced stress transfer to the filler particles. Mir *et al.* [71] observed similar G" values of non-crosslinked components compared to crosslinked components at higher frequency region. As mentioned earlier, the melt shearing conditions may also lead to a fraction of the surface adsorbed polymer chains to dissociate from the filler surface, thus, resulting in loss of interfacial contact between the polymer and filler phases. Comparing also the composites with different modified sepiolite minerals, S-PEG and S-PVP containing composites exhibited the highest storage modulus, thus, also corroborating the earlier findings from tensile analysis. It also confirmed that the nature of polymer-filler interface generated due to the type of filler surface modification contributes significantly towards the composite properties. The loss modulus of the composites was also observed to be higher than the storage modulus for the complete angular frequency range, thus, indicating the dominance of viscous nature at the melt shearing conditions.

The morphology of the S-PEG and S-PVP composites was also studied further through microscopy. As observed in the optical micrographs of the composites in Figure 7.12, the filler particles (observed as black domains) were uniformly distributed in the polymer matrix and no large filler aggregates were visible. It, thus, confirmed the earlier findings of enhanced composite properties resulting due to the effective polymer-filler interactions leading to enhanced degree of filler dispersion. Figure 7.13 demonstrates the TEM images of

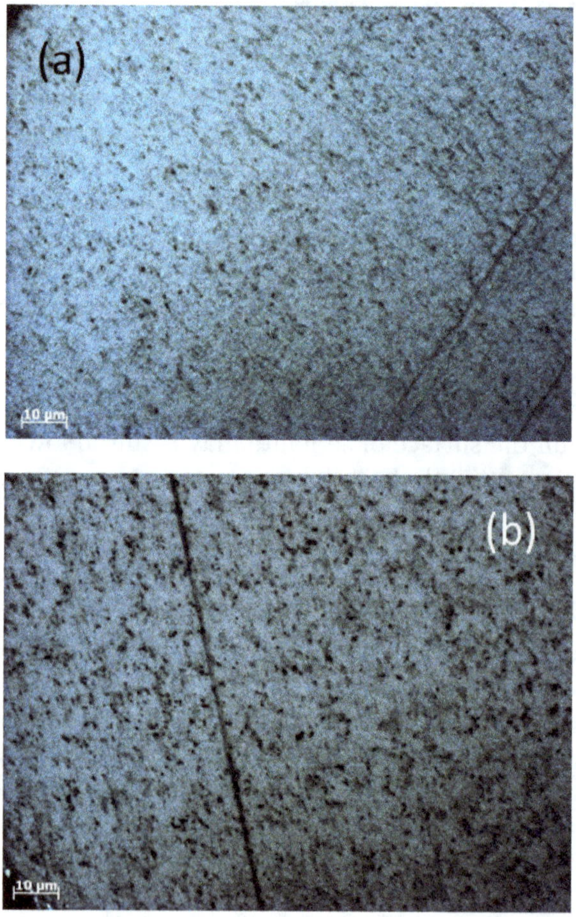

Figure 7.12 Optical micrographs of the (a) HDPE/S-PEG and (b) HDEP/S-PVP nanocomposites.

the composites with S-PEG and S-PVP at two different magnifications, in comparison with the composite with S-P. Though some degree of filler delamination was observed in the case of S-P containing composite, however, it also exhibited the presence of occasional large filler aggregates, as shown in Figure 7.13 a. On the other hand, the composites with both modified sepiolite fillers exhibited large extent of nanoscale filler delamination in the matrix, therefore, confirming the findings from optical microscopy. The enhanced interfacial contacts achieved owing to the observed filler delamination, thus, resulted in superior heat and stress transfer from the polymer chains to the filler

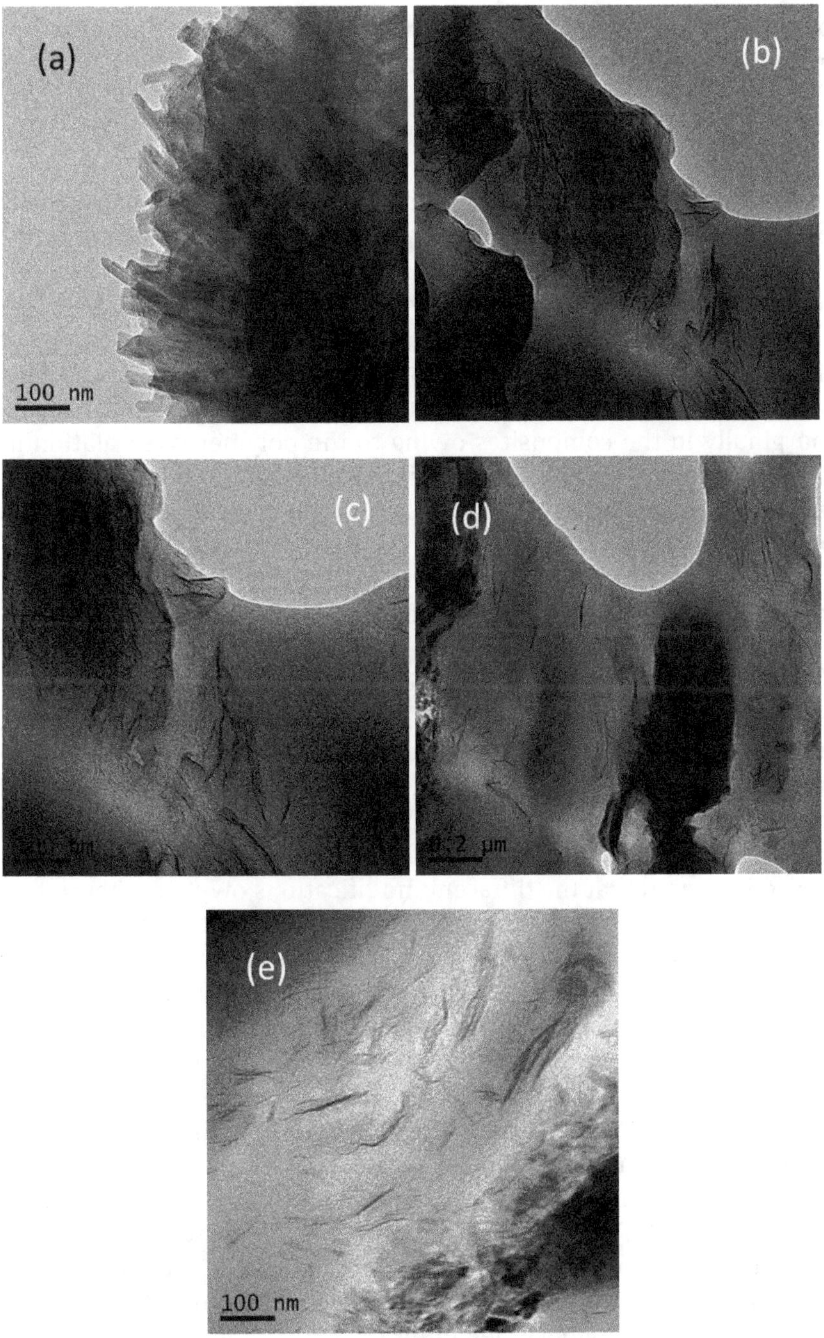

Figure 7.13 TEM images of (a) HDPE/S-P, (b,c) HDPE/S-PEG and (d,e) HDPE/S-PVP.

particles, thereby, resulting in significant improvement of the properties in the S-PEG and S-PVP containing composites.

7.4 Conclusions

In the current study, surface modification of 1-D sepiolite nano-fibrils with a variety of polymers was demonstrated using physical adsorption method. The modification of sepiolite through van der Waals interactions of the polymer chains with the inorganic surface was also confirmed through shifts in IR bands. In addition, the morphology of the sepiolite powder was also confirmed to be affected after surface modification. The basal plane spacing of the modified filler increased marginally in the composites owing to the polymer intercalation in the filler interlayers. In addition, increased extent of filler exfoliation due to the incorporation of modified filler also resulted in efficient heat transfer from the polymer chains to the inorganic filler particles, as demonstrated by the increment in the thermal stability of the polymer in the composites. Composites with S-PEG and S-PVP were observed to have largest increased in the thermal stability. The composites with S-P, S-PEG, S-PVP and S-PU exhibited a significant increase in the degree of crystallinity of the polymer, though the peak melting point of the polymer was observed to be almost constant in the different nanocomposites. S-PEG and S-PVP modified sepiolite also exhibited largest increment in the tensile modulus, thus, indicating good degree of phase intermixing of the modified filler and polymer. It also indicated that the different modifications owing to their interactions with the polymer chains had different effect on overall composite properties. Also, in the absence of any thermodynamic interaction between the polymer and modifier molecules, the effect of shear forces in the compounder as well as changes in crystallinity on the composite properties can also be expected to be significant. The storage modulus of the composites with modified sepiolite was observed to be lower than the composite with unmodified sepiolite owing to chain slipping at the interface and corresponding reduction in the stress transfer at the melt shearing conditions. The morphological analysis of the S-PEG and S-PVP containing composites also revealed uniform dispersion of sepiolite in the polymer matrix, thus, corroborating the earlier findings of enhanced mechanical and thermal properties owing to polymer-filler interactions leading to enhanced degree of filler dispersion. Overall, it could be confirmed that the surface tuning of the sepiolite surface was able to eliminate the need of an external compatibilizer. In specific, the S-PEG and S-PVP containing nanocomposites represent advanced materials with optimized polymer-filler interactions.

References

1. Mittal, V. (2009) Polymer layered silicate nanocomposites: A review. *Materials*, **2**, 992-1057.
2. Mittal, V. (2014) Functional polymer nanocomposites with graphene: A review. *Macromolecular Materials and Engineering*, **299**, 906-931.
3. Reichert, P., Nitz, H., Klinke, S., Brandsch, R., Thomann, R., and Mulhaupt, R. (2000) Poly(propylene)/organoclay nanocomposite formation: Influence of compatibilizer functionality and organoclay modification. *Macromolecular Materials and Engineering*, **275**, 8-17.
4. Uthirakumar, P., Kim, C. J., Nahm, K. S., Hahn, Y. B., and Lee, Y. S. (2004) Preparation and characterization of new difunctional cationic radical initiator-montmorillonite hybrids. *Colloids and Surfaces A: Physicochemical and Engineering Aspects*, **247**, 69-75.
5. Fan, X., Xia, C., Fulghum, T., Park, M. K., Locklin, J., and Advincula, R. C. (2003) Polymer brushes grafted from clay nanoparticles adsorbed on a planar substrate by free radical surface-initiated polymerization. *Langmuir*, **19**, 916-923.
6. Meier, L. P., Shelden, R. A., Caseri, W. R., and Suter, U. W. (1994) Polymerization of styrene with initiator bound to high surface area mica: Grafting via an unexpected mechanism. *Macromolecules*, **27**, 1637-1642.
7. Velten, U., Shelden, R. A., Caseri, W. R., Suter, U. W., and Li, Y. (1999) Polymerization of styrene with peroxide initiator ionically bound to high surface area mica. *Macromolecules*, **32**, 3590-3597.
8. Prucker, O., and Ruhe J. (1998) Mechanism of radical chain polymerizations initiated by azo compounds covalently bound to the surface of spherical particles. *Macromolecules*, **31**, 602-613.
9. Mittal, V. (2007) Polymer chains grafted "to" and "from" layered silicate clay platelets. *Journal of Colloid and Interface Science*, **314**, 141-151.
10. Moet, A. S., and Akelah, A. (1993) Polymer-clay nanocomposites: Polystyrene grafted onto montmorillonite interlayers. *Materials Letters*, **18**, 97-102.
11. Lofas, S., and Johnsson, B. (1990) A novel hydrogel matrix on gold surfaces in surface plasmon resonance sensors for fast and efficient covalent immobilization of ligands. *Journal of the Chemical Society, Chemical Communications*, 1990, 1526.

12. Wilner, I., Riklin, A., Shoham, B., Rivenson, D., and Katz, E. (1993) Development of novel biosensor enzyme electrodes: Glucose oxidase multilayer arrays immobilized onto self-assembled monolayers on electrodes. *Advanced Materials*, **5**, 912.
13. Pan, S., Castner, D. G., and Ratner, B. D. (1998) Multitechnique surface characterization of derivatization efficiencies for hydroxyl-terminated self-assembled monolayers. *Langmuir*, **14**, 3545-3550.
14. Persson, H. H. J., Caseri, W. R., and Suter, U. W. (2001) Versatile method for chemical reactions with self-assembled monolayers of alkanethiols on gold. *Langmuir*, **17**, 3643-3650.
15. Guiomar, A. J., Guthrie, J. T., and Evans, S. D. (1999) Use of mixed self-assembled monolayers in a study of the effect of the microenvironment on immobilized glucose oxidase. *Langmuir*, **15**, 1198-1207.
16. Mittal, V. (2007) Esterification reactions on the surface of layered silicate clay platelets. *Journal of Colloid and Interface Science*, **315**, 135-141.
17. Greenland, D. J., Laby, R. H., and Quirk, J. P. (1965) Adsorption of amino-acids and peptides by montmorillonite and illite. Part 2. Physical adsorption. *Transactions of the Faraday Society*, **61**, 2024-2035.
18. Zhou, Q., Frost, R. L., He, H., Xi, Y., and Liu, H. (2007) Adsorbed para-nitrophenol on HDTMAB organoclay - A TEM and infrared spectroscopic study. *Journal of Colloid and Interface Science*, **307**, 357-363.
19. Zhou, Q., He, H., Frost, R. L., and Xi, Y. (2007) Adsorption of p-nitrophenol on mono-, di-, and rrialkyl surfactant-intercalated organoclays: A comparative study. *Journal of Physical Chemistry C*, **111**, 7487-7493.
20. Mittal, V., and Herle, V. (2008) Physical adsorption of organic molecules on the surface of layered silicate clay platelets: A thermogravimetric study. *Journal of Colloid and Interface Science*, **308**, 295-301.
21. Mittal, V. (2010) Clay exfoliation in polymer nanocomposites: Specific chemical reactions and exchange of specialty modifications on clay surface. *Philosophical Magazine*, **90**(17), 2489- 2506.
22. Chaudhry, A. U., and Mittal, V. (2013) High-density polyethylene nanocomposites using masterbatches of chlorinated polyethylene/graphene oxide. *Polymer Engineering and Science*, **53**, 78-88.
23. Mahboube, M., Hamid, G., and Fatemeh, K. (2014) Structural characterization of high-density polyethylene nanocomposites: effect of compatibilizer type and high aspect ratio organoclay platelets. *Polymer Bulletin*, **71**, 2709-2731.
24. Zebarjad, S. M., and Sajjadi, S. A. (2008) On the strain rate sensitivity of HDPE/$CaCO_3$ nanocomposites. *Materials Science and Engineering A*, **475**, 365-367.
25. Singh, V. P., Vimal, K. K., Kapur, G. S., Sharma, S., and Choudhary, V.

(2016) High-density polyethylene/halloysite nanocomposites: morphology and rheological behaviour under extensional and shear flow. *Journal of Polymer Research*, **23**, 43.

26. Doriga, A., D'Amato, M., and Pegoretti, A. (2012) Thermo-mechanical properties of high density polyethylene – fumed silica nanocomposites: effect of filler surface area and treatment. *Journal of Polymer Research*, **19**, 9889.
27. Chrissafis, K., Paraskevopoulos, K. M., Pavlidoua, E., and Bikiaris, D. (2009) Thermal degradation mechanism of HDPE nanocomposites containing fumed silica nanoparticles. *Thermochimica Acta*, **485**, 65-71.
28. Liao, C. Z., and Tjong, S. C. (2013) Mechanical and thermal performance of high-density polyethylene/alumina nanocomposites. *Journal of Macromolecular Science, Part B: Physics*, **52**, 812-825.
29. Oliveira, P. L., Araujo, R. S., Oliveira, V. R., Morais, J. S., and Marques, M. F. V. (2016) Nanocomposites of polyethylene blends using organomica. *Macromolecular Symposia*, **367**, 143-150.
30. Rezanavaz, R., Razavi Aghjeh, M. K., and Babaluo, A. A. (2010) Rheology, morphology, and thermal behavior of HDPE/clay nanocomposites *Polym Composites*, **31**, 1028-1036.
31. Lee, Y. H., Park, C. B., Sain, M., Kontopoulou, M., and Zheng, W. (2007) Effects of clay dispersion and content on the rheological, mechanical properties, and flame retardance of HDPE/clay nanocomposites. *Journal of Applied Polymer Science*, **105**, 1993-1999.
32. Sanchez-Valdes, S. (2016) Influence of maleated elastomer on filler dispersion, mechanical and antimicrobial properties of hybrid HDPE/clay/silver nanocomposites. *Journal of Adhesion Science and Technology*, **30**, 1006-1016.
33. Mokhena, T. C., and Luyt, A. S. (2014) Investigation of polyethylene/sisal whiskers nanocomposites prepared under different conditions. *Polymer Composites*, **35**, 2221-2233.
34. Lei, Y., Wu, Q., Yao, F., and Xu, Y. (2007) Preparation and properties of recycled HDPE/natural fiber composites. *Composites, Part A*, **38**, 1664-1674.
35. Lee, M.-G., and Nho, Y. C. (2001) Electrical resistivity of carbon black-filled high-density polyethylene (HDPE) composite containing radiation crosslinked HDPE particles. *Radiation Physics and Chemistry*, **61**, 75-79.
36. Jeon, K., Lumata, L., Tokumoto, T., Steven, E., Brooks, J., and Alamo, R. G. (2007) Low electrical conductivity threshold and crystalline morphology of single-walled carbon nanotubes e high density polyethylene nanocomposites characterized by SEM, Raman spectroscopy and AFM. *Polymer*, **48**, 4751-4764.
37. Xiang, D., Jones, E. H., and Linton, D. (2015) Characterization and structure-property relationship of melt-mixed high density

PE/multi-walled carbon nanotube composites under extensional deformation. *RSC Advances*, **5**, 47555-47568.

38. Jiang, X., and Drzal, L. T. (2012) Multifunctional high-density polyethylene nanocomposites produced by incorporation of exfoliated graphene nanoplatelets 2: Crystallization, thermal and electrical properties. *Polymer Composites*, **33**, 636-642.
39. Fim, F. C., Basso, N. R. S., Graebin, A. P., Azambuja, D. S., and Galland, G. B. (2013) Thermal, electrical, and mechanical properties of polyethylene-graphene nanocomposites obtained by in-situ polymerization. *Journal of Applied Polymer Science*, **128**, 2630-2637.
40. Ray, S. S. (2013) *Clay-Containing Polymer Nanocomposites*, Elsevier, USA.
41. Shiraz, N. Z., Enferad, E., Monfared, A., and Mojarrad, M. A. (2013) Preparation of nanocomposite based on exfoliation of montmorillonite in acrylamide thermosensitive polymer. *ISRN Polymer Science*, Article ID 280897.
42. Sapalidis, A. A., Katsaros, F. K., Steriotis, T. A., and Kanellopoulos, N. K. (2012) Properties of poly(vinyl alcohol)-bentonite clay nanocomposite films in relation to polymer-clay interactions. *Journal of Applied Polymer Science*, **123**, 1812-1821.
43. Velasco, J. I., Antunes, M., Ayyad, O., Saiz-Arroyo, C., Rodriguez-Perez, M. A., Hidalgo, F., and de Saja, J. A. (2007) Foams based on low density polyethylene/hectorite nanocomposites: Thermal stability and thermo-mechanical properties. *Journal of Applied Polymer Science*, **105**, 1658-1667.
44. Bilotti, E., Duquesne, E., Deng, H., Zhang, R., Quero, F., Georgiades, S. N., Fischer, H. R., Dubois, P., and Peijs, T. (2014) In situ polymerised polyamide 6/sepiolite nanocomposites: Effect of different interphases. *European Polymer Journal*, **56**, 131-139.
45. Tartaglione, G., Tabuani, D., Camino, G., and Moisio, M. (2008) PP and PBT composites filled with sepiolite: Morphology and thermal behaviour. *Composites Science and Technology*, **68**(2), 451-460.
46. Vengatesan, M. R., Singh, S., Stephen, S., Prasanna, K., Lee, C. W., and Mittal, V. (2017) Facile synthesis of thermally reduced graphene oxide-sepiolite nanohybrid via intercalation and thermal reduction method. *Applied Clay Science*, **135**, 510-515.
47. Chen, H., Zheng, M., Sun, H., and Jia, Q. (2007) Characterization and properties of sepiolite/polyurethane nanocomposites. *Materials Science and Engineering A*, **445-446**, 725-730.
48. Yazdi, Z. T., and Ataei, S. M. (2015) Poly (ether-imide) and related sepiolite nanocomposites: investigation of physical, thermal, and mechanical properties. *Polymers for Advanced Technologies*, **26**, 308-314.
49. Liu, Y., Zhao, J., Deng, C.-L., Chen, L., Wang, D.-Y., and Wang, Y.-Z. (2011) Flame-retardant effect of sepiolite on an intumescent flame-

retardant polypropylene system. *Industrial & Engineering Chemistry Research*, **50**, 2047-2054.

50. Mittal, V., Chaudhry, A. U., Cole, M., and Matsko, N. (2017) Polyethylene nanocomposites with psyllium husk, graphene, sepiolite, and their hybrids. In: *Functional Polymer Blends and Nanocomposites*, Mittal, V. (ed.), Central West Publishing, Australia, pp. 165-194.
51. Nohales, A., Munoz-Espi, R., Felix, P., and Gomez, C. M. (2011) Sepiolite-reinforced epoxy nanocomposites: Thermal, mechanical, and morphological behavior. *Journal of Applied Polymer Science*, **119**, 539-547.
52. Pinnavaia, T. J., and Beall, G. W. (2000) *Polymer-Clay Nanocomposites*, Wiley, USA.
53. Lopez, D. G., Fernandez, J. F., Merino, J. C., Santaren, J., and M.Pastor, J. (2010) Effect of organic modification of sepiolite for PA 6 polymer/organoclay nanocomposites. *Composites Science and Technology*, **70**, 1429-1436.
54. Lemic, J., Tomasevic-Canovic, M., Djuricic, M., and Stanic, T. (2005) Surface modification of sepiolite with quaternary amines. *Journal of Colloid and Interface Science*, **292**, 11-19.
55. Ozdemir, O., Cinar M., Sabah, E., Arslan, F., and Celik, M. S. (2007) Adsorption of anionic surfactants onto sepiolite. *Journal of Hazardous Materials*, **147**, 625-632.
56. Mejia, A., Garcia, N., Guzman, J., and Tiemblo, P. (2014) Surface modification of sepiolite nanofibers with PEG based compounds to prepare polymer electrolytes. *Applied Clay Science*, **95**, 265-274.
57. Alan, N., and Isci, S. (2014) Surface modification of sepiolite particles with polyurethane and polyvinyl alcohol. *Progress in Organic Coatings*, **77**, 444-448.
58. Turhan, Y., Turan, P., Dogan, M., Alkan, M., Namli, H., and Demirbas, O. (2008) Characterization and adsorption properties of chemically modified sepiolite. *Industrial & Engineering Chemistry Research*, **47**, 1883-1895.
59. Vengatesan, M. R., and Mittal, V. (2015) Surface modification of nanomaterials for application in polymer nanocomposites: An overview. In: *Surface Modification of Nanoparticle and Natural Fiber Fillers*, Mittal, V. (ed.), Wiley VCH, Germany, pp. 1-28.
60. Isci, S. (2015) Surface treatment of sepiolite particles with polymers. In: In: *Surface Modification of Nanoparticle and Natural Fiber Fillers*, Mittal, V. (ed.), Wiley VCH, Germany, pp. 151-170.
61. Mittal, V. (2010) Polymer nanocomposites: Synthesis, microstructure, and properties. In: *Optimization of Polymer Nanocomposite Properties*, Mittal, V. (ed.), Wiley VCH, Germany, pp. 1-20.
62. Chivrac, F., Pollet, E., Schmutz, M., and Averous, L. (2010) Starch nano-biocomposites based on needle-like sepiolite clays. *Carbohydrate Polymers*, **80**, 145-153.

63. Roy, N., and Bhowmick, A. K. (2010) Novel in-situ polydimethylsiloxane-sepiolite nanocomposites: Structure-property relationship. *Polymer*, **51**, 5172-5185.
64. Bidsorkhi, H. C., Soheilmoghaddam, M., Pour, R. H., Adelnia, H., and Mohamad, Z. (2014) Mechanical, thermal and flammability properties of ethylenevinyl acetate (EVA)/sepiolite nanocomposites. *Polymer Testing*, **37**, 117-122.
65. Joshi, M., Butola, B. S., Simon, G., and Kukaleva, N. (2006) Rheological and viscoelastic behavior of HDPE/octamethyl-POSS nanocomposites. *Macromolecules*, **39**, 1839-1849.
66. Nanjegowda, L. M., Bommulu, R., Juikar, V., and Hatna, S. (2013) Investigation on the influence of different compatibilizers on polycarbonate and high density polyethylene blends: Mechanical properties, thermal properties, morphology, and chemical resistance. *Industrial & Engineering Chemistry Research*, **52**, 5672-5682.
67. Zhu, L., Fan, H.-N., Yang, Z.-Q., and Xu, X.-H. (2010) Evaluation of phase morphology, rheological, and mechanical properties based on polypropylene toughened with poly(ethylene-co-octene). *Polymer-Plastics Technology and Engineering*, **49**, 208-217.
68. Wu, D., Wang, X., Song, Y., and Jin, R. (2004) Nanocomposites of poly(vinyl chloride) and nanometric calcium carbonate particles: Effects of chlorinated polyethylene on mechanical properties, morphology, and rheology. *Journal of Applied Polymer Science*, **92**, 2714-2723.
69. Kumar, M., Priya, N. S., Kanagaraj, S., and Pugazhenthi, G. (2016) Melt rheological behavior of PMMA nanocomposites reinforced with modified nanoclay. *Nanocomposites*, **2**, 109-116.
70. Ahmad, A. A., Al-Juhani, A. A., Thomas, S., De, S. K., and Atieh, M. A. (2013) Effect of modified and nonmodified carbon nanotubes on the rheological behavior of high density polyethylene nanocomposite. *Journal of Nanomaterials*, Article ID 731860.
71. Mir, S., Yasin, T., Halley, P. J., Siddiqi, H. M., Ozdemir, O., and Nguyen, A. (2013) Thermal and rheological effects of sepiolite in linear low-density polyethylene/starch blend. *Journal of Applied Polymer Science*, **127**, 1330-1337.

8

Polyethylene/Graphene Nanocomposites: A Review of Morphological, Mechanical and Thermal Properties

8.1 Introduction

Polymer nanocomposites, with at least one of the dimensions of the dispersed phase at the nanoscale (<100 nm), are an important area of research because of their potential use in a wide range of applications. There are a variety of uses in industry as well as in daily life that devices made out of neat polymers or metals cannot be put to because they do not meet the requisite thermal, mechanical and electrical properties. The compounding of polymers with nanomaterials, most of which have been found to have excellent mechanical, thermal and electrical properties, has enabled the production of nanocomposites whose properties are significantly better than those of the neat polymers due to the synergetic interaction between the polymers and nanomaterials. Nanocomposites have, due to these improved properties, strong potential of application in various commodity and engineering sectors. Out of various reinforcement materials used to develop nanocomposites, silica, alumino-silicate clays, carbon nanotubes and graphene have been extensively reported to impart significant improvement in a variety of polymer properties. For instance, Figure 8.1 shows the height, amplitude, and phase images of the compatibilized polyethylene (PE) nanocomposites, using silica and alumino-silicate as fillers and ethylene vinyl acetate-graft-maleic anhydride (EVA-*g*-MA) as compatibilizer. In the phase images, the filler particles were observed to be white in color, thus, signifying harder regions, whereas the polymer matrix was noted as the dark phase. The phase of the polymer near the interface with the filler surface as well as in the bulk was similar, thus, indicating that the filler did not have any specific effect on the crystallization process. The filler-polymer interface in the micrographs also indicated that the compatibility between the phases was significantly improved in the presence of EVA-*g*-MA as compared to the uncompatibilized composites, which correspondingly impacted the mechanical and thermal properties of

Baron Rubahamya and Vikas Mittal
Khalifa University of Science and Technology, Abu Dhabi, UAE

the developed nanocomposite.

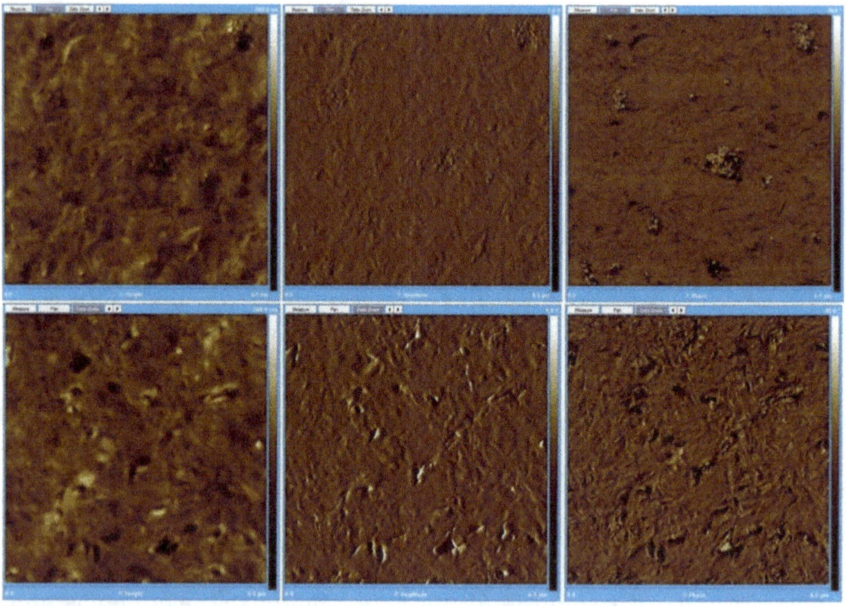

Figure 8.1 AFM height, amplitude and phase images of PE+EVA-*g*-MA+Silica (upper row) and PE+EVA-*g*-MA+Silicate (lower row) composites at lower magnification. Reproduced from Reference 1 with permission from Sage Publications.

Graphene has, since its discovery, gained significant interest from researchers because of its superior properties [2]. Graphene is a two-dimensional nanomaterial made up of sp^2 carbon atoms, has the thickness of one atom and a honeycomb structure. It is the strongest single layer material that has ever been measured having a Young's modulus of 1 TPa and a tensile strength of 130 GPa [3]. Graphene also shows excellent electrical and thermal conductivity values, being 6000 S/cm and 5000 W/(m.K) respectively [4]. These properties coupled with its vast surface area make graphene a nanomaterial of great interest for research concerning its application in making nanocomposites that leverage its individual properties to produce materials of greatly improved mechanical, thermal and electrical properties compared to the neat polymers. This increased interest in graphene can be evidenced by the great volume of research studies on polymer nanocomposites incorporating graphene reported in the recent years [2,5].

The synthesis of graphene has been achieved through various routes including bottom-up as well as top-down processes. Some of the top-down processes include the exfoliation of graphite and the derivatives of graphite, most commonly graphite oxide, into sheets of graphene or modified graphene [6]. The exfoliation route of graphene synthesis by directly exfoliating graphite has transitioned from the earlier techniques of micromechanically breaking down graphite [7] to the recent techniques of exfoliating graphite into single layer and multiple layer graphene by dissolving it in superacids [8], electrochemically functionalizing it using ionic liquids [9] as well as the use of sonication [10]. On the other hand, the exfoliation of graphite into sheets of graphite oxide and their subsequent reduction, provides a great potential for the production of graphene on a large scale. The most commonly used methods for the production of graphene from graphite oxide are the Hummer's method [11] and a number of variations of this method, in which graphite is subjected to a process of oxidation in the presence of nitric acid using strong oxidants such potassium permanganate. Bottom-up methods though lead to better control on the structure and properties of graphene, however, these have not been as widely used as top-down methods because of the costly process and the fact that these lead to production of low amounts of graphene. Some of the methods that have been used include chemical vapor deposition [12], surfactant self-assembly [13], reduction of carbon monoxide [14] and arc discharge [15].

To be able to obtain the desired properties that result from compounding polymers with nanomaterials such as graphene, the dispersion of the nanomaterial in the polymer matrix is at nanoscale is of great importance. Thus, a higher degree of graphene dispersion in the polymer matrix has been commonly reported to result in greater improvement of properties [16]. One important advantage of the nanofillers such as graphene is the possibility to achieve improvements in the properties of the polymers with low amounts of filler loading due to the high surface areas and high aspect ratios of these nano-fillers [17]. The interfacial interaction between the polymer matrix and graphene nano-sheets is also influenced by various factors including surface reactive groups, hydrophobicity and molecular weight of the polymer chains, functional groups in the polymer chains, among others. In addition, the dispersion of graphene in the polymer matrices is also influenced by the method of compounding of the polymer with graphene [4] and compounding techniques include, but are not limited to, the following commonly used ones [18]:

In-situ intercalative polymerization: In this technique, graphene or its modified version is simultaneously compounded with the polymer matrix during the process of polymerization. This is performed by mixing graphene with the monomer in the presence of an initiator and subsequently providing the optimum conditions for polymerization.

Intercalation with the polymer or pre-polymer in solution: In this method, a solvent in which the polymer or a pre-polymer (such as an oligomer) can dissolve and in which graphene or its modified form can swell is used. The dissolved polymer and swollen graphene are mixed in solution allowing for the graphene sheets to be intercalated by the polymer chains and a nanocomposite is achieved when the solvent is removed, by means of evaporation. Even though the entropy of the system is reduced due to the restraining of the polymer chains within the graphene sheets, this process is thermodynamically possible due to the increase in entropy achieved by the desorption of solvent molecules from the surface of the graphene sheets.

Melt intercalation: In this method, the polymer forms a melt at a raised temperature and in the molten state is mixed with graphene or modified graphene under shear, such as by using a rotor. The mixture is then cooled to obtain a nanocomposite in which the graphene sheets are exfoliated and intercalated by the polymer chains. The temperature of compounding should be carefully chosen to achieve thorough mixing without degradation of the polymer as well to achieve the optimum viscosity at which the polymer melt can withstand the shear stress in the compounder. This method has been widely used to synthesize nanocomposites of graphene with thermoplastic polymers [5].

As mentioned earlier, the interfacial interaction between the polymer matrix and graphene plays an important role in the extent of dispersion of the graphene sheets. The graphene sheets have the tendency to agglomerate by means of van der Waals forces and through π-π stacking, especially when added in large amounts [19]. This phenomenon is usually irreversible to a large extent and the agglomerates formed lead to significant reduction in the surface area and aspect ratio. To achieve improved interfacial interactions between graphene and polymer matrices, and consequently improved dispersion and nanocomposite properties, graphene sheets' surfaces are usually chemically functionalized. For the dispersion of graphene in polar polymers such as polyurethane, polyamide and poly (vinyl alcohol), the surfaces of the nano-sheets are modified with polar groups including epoxies, hydroxyl groups and carboxylic acid groups [20-22], while for dispersion in non-polar polymer matrices such as polyolefins, the surface is usually rendered organophilic by the incorporation of groups including maleic acid grafted polyolefins, alkylamines, phenyl isocyanate, silanes, and others [23,24].

PE is a widely used thermoplastic polyolefin due to its large number of beneficial properties. The most common forms of PE include low density polyethylene (LDPE), high density polyethylene (HDPE), ultra-high molecular weight polyethylene (UHMWPE) and linear low density polyethylene (LLDPE). LDPE has a melting temperature between 100 and 120 °C, a specific gravity ranging from 0.91 to 0.94 and has partial crystallinity ranging between 50 to 70%. LDPE exhibits a greater level of branching compared to HDPE, with 10 to 20 branches for every 1000 carbon atoms compared to HDPE's average 1 branch for every 200 carbon atoms in the main chain. Existence of branches in the carbon chains is a defect and affects the degree of crystallinity, which along with molecular weight and molecular weight distribution determines the structural properties of polyethylene. Mechanical properties including tensile strength, rigidity and others are observed to be directly proportional to crystallinity, and consequently inversely proportional to the amount of branching within the chain. Toughness of polyethylene, however, decreases with increasing degree of crystallinity [25]. Polyethylene finds application in a variety of areas including electrical appliances, industrial equipment, construction, body part replacements, packaging, transportation and agriculture, among others [25]. The wide range of applications of polyethylene, therefore, makes it a polymer of interest for further developing its nanocomposites, thus, expanding its spectrum of applications further.

This study aims to review recent studies reporting the development of polyethylene/graphene nanocomposites with a focus on the composite properties, especially morphological, mechanical and thermal properties. In addition, these properties have also been correlated with the synthesis methods as well as filler type, content, compatibilization and surface functionalization. Specifically, various mechanical properties such as Young's modulus, yield stress, yield strain, tensile strength, elongation at break, impact strength and dynamic mechanical response have been reviewed. On the other hand, thermal properties such as glass transition temperature, melting temperature, degree of crystallinity, thermal degradation, etc., have been studied. In addition, some insights into the rheological performance of nanocomposites have also been provided.

8.2 Review of Nanocomposite Properties

Along with other applications, UHMWPE has been extensively studied for its use in replacement of human body parts such as the joints of knee, shoulder, hip and spine. This requires for it to have specific mechanical properties such as high tensile strength to ensure increased wear resistance and prevention of early failure as well as high yield stress to ensure that the material does not deform easily. To this end, attempts have been made to improve the inherent mechanical properties of UHMWPE for it to be able to meet the requirements of this application and some of the techniques that have been used include increasing its degree of crosslinking. UHMWPE/graphene nanocomposites present a potential alternative for meeting these goals. A range of synthesis methods have been reported to generate these UHMWPE nanocomposites in the recent years. For instance, Lahiri *et al.* [26] synthesized a nanocomposite of UHMWPE and graphene nano-platelets (GNP) by an electrostatic spraying procedure. In this procedure, GNPs were uniformly dispersed in acetone using ultrasonication and UHMWPE powder was added subsequently with further ultrasonication. It was followed by removal of the solvent by means of evaporation. Electrostatic spraying was then used to deposit films of the nanocomposite on non-stick substrate coated with Teflon. The films were then cured, cooled and peeled off the substrate. The compositions of the synthesized films were neat UHMWPE, UHMWPE with 0.1 wt% as well as 1 wt% of GNPs. Analysis of the mechanical properties of these films exhibited that the elastic modulus of neat UHMWPE increased from 0.53 GPa to 0.69 GPa for its nanocomposite containing 0.1 wt% GNPs and increased further to 1.19 GPa when the GNPs concentration was 1 wt%. The same trend was observed for yield stress which increased from 22 to 35 and then to 42 MPa respectively. Tensile strength and toughness, on the other hand, showed different trends with the tensile strength increasing from 45 MPa for neat UHMWPE to 77 MPa when the nanocomposite contained 0.1 wt% GNPs, but then reduced to 68 MPa when the GNPs content was 1 wt%. Toughness, on the other hand, was initially 98 $J.m^{-3}$, increasing to 151 $J.m^{-3}$ before reducing to 30 $J.m^{-3}$ respectively.

These trends were explained by the fact that the stress transfer from UHMWPE to the platelets resulted in the continuous increase in elastic modulus and yield stress with increase in weight percentage of the platelets. However, due to the agglomeration that resulted from increased amounts of GNPs, the interfacial interactions between UHMWPE and the platelets deteriorated making the nanocomposites brittle with reduced tensile strength and toughness. In another study, Suner *et al.* [27] prepared nanocomposites of UHMWPE with graphene oxide (GO) using a ball milling technique in which a slurry of UHMWPE powder and GO powder in ethanol was mixed using a planetary ball mill. Afterwards, the solvent was evaporated and the nanocomposite sheets were molded from the powders using a hot press. The nanocomposites were prepared with GO content of 0.1, 0.3, 0.5, 0.7, 1 and 2 wt%. The mechanical properties exhibited the same trends as noted by Lahiri *et al.* [26] where improved properties as compared to neat UHMWPE were achieved with optimum loading of 0.5 wt% of GO and higher amounts of loading resulting in decrease in the values. Thus, the toughness, fracture stress and fracture strain of the nanocomposite with 2 wt% GO were observed to be lower than those of neat UHMWPE. The melting temperature, temperature of crystallization and degree of crystallization values were constant with increasing GO content and were observed to be same as neat UHMWPE. The nanocomposites, however, exhibited improved thermal stability where the oxidation temperature, the linear weight loss initiation temperature and the temperature of complete volatilization increased with increasing GO content.

Sturzel *et al.* [28] synthesized UHMWPE/graphene nanocomposite via an *in-situ* polymerization route (Figure 8.2). GO was prepared

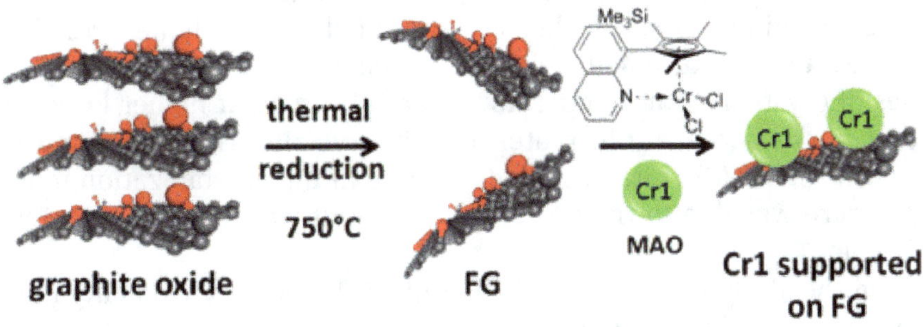

Figure 8.2 Synthesis of UHMWPE/graphene nanocomposite via *in-situ* polymerization. Reproduced from Reference 28 with permission from American Chemical Society.

from graphite by means of a modified Hummer's method before being thermally reduced to obtain functionalized graphene (FG). By dispersing FG/methylaluminoxane (MAO) in n-heptane, the authors immobilized catalyst sites composed of MAO-activated chromium on the FG to enable the dispersion of the nano-sheets in UHMWPE to happen simultaneously with the polymerization process under optimum conditions. Nanocomposite samples were obtained using hot press molding with varying FG compositions of 0.5, 1, 2.5, 5 and 10 wt%. The Young's modulus increased with increasing FG content of the nanocomposite, increasing from 420 MPa for the commercial UHMWPE used as the reference point through 630 MPa for the composite with 0.5 wt% FG content up to 991 MPa for 10 wt% FG content. The elongation at break and tensile strength, on the other hand, increased with increasing FG content until 1 wt% FG content, but then reduced as the content of filler increased, with the 10 wt% FG content nanocomposites exhibiting even lower properties than the commercial UHMWPE. This observation further evidenced that maximization of the mechanical properties of UHMWPE/graphene nanocomposites can be attained at low graphene loadings and deterioration happens with large loading amounts of the nanomaterial. It should also be noted that Sturzel *et al.* [28] achieved greater values of tensile strength for the nanocomposites at the same graphene loading as Lahiri *et al.* [26] and Suner *et al.* [27]. This could be due to the use of FG and *in-situ* polymerization technique that improved interfacial interaction between the polymer matrix and filler phases as well as reduced the agglomeration tendency of the filler nano-sheets. In addition, the authors indicated that the melting temperature of the nanocomposites was almost constant with increasing filler content with a difference of only a few degrees (~3 °C), while the temperature of crystallization increased when FG was added to the polymer with the highest temperature being observed at 0.5 wt% fraction. Lower values of crystallization temperature were observed at higher FG content, but these were still greater than the crystallization temperature of neat UHMWPE. The observed increase in the crystallization temperature was due to the nucleation sites that were provided by the FG filler.

In the study reported by Chen *et al.* [29], powdery GO generated using modified Hummer's method from graphite was uniformly dispersed by means of ultrasonication in alcohol to form a suspension to which UHMWPE powder was added and the solvent evaporated after further ultrasonication. Nanocomposite samples of 0, 0.1, 0.3, 0.5 and

1 wt% GO content were then molded from the solid using hot pressing. The yield strength was observed to increase with increasing GO content while the ultimate tensile strength and elongation at break both reached maximum at 0.5 wt% GO content before showing a tremendous decrease for 1 wt% GO composite. These findings were also similar to the observation reported by Suner *et al.* [27] for the composites of UHMWPE with derivative of graphene (GO) using the similar route of nanocomposite synthesis. The authors also reported that the melting point of the nanocomposites was constant, while the degree of crystallization increased with addition of GO with the highest observed degree of crystallization obtained at 0.1 wt% GO content. It was observed to subsequently reduce as more GO was added. The filler provided nucleation sites, but high filler content led to agglomeration, thus, resulting in poor dispersion.

Kuila *et al.* [30] functionalized graphite oxide (GO) using dodecylamine (DA) through the nucleophilic substitution of the DA moieties onto the epoxy groups of GO and subsequently reduced the product and dried it. The functionalized GO was then compounded with linear low density polyethylene (LLDPE) by means of solution dispersion using xylene as the solvent. Nanocomposite samples with varying amounts of filler were compression molded. Analysis of the dynamic mechanical properties of the nanocomposites indicated that the storage modulus increased with increasing amount of functionalized GO in the polymer matrix, whereby the nanocomposite sample was observed to have a storage modulus 118% higher than that of virgin LLDPE. The observed improvement in the properties was attributed to the filler decreasing the mobility of the LLDPE chains. Loss modulus measurements were not reported by the authors, but the loss factor at 80 °C showed a general decrease with increasing filler content in the nanocomposite. The nanocomposites exhibited constant melting temperature, while the crystallization temperature increased with increasing graphene content. It was observed to reduce at 8 wt% of functionalized reduced graphene oxide content, yet was still greater than that of neat LLDPE. TGA also revealed that the onset of degradation occurred at lower temperatures for the nanocomposites than for neat LLDPE because of the degradation of the dodecylamine. However, the thermal stability of the nanocomposites increased with increasing filler content as was seen based on the increasing temperatures required for 50% and 80% weight loss as the filler content increased. This was attributed to the filler nano-sheets reducing the rate of degradation by preventing the movement of radicals and flow of oxygen through the polymer matrix, thereby, leading to a shielding effect.

In another study, Yang *et al.* [19] studied the influence of different chain length alkylamine functionalizing agents on the properties of the nanocomposites of functionalized graphene oxide and LDPE. Octadecylamine (ODA), dodecylamine (DDA) and octylamine (OTA) were the three different alkylamines utilized to functionalize reduced graphene oxide, which was used to make nanocomposites with LDPE by means of solution dispersion. The nanocomposites with 0.5 wt% filler loading of reduced graphene oxide functionalized with each of the different chain length alkylamines were analyzed for their properties. It was observed that in each of the three cases, the Young's modulus and tensile strength were higher for the nanocomposite than that of neat LDPE, whereas the elongation at break reduced. Good filler dispersion and improved interfacial interaction were the reasons for stress transfer from the polymer matrix to the filler and resulted in improved mechanical properties. However, the reduction in elongation at break was attributed to certain extent of agglomeration. The filler modified with DDA had higher tensile strength than the other two nanocomposites at the same filler loading, with the nanocomposite with OTA functionalization leading to the least tensile strength. Characterization of the functionalized reduced graphene oxide showed that DDA functionalized reduced graphene oxide contained more primitive graphene than ODA which was reported to be the reason for the higher tensile strength. The temperature of maximum weight loss of the nanocomposites was reported to be greater than that of neat LDPE owing to the barrier effect and the slowing down of oxygen permeation. These results were also similar to those reported by Kuila *et al.* [30].

Mittal and Chaudhry [31] studied the effect of amphiphilic compatibilizers on the properties of nanocomposites of high density polyethylene and thermally reduced graphene. Nanocomposites were prepared by melt mixing the polymer and various types of compatibilizers with thermally reduced graphene with a composition of 5 wt% compatibilizer and 1 wt % thermally reduced graphene. Different compatibilizers resulted in different interactions between the polymer matrix and filler phases. Specifically, the molecular weight, extent of grafting, and polarity in the compatibilizers were the factors which determined their interactions with the filler surface. It was observed that the highest improvements in tensile modulus occurred in the nanocomposites generated using 25% chlorinated polyethylene

(CPE25), ethylene-α-olefin copolymer (E-AO) and ethylene acrylic acid copolymer (EAA), while 35% chlorinated polyethylene (CPE35) resulted in increased plasticization because of its amorphous nature. Dynamic mechanical properties also exhibited improvement for the nanocomposites whereby storage modulus was reported to increase to a varying degree. To attain an optimum amount of compatibilizer in the nanocomposites for property enhancement, the amount of best performing compatibilizer EAA was varied in the composites. As observed in Figure 8.3, the extent of filler exfoliation was observed to increase with increasing EAA content, thus, confirming positive interactions between EAA and thermally reduced graphene, though no specific chemical interactions could be detected. The observed

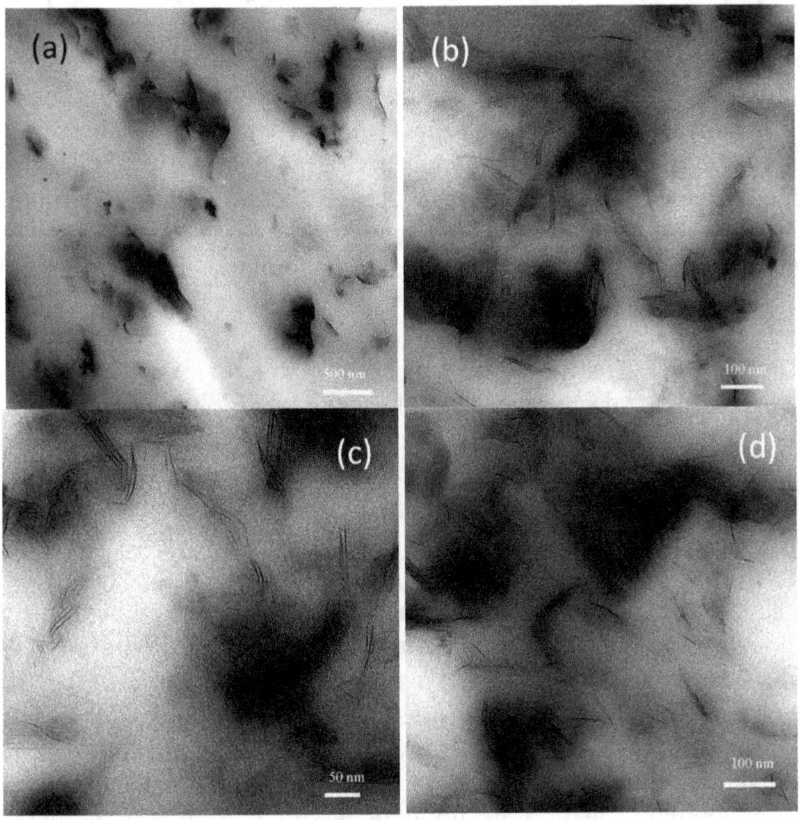

Figure 8.3 TEM micrographs of (a) and (b) PE/G/5%EAA nanocomposite at different magnifications; (c) PE/G/10% EAA nanocomposite and (d) PE/G/15%EAA nanocomposite. Reproduced from Reference 31 with permission from Wiley.

properties from these composites revealed that 7.5 wt% was the optimum amount of EAA in the composites as the mechanical and rheological properties exhibited improvement due to filler exfoliation. Beyond 7.5 wt% content of EAA in the composites, the extent of filler exfoliation though was observed to increase, the performance of the nanocomposites deteriorated due to matrix plasticization. The authors also observed that the compatibilized nanocomposites had higher thermal stability than the non-compatibilized HDPE/G nanocomposite, which was itself more thermally stable than neat HDPE. Nanocomposites synthesized using ethylene acrylic acid copolymer (EAA) as compatibilizer were reported to have the highest thermal stability while those synthesized using 25% chlorinated polyethylene (CPE25) showed increased melt enthalpy because of its semi-crystalline nature. Other nanocomposites, including non-compatibilized HDPE/G nanocomposite, were reported to show lower melt enthalpy than neat HDPE because of reduced crystallinity. Nanocomposites synthesized using ethylene methacrylic acid copolymer with zinc ion, ethylene butyl acrylate copolymer and ethylene methyl acrylate copolymer had dual peak in DSC micrographs due to possible incompatibility, similar to the composites generated using ethylene-*g*-maleic anhydride copolymer, ethylene-α-olefin copolymer and ethylene vinyl acetate-*g*-maleic anhydride due to lack of miscibility on the molecular level.

Park *et al.* [32] studied the effect of functionalization of graphene on the mechanical properties of polyethylene/graphene nanocomposites. Unfunctionalized graphene oxide (GO), thermal reduced graphene oxide (TRG) and 3-isopropenyl-α,α-dimethylisocyanate (TMI)/octadecylamine functionalized graphene oxide (FGO) were dispersed in HDPE, LLDPE and oxidized polyethylene (OPE) by means of solvent blending. The tensile modulus of all polyethylene types was enhanced upon incorporation of FGO. The FGO/HDPE nanocomposite exhibited the highest tensile modulus, whereas TRG and GO showed less increase and a slight decrease was reported for the OPE/graphene nanocomposites. Due to the low molecular weight, viscosity and oxide groups of OPE, formation of unstable heterogeneous nanocomposites was suggested, which led to the observed lack of improvement in the modulus upon dispersion of graphene in the matrix. FGO and TRG were well dispersed in polyethylene whereas GO was poorly dispersed, therefore, inconsistent increase in normalized modulus was reported for polyethylene/GO nanocomposites. In another study, Kim *et al.* [33] also reported the effect of polyethylene functionalization and blending methods on the properties of polyethylene/graphene nanocomposites. TRG was blended with linear low density polyethylene (LLDPE) and its amine, nitrile, and isocyanate functionalized analogs using both melt and solution blending. Graphene platelets were well exfoliated in functionalized LLDPE matrices, whereas phase separation was observed for the unmodified LLDPE. Solvent based blending was also observed to be more effective in dispersing the graphene platelets as compared to melt compounding.

Jiang and Drzal [34] also reported that the thermal stability and thermal conductivity of PE/exfoliated graphene nano-platelets based nanocomposites enhanced with increasing graphene concentration. In another study, Jiang and Drzal [35] coated the graphene nanoplatelets (GNPs) with paraffin wax before dispersing in HDPE by means of melt extrusion and injection-molding to make the test sample coupons. A nanocomposite of graphene without paraffin wax coating was generated by the same method to be used as a reference and samples with varying paraffin wax to GNP weight ratio were prepared. For all the nanocomposites, a 5 vol% filler loading was used. The flexural strength and flexural modulus were observed to initially increase with increased paraffin wax to GNPs ratio, but then reduced when the paraffin wax content of the nanocomposites exceeded 5 wt%. Further tests to affirm the reason for this occurrence attributed the observed effect to the presence of excess low molecular weight paraffin wax in the nanocomposites, thus, leading to the deterioration of the mechanical properties after the 5 wt% threshold.

El Achaby *et al.* [36] melt-compounded reduced graphene oxide nano-sheets (GNs) with HDPE in a twin screw mixer without modifying the surface of the GNs and obtained the test samples by means of hot press molding. The nanocomposites had filler loadings of 0.5, 1 and 3 wt%. Dynamic mechanical analysis showed that both storage and loss moduli of the nanocomposites increased with increasing filler loading and that the storage modulus increased more than the loss modulus. The tensile strength and Young's modulus were also observed to increase with filler loading. The tensile strength was observed to increase by 77% and the Young's modulus by 87% as compared to neat HDPE when the polymer was compounded with 3 wt% GNs. The nanocomposites also exhibited increasing thermal stability with increasing filler content. TGA revealed that the initiation of weight loss occurred at a higher temperature and the temperature required to degrade 5% of polymer weight significantly increased from 431 °C for neat HDPE to 466 °C for the nanocomposite with 3 wt% GNs content.

Mittal and Chaudhry [37] studied the effect of graphene content on the properties of PE/graphene nanocomposites. Nanocomposite with 10% filler content had significantly low elongation indicating the presence of filler aggregates causing stain hardening of polymer chains. The tensile modulus exhibited gradual enhancement with filler fraction and at 10% filler content, an increment of 62% in tensile modulus as compared to pure polymer was observed. The yield stress was observed to remain constant with filler fraction except at 10% content where a small decrease was observed. The yield strain remained unchanged till 3% filler content and exhibited significant reduction on further enhancing the filler content and attained a plateau value at 7% graphene oxide amount. Peak stress was also observed to increase as a function of filler fraction, except for 10% filler composite. The enthalpy of HDPE in the composites (normalized to 100% polymer) was observed to decrease gradually with increasing filler content indicating that the filler stacks hindered the polymer crystallization. As shown in Figure 8.4, the peak melting point was observed to remain largely unchanged. Due to hindrance to crystallization on graphene addition, the peak crystallization temperature was also observed to decrease by 2-3 °C as a function of filler content. The onset and peak degradation temperatures were observed to increase with filler content confirming the efficient heat transfer from polymer chains to graphene.

Chaudhry and Mittal [38] also reported graphene-based nanocomposites generated by solution mixing of the graphene filler with chlorinated polyethylene (CPE) compatibilizers (of different chlorination levels of 25% and 35%, thus, denoted as CPE25 and CPE35) followed by melt mixing of the solution mixed masterbatches with high density polyethylene (Figure 8.5). The CPE35 masterbatch was observed to have better graphene dispersion as compared with CPE25 containing masterbatch. Though complete nanoscale delamination of the graphene platelets was not observed in the composites, however, the composites with CPE35 compatibilizer also had much better filler dispersion as compared with corresponding CPE25 nanocomposites. Higher extent of chlorination in the matrix resulted in higher magnitude of interfacial interactions between the polymer and the filler surface, resulting in increased extent of filler delamination. Peak melting point as well as overall crystallinity was affected by graphene oxide as well as compatibilizers. The mechanical performance of the

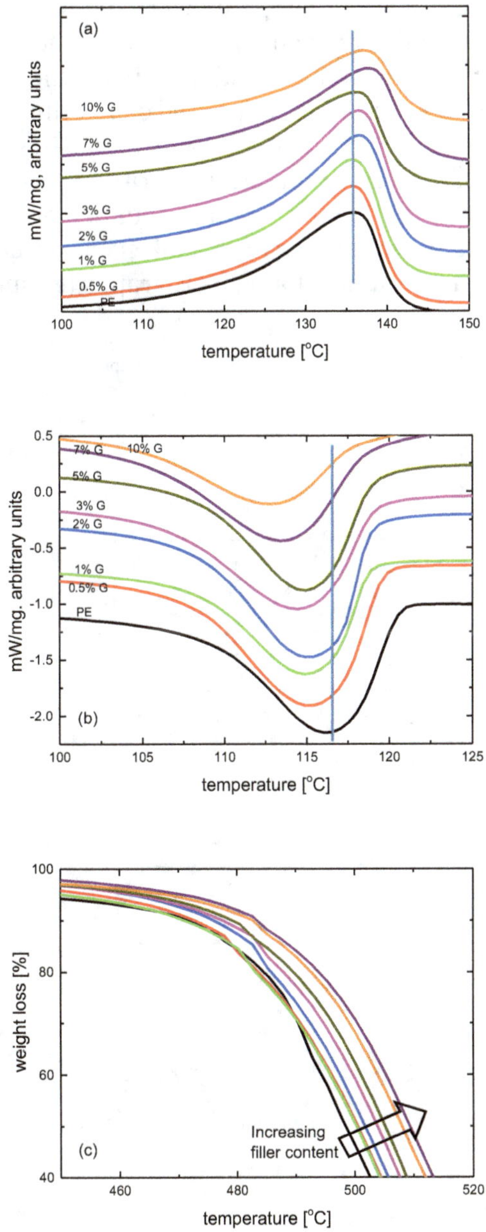

Figure 8.4 DSC (a) melting and (b) crystallization thermograms of HDPE and its nanocomposites with increasing filler content; (c) TGA thermograms of the polymer and nanocomposites. Reproduced from Reference 37 with permission from Wiley.

nanocomposites was affected by the factors like filler delamination, plasticization of the matrix, compatibilizer content as well as their chlorination level and amorphous/semi-crystalline nature. An increase of 16% in tensile modulus and 21% in tensile strength of the composites with 5 wt% of 25% chlorinated compatibilizer at 0.5 wt% graphene content was observed. The nanocomposites were also observed to have reduced low-temperature brittleness on the addition of CPE (especially CPE35) which was also retained even on graphene addition [39]. The height corrugation analysis at different places of the scanned area in the atomic force micrographs (25 mm^2) indicated

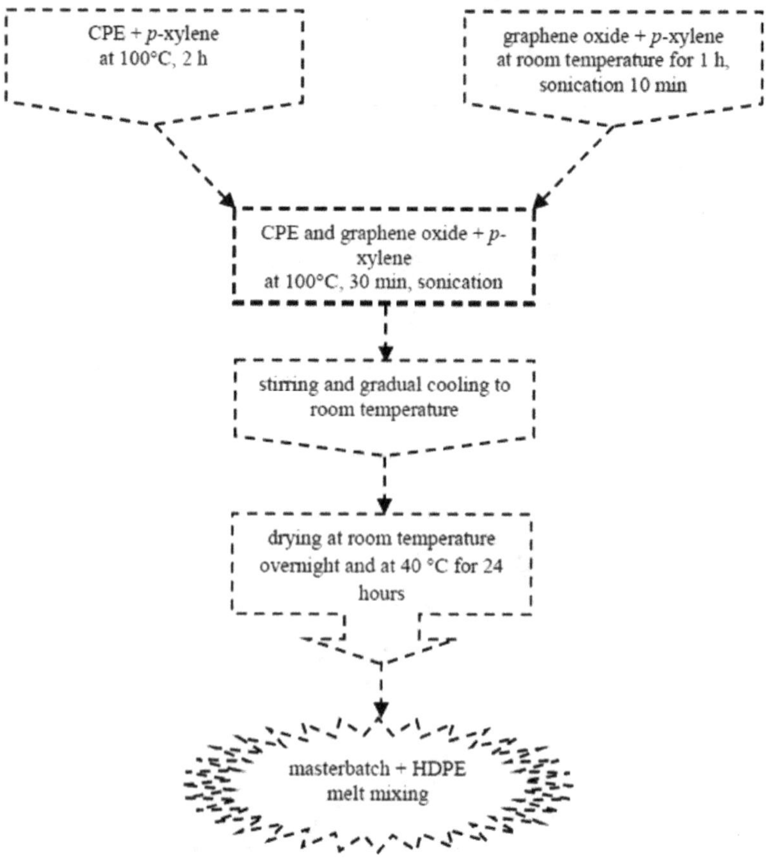

Figure 8.5 Schematic representation of the synthesis of CPE-graphene oxide masterbatches by solution mixing and subsequently HDPE nanocomposites by melt mixing. Reproduced from Reference 38 with permission from Wiley.

that the composites followed similar height variation trend as the PE-CPE blends, however, the height variation was even higher in the composite samples. For example, for PE-G sample, a value of 113 nm for surface variation was calculated which was enhanced markedly to 198 nm for PE-G-CPE25 nanocomposite and 564 nm for PE-G-CPE35 nanocomposite. The samples of atomic force microscopy were prepared at -140 °C, whereas the analysis was performed at room temperature.

In other studies reporting the use of *in-situ* polymerization, Fim *et al.* [40] synthesized HDPE/graphene nanocomposites using graphene nano-sheets (GNSs) with methylaluminoxane as cocatalyst and Cp_2ZrCl_2 as catalyst. Dispersion of the GNSs took place simultaneously along with the polymerization process. Dynamic mechanical analysis exhibited that both storage and loss moduli were higher for the nanocomposites than neat HDPE. The tensile strength and elongation at break values of the nanocomposites were, however, lower than those of neat HDPE. Since the filler loadings used in this study were above 1 wt%, the observation about reduction of tensile strength and elongation at break also mirrored the results of Sturzel *et al.* [28], where *in-situ* polymerization too was used for the composites generation and the tensile strength and elongation at break reduced beyond 1 wt% filler loading due to filler agglomeration. Elastic modulus, on the other hand, was reported to increase with increasing graphene content, as also reported by Sturzel *et al.* [28]. The authors also reported an increase in the thermal stability of the nanocomposites with increase in content of graphene nano-sheets (GNSs). This was indicated by an increase in the temperatures for weight loss onset and maximum weight loss. Comparing different synthesis methods, the observed thermal findings indicated that increased thermal stability was observed in the nanocomposites, irrespective of the different synthesis methods. Synthesis of polyethylene/graphene nanocomposites by *in-situ* polymerization was also reported by Zhang *et al.* [41,42]. In both cases, filler/$MgCl_2$-supported titanium-based Ziegler-Natta catalysts prepared by a facile spray-coagglomeration method were used to synthesize polyethylene and distribute the filler within the polymer matrix. Polyethylene/graphene and polyethylene/graphene oxide nanocomposites prepared by this method [41] exhibited improvement of tensile strength, modulus and elongation at break with increasing filler content. Low amounts of filler (up to 0.39 wt% for graphene and 0.37 wt% for graphene oxide) were used in the study. The results indicated a good stiffness-toughness balance for polyethylene/graphene nanocomposites. Similar improvements in tensile strength and modulus upon addition of graphene were reported by the authors in the other study [42]. Though tensile strength and modulus increased with increasing filler content among the nanocomposites, elongation at break reduced with increasing filler content, as was reported by Fim *et al.* [40]. The authors reported that melting temperature and degree of crystallization reduced with increase in filler content, while crystallization temperature increased [41,42]. Reduction in melting temperature and degree of crystallization was attributed to the hindrance in crystallization due to the presence of the filler particles resulting in faults and imperfections. On the other hand, the increase in crystallization temperature was attributed to the filler acting as a nucleating agent. Thermal stability of the nanocomposites was reported to increase with increase in filler content due to the filler particles acting as barrier to the transport of heat, oxygen and pyrolysis products through the nanocomposite matrix. Comparing polyethylene/graphene oxide (PE/GO) and polyethylene/graphene (PE/G) nanocomposites, it was reported that PE/GO had higher thermal stability than PE/G nanocomposites [41].

In another study, Mittal *et al.* [43] synthesized three different polyethylene samples (molecular weights PE1 < PE2 < PE3) using coordination polymerization, in order to study the effect of the variation in molecular weight on the properties of the nanocomposites, using thermally reduced graphene as filler. Due to efficient heat transfer to the filler platelets owing to their better dispersion in PE1, the highest extent of improvement in the initiation of degradation temperatures was observed in this case. For instance, the PE1 composite with 4 wt% filler fraction was observed to be have initiation delayed by ~40 °C as compared to pure polymer. Due to the higher degree of crystallinity, PE1 had higher tensile modulus of 1334 MPa, which was improved to the maximum extent of 35% in the composite with 4 wt% filler content. Thus, the mechanical properties were enhanced due to the combination of crystallinity and filler effect. The peak stress as well as elongation of the composites was improved in most cases as compared to pure polymers. As shown in Figure 8.6, the PE1 composites also exhibited lowest melt viscosity among the composites probably due to the lower degree of entanglements, which indicated ease of processing similar to the pure polymer. The corresponding PE2 and PE3 composites exhibited an order of magnitude higher viscosity at lower range of angular frequency. For PE1, the degree of crystallinity enhanced on adding 2 and 4 wt% of graphene due to nucleation

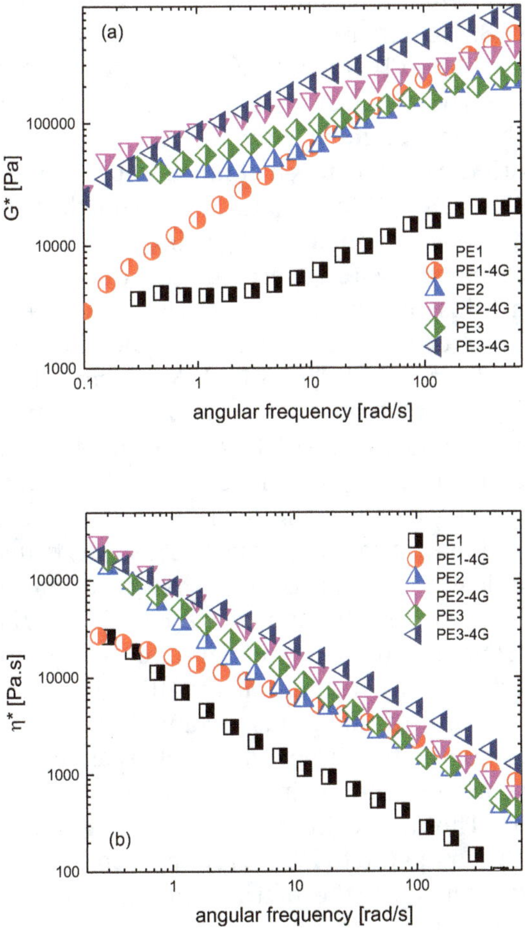

Figure 8.6 (a) Complex modulus and (b) complex viscosity of the pure polymers and nanocomposites (4% graphene content) as a function of angular frequency. Reproduced from Reference 43 with permission from Springer.

induced by the graphene nano-platelets, however, the crystallinity was observed to decrease in PE2 and PE3.

Mittal and Chaudhry [44] also compared the two main routes of nanocomposite generation, viz., the masterbatch method and direct melt mixing, in relation with the properties of the polyethylene-thermally reduced graphene nanocomposites compatibilized with functional polymers. Filler dispersion was observed to improve in the masterbatch method due to preferential interactions of compatibilizer with the filler, before melt mixing with the matrix polymer. For instance, Figure 8.7 compares the morphology of the HDPE/G, HDPE/G/5 EAA/MM, and HDPE/G/5 EAA/MB composites at different magnifications (where EAA, MM and MB denote ethylene-co-acrylic acid copolymer, melt mixing method and masterbatch method respectively). Both the composites had much improved filler dispersion as compared to HDPE/G composite. Comparing the two, the composite generated with direct melt mixing exhibited less optimal filler dispersion than the composite generated with the masterbatch approach. Filler tactoids of higher thickness were observed for the HDPE/G/5 EAA/MM composite, whereas the tactoid thickness was relatively reduced for the HDPE/G/5 EAA/MB composite. No single exfoliated graphene platelets were observed for the HDPE/G/5 EAA/MM composite, whereas the HDPE/G/5 EAA/MB composite exhibited presence of single graphene layers. Also, the extent of intercalation was observed to be qualitatively higher in the composite generated with the masterbatch approach. This also translated into improved mechanical properties as more than two times increment in tensile modulus was observed in the HDPE/G/5 EAA/MB composite (41%, masterbatch method) as compared to HDPE/G/5 EAA/MM (20%, melt mixing method). In addition, the HDPE/G/5 EAA/MB and other MB based composites did not show any decrease in degree of crystallinity as compared to pure polymer, whereas the corresponding composites generated with melt mixing exhibited slight decrease in the crystallinity. The composites were observed to have enhanced thermal stability as compared to HDPE. Five percent EAA-based composites exhibited even higher thermal stability than HDPE/G composite due to improved filler dispersion.

In the work reported by Wang *et al.* [45], reduced graphene oxide was functionalized with vinyl triethoxysilane (VTES) and compounded with LDPE by means of solution dispersion. The tensile strength and Young's modulus of the nanocomposites increased with increasing filler content, showing an increment of 27% and 92.8% respectively for the nanocomposite with 7 wt% filler content, as compared to neat LDPE. The elongation at break increased until a peak was achieved at filler content of 3 wt% and then started reducing with further increase in graphene content. The nanocomposites of LDPE with un-functionalized graphene had lower elongation at break than neat LDPE even with 1 and 3 wt% filler loading. The chemical bonds formed as a result of reaction of the vinyl groups in VTES with

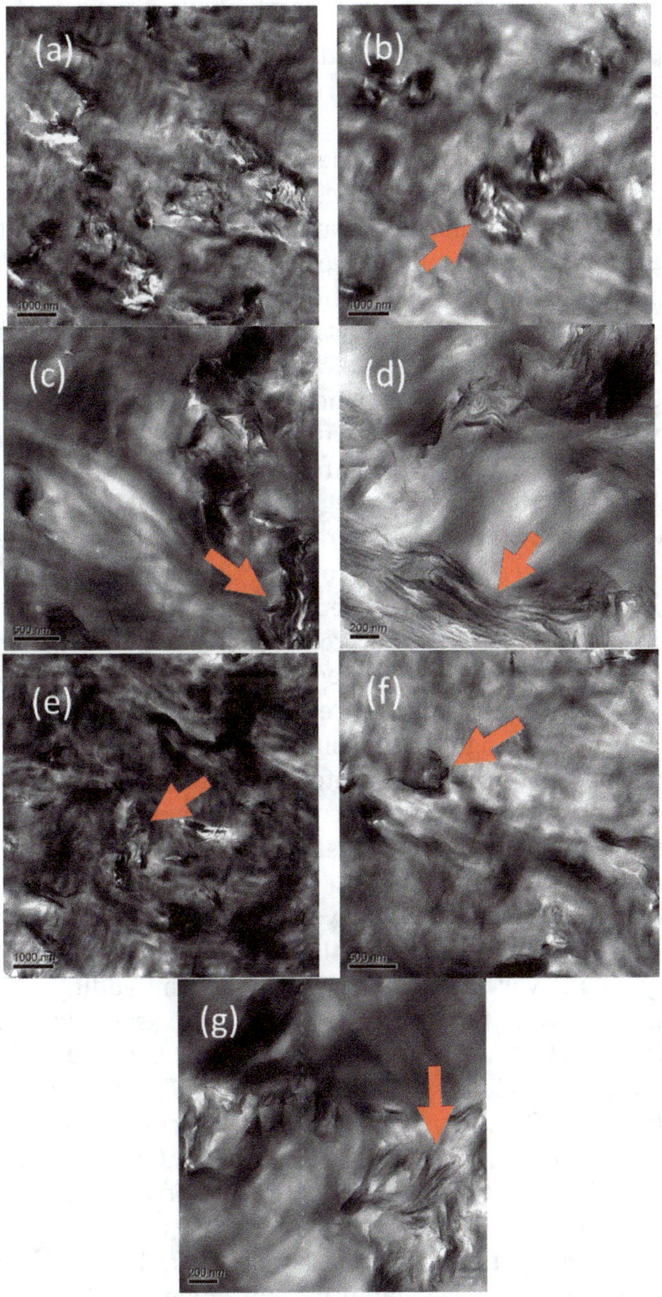

Figure 8.7 TEM micrographs of (a) HDPE/G; (b-d) HDPE/G/5 EAA MM and (e-g) HDPE/G/5 EAA MB composites. Reproduced from Reference 44 with permission from Springer.

the radicals in the LDPE improved the interfacial interaction of graphene with the polymer matrix and, hence, resulted in enhanced mechanical properties. Improved thermal stability of LDPE was also reported by the authors, as the weight loss initiation temperature increased with addition of the graphene filler. In another study, Huang *et al.* [46] prepared nanocomposites with varying fractions of octadecylamine (ODA) functionalized graphene oxide nano-sheets (GONs) and LDPE by means of solution coagulation. Improvement in yield strength upon addition of the functionalized GONs was reported to be limited, while the Young's modulus increased significantly with an increase of up to approximately 59% being observed on the addition of 4 wt% GONs. The observed enhancement was attributed to interfacial transfer of stress that resulted from the fully exfoliated uniformly dispersed GONs and their strong interfacial interaction with the LDPE matrix. However, the increase in rigidity resulted in the loss of ductility as evidenced by the decrease in elongation at break of the nanocomposites with increasing filler content. The results were also similar to the nanocomposites reported by Wang *et al.* [45]. Crystallization peak temperature was reported to increase with enhancing the filler content due to the heterogeneous nucleation effect. The thermal stability was also observed to increase with filler content due to the barrier effect. No change in the melting temperature and melting enthalpy was observed, which indicated that the filler only acted as nucleating agent, thus, accelerating crystallization kinetics but did not have significant effect on the crystalline structure and crystallinity of LDPE nanocomposites.

Patwary and Mittal [47] also developed polyethylene nanocomposites with 5 wt% silica, alumino-silicate and thermally reduced graphene by adding varying amounts of pro-oxidant additive. The graphene particles were observed to be uniformly dispersed in the polymer matrix. In the graphene composite with 2.5 wt% additive content, the tensile modulus was increased by 1.95 times that of pure polymer (Figure 8.8). Increasing the additive content was also observed to enhance the mechanical performance. Graphene nanocomposite with 1% additive content had also 40% and 33% increment in storage modulus at 50 °C and 70 °C respectively as compared to pure PE. Degree of embrittlement of the samples with 2.5% additive was calculated from the carbonyl index measurements as a function of exposure time adjusted to real environment in a constant temperature of 30 °C in sunlight. PE sample containing only 2.5 wt% additive exhibited the fastest rate of degradation reaching 100% embrittlement after ~795 accelerated ageing hours (equivalent to around 8.7 months at 30 °C). Graphene based composite did not reach 100 % embrittlement during the test period, but was still ~48.2% embrittled at the end of the test. Graphene platelets slowed the photo-oxidation due to known property of strong UV absorption by the graphene platelets. However, the observed results confirmed that the addition of pro-oxidant in the composites was necessary to attain their

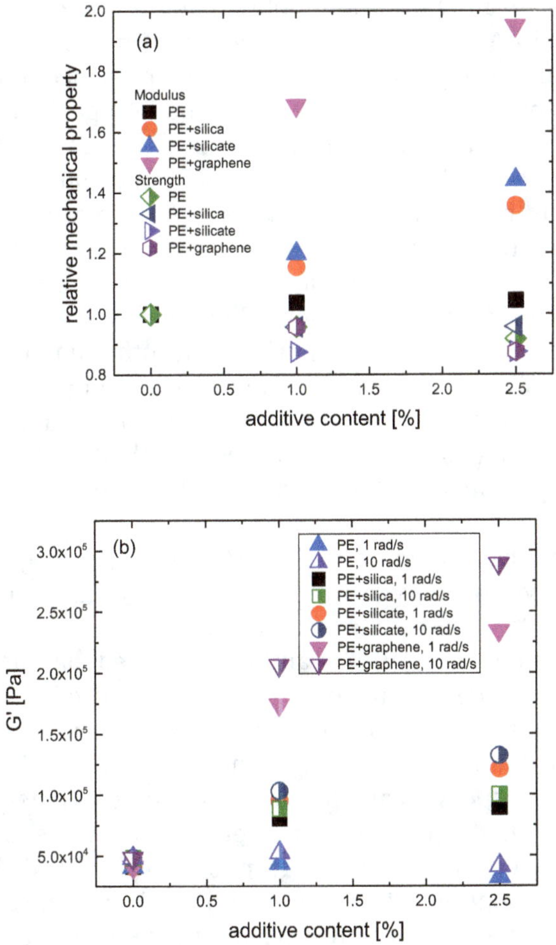

Figure 8.8 (a) Relative tensile modulus and strength of the materials plotted as a function of additive content; (b) comparison of storage modulus in the nanocomposites as a function of additive content at an angular frequency of 1 rad/s (solid dots) and 10 rad/s (half solid dots). Reproduced from Reference 47 with permission from Elsevier.

degradation and a controlled and fairly fast degradation could still be achieved in varying degrees even in thick composite plaques with 5 wt% filler content.

In another study, Mittal and Krauss [48] also reported the mapping of the component distribution in the nanocomposites by using hyperspectral imaging. Ethylene-co-acrylic acid (EAA) and ethylene-comethacrylic acid ionomer (EMAZ) copolymers were used as compatibilizers for polyethylene-graphene nanocomposites generated by melt mixing. Figure 8.9 presents the mapping of the distribution of graphene and EAA at two different locations in the PE-5%EAA-G nanocomposite. Pixels in red represented the presence of graphene peak, whereas pixels in green represented the EAA peak in the nanocomposite. The dark area in the micrographs without any colored signal corresponded to PE. EAA in the PE-EAA-G nanocomposite sample was observed to bond/mix well with the graphene causing spectral shifting and mixing. Several areas were observed to contain characteristic peaks of both EAA and graphene in the nanocomposite. The EAA compatibilizer was, thus, observed to be effective in dispersing the graphene throughout the PE matrix. On the other hand, nanocomposite with EMAZ had no spectral mixing as the components did not mix optimally with each other. The hyperspectral imaging and mapping was, thus, demonstrated to be a useful method for the determination of component distribution in complex nanocomposite systems.

Recently, different chemical routes to functionalize graphene with short-chain polyethylene were reported which influenced the mechanical properties of graphene-based high density polyethylene (HDPE) nanocomposites [49]. The authors observed that the polyethylene grafting on the graphene through thiol-yne click chemistry showed a significant effect on the Young's modulus and elongation at break of the resultant composites. In another study, Vasileiou *et al.* also studied the dispersion and properties of polyethylene-graphene composite using maleated linear low-density polyethylene (LLDPE) derivatives and thermally reduced graphene oxide (TRG) through a non-covalent compatibilization approach [50]. It was observed that the pyridine functionalized polyethylene matrix exhibited a better dispersion with TRG due to the π-π interactions, thereby resulting in significantly improved Young's moduli without reduction of tensile stress and elongation at the break.

In another study, PE based blend nanocomposites with ethylene acrylic acid (EAA) copolymer (40:60) and graphene (1 wt%) could be

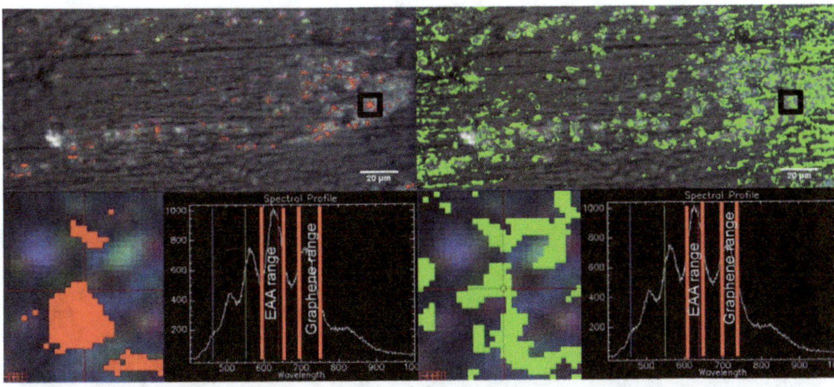

(a)

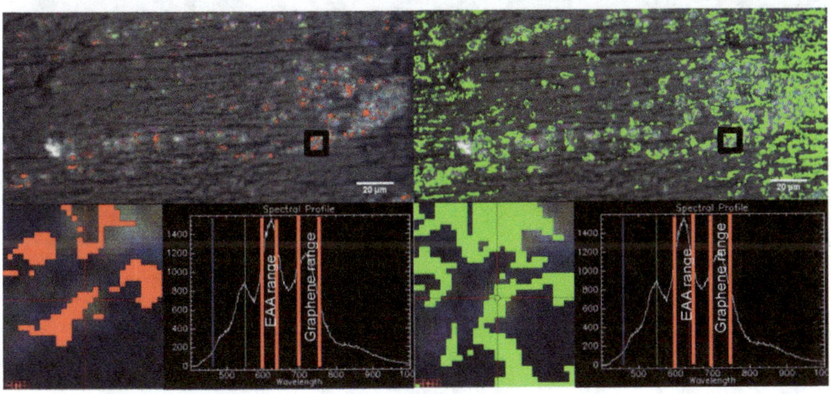

(b)

Figure 8.9 Mapping of distribution of graphene and EAA at two different locations in PE-5%EAA-G nanocomposite. Pixels in red represent the presence of graphene peak, whereas pixels in green represent the presence of EAA peak in the nanocomposite. Reproduced from Reference 48 with permission from Springer.

generated by melt mixing. Figure 8.10 presents the morphological, thermal, rheological and mechanical properties of the composite with 1 wt% filler. Phase immiscibility was observed in the PE and EAA phases from the microscopy analysis. Good degree of graphene distribution and delamination was also observed, as shown in Figure 8.10 a and b. In addition, graphene was observed to be dispersed in both PE and EAA phases. The DSC thermogram of the polymer blend exhibited individual melt transitions of PE and EAA, thus, confirming the phase immiscibly (Figure 8.10 c). The peak melting points of both phases were enhanced in the nanocomposite due to the efficient heat

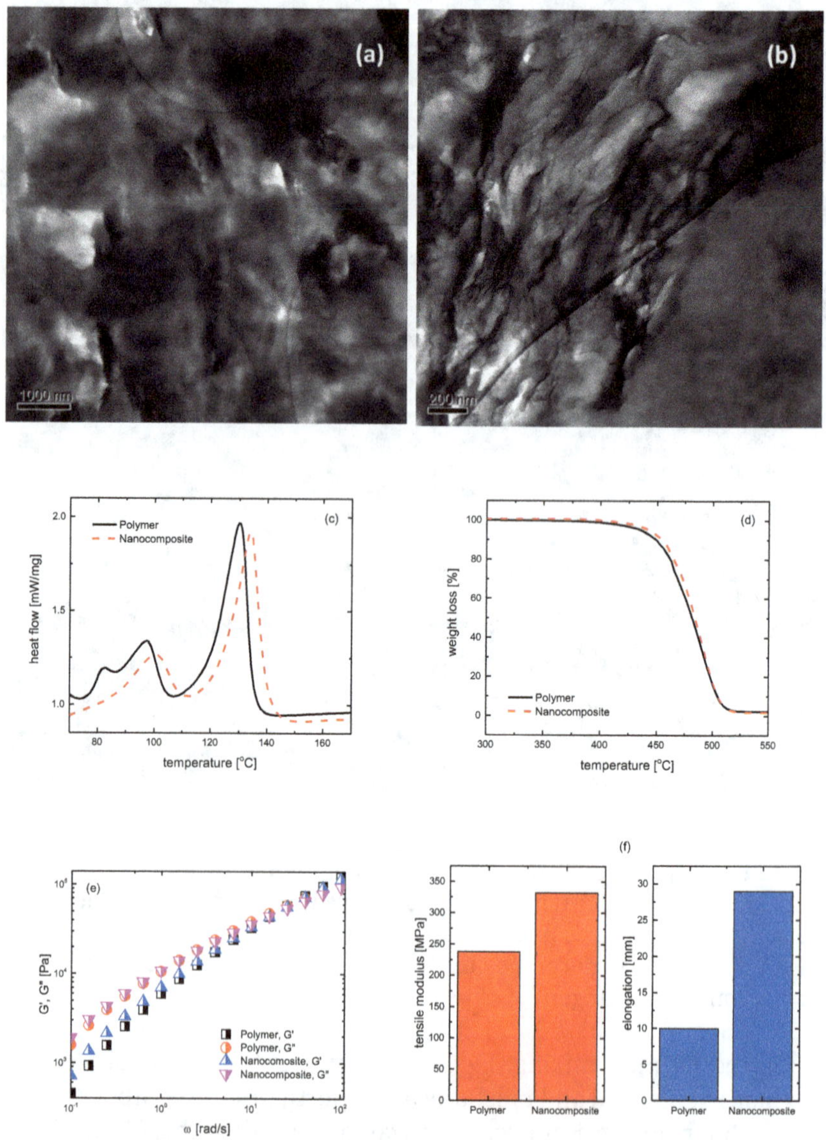

Figure 8.10 Characterization of the morphology as well as thermal, rheological and mechanical properties of the 40%PE+60%EAA+1%G nanocomposite: (a,b) TEM micrographs at different magnifications, (c) DSC thermograms, (d) TGA thermograms, (e) storage and loss modulus and (f) tensile modulus and elongation of the composite in comparison with pure polymer blend.

transfer to the graphene platelets dispersed in both phases. Similarly, the thermal degradation behavior of the nanocomposites also exhibited slight improvement in the onset and peak degradation temperatures. The storage modulus of the composite was observed to be better than the pure polymer, though the loss modulus was higher than the storage modulus at lower angular frequencies. The tensile behavior also exhibited increased tensile modulus of 332 MPa with incorporation of 1 wt% graphene, in comparison to 238 MPa for the polymer blend, thus, indicating effective stress transfer to the uniformly dispersed graphene platelets. As a result, the elongation of the nanocomposite was also enhanced significantly as compared to the PE/EAA blend, thus, indicating synergistic enhancement of toughness and stiffness of the polymer.

Tarani *et al.* [51] also studied the effect of the size of graphene platelets (GNPs) on HDPE melting behavior and crystallization kinetics. It was reported that the crystallization temperature decreased with increase in the diameter of the GNPs. This was reported to be due to the hindrance to lamella formation by the higher diameter GNPs. Higher diameter GNPs were also observed to be rejected from the crystalline phase during the segregation that took place in the process of crystallization, thus, resulting in the difficulty in packing of the HDPE polymer chains. The authors also reported reduction in melting enthalpy and crystallinity attributed to restriction of HDPE chain mobility by the GNPs during the generation of an ordered crystal structure. Melting enthalpy was reported to reduce with increase in GNP diameter.

In another study, Cheng *et al.* [52] investigated the crystallization of PE induced by reduced graphene oxide (RGO) (Figure 8.11) and reported increase of both thermal stability and rate of crystallization in the presence of RGO. The synthesis method involved the dispersion, in two separate vessels, of low amounts of PE and RGO in dichlorobenzene (DCB) (0.05 wt% and 0.01 wt%, respectively) followed by crystallization, mixing the product with a concentrated solution of PE in DCB and then subsequent crystallization to obtain the nanocomposite. Non-isothermal DSC revealed that the melting temperature of the nanocomposites remained the same with increasing RGO content, whereas the crystallization temperature and degree of crystallization increased with increasing the RGO content in the nanocomposites due to the nucleation effect provided by the RGO nano-sheets. Isothermal crystallization analysis exhibited that the time required to complete 50% of the crystallization process reduced with increasing the filler content and the crystallization rate constant as calculated based on the Avrami equation [53] increased with increasing the filler content. Both observations indicated that the rate of crystallization increased with the addition of graphene.

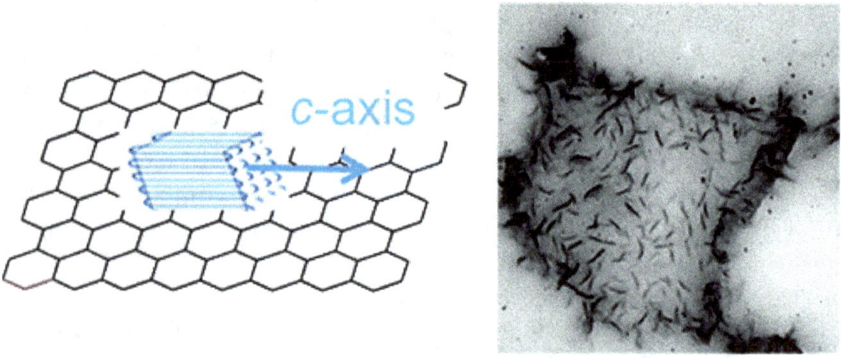

Figure 8.11 PE/RGO nanocomposites (PGNs) fabricated through solution crystallization/precipitation using the PE-decorated RGO hybrid. Reproduced from Reference 52 with permission from American Chemical Society.

Hu *et al.* [54] prepared hyperbranched flame retardant functionalized GO (FGO) using N-aminoethylpiperazine, hydroxyethyl acrylate, methyldichlorophosphate and GO via *in-situ* polymerization technique. FGO was incorporated into cross-linked PE matrix and its flammability and thermal stability were studied. It was observed that the incorporation of FGO in to the cross-linked PE endowed a highly flame retardant and thermally stable PE composite. In a similar work, Han *et al.* [55] studied the flame retardancy behavior of polyethylene/alumina trihydrate (PE/ATH) composites with the effect of incorporation of graphene nano-platelets (GNPs). The composites were prepared through melt blending method. The incorporation of 0.2 wt% of GNPs into the PE/ATH composite decreased the heat release peak by 18%.

Ran *et al.* [56] developed high density polyethylene/brominated polystyrene/graphene nano-platelet (GNPs) composites *via* melt blending and studied the effect of Friedel-Crafts reaction on their thermal stability and flammability. Anhydrous aluminium trichloride ($AlCl_3$) was used as a Lewis catalyst and was added to initiate the Friedel-Crafts reaction for effective dispersion of GNPs into the polymer blends. The results revealed that the GNPs were partly unfolded and the domains became smaller in the presence of $AlCl_3$. The incorporation of $AlCl_3$ into polymer-graphene nanocomposites reduced the flammability and slowed down the heat release rate.

In another study, Carotenuto *et al.* [57] fabricated an electrically conductive polyethylene supported graphene film based on simple mechanical approach. The film exhibited a semiconducting nature at lower temperature. It was observed that the film exhibited good conductivity with increasing graphene content, but it also lowered the optical transparency. Wang *et al.* [58] developed electrically conductive ultrahigh molecular weight polyethylene (UHMWPE)/graphene nanocomposites with segregated structure. The authors found that the UHMWPE powder could generate a static electricity after high speed mechanical friction and adsorb graphene on the surface. The composite exhibited high electrical conductivity with percolation threshold of 0.1 vol%. In another study, Du *et al.* [59] prepared multiwalled carbon nanotube (MWCNT) and graphene nano-sheets (GNS) incorporated HDPE nanocomposites with segregated network structure and compared their electrical properties. HDPE/ MWCNT composites exhibited higher conductivity than that of HDPE/GNS composites at the same filler content. This was due to different conductivity mechanism between the two composites, while the HDPE/MWCNT composites possessed three-dimensional conductive system, the HDPE/GNS composites exhibited a two-dimensional conductive system.

8.3 Conclusions

The excellent mechanical and thermal properties of graphene, along with its large surface area have made it a nanomaterial of wide research interest. The incorporation of graphene or its derivatives in polymer matrices enables the synergetic combination of its properties with those of the polymers to produce nanocomposites that are suitable for applications for which the neat polymers are not useful. One such polymer that is used widely is polyethylene (of different densities) and the mechanical and thermal properties of polyethylene/graphene nanocomposites comprise some of the areas of interest of recent research.

In this review, the properties of polyethylene/graphene nanocomposites reported in recent studies have been summarized. The incorporation of graphene to the polymer was observed to result in the improvement of mechanical and thermal properties as compared to pure PE, irrespective of the synthesis method of the composites. The review summarized the studies reporting the generation of the nanocomposites using different synthesis methods, such as solution dispersion, melt mixing and *in-situ* polymerization. The mechanical properties studied in the review included Young's modulus, yield stress, yield strain, elongation at break, tensile strength and dynamic mechanical properties. The Young's modulus, yield stress and yield strain increased in the reported studies with increasing graphene content in the nanocomposites. This was attributed to the effective stress transfer from the polyethylene chains to the graphene platelets, which resulted in the improved properties. The elongation at break, however, exhibited an increase with increasing graphene content until a certain optimum loading, after which a decrease was observed. This was attributed to the agglomeration of the graphene nano-sheets when added in large contents, hence reducing the effective aspect ratio as well as elongation at break. The dynamic mechanical properties such as storage modulus and loss modulus were also observed to increase with increasing graphene content. In the cases when the graphene functionalization was performed, the properties such as tensile strength were reported to be superior than the composites with un-functionalized graphene due to interfacial interaction and effective filler dispersion. Similar to the mechanical properties, the thermal properties of the nanocomposites were also reported to be better than those of neat polyethylene. The melting temperature of the polymer in nanocomposites was reported to remain constant irrespective of the filler content. The crystallization temperature and degree of crystallinity, on the other hand, were generally reported to increase with increasing graphene content in the nanocomposites because of the nucleation effect of the filler platelets. The rate of crystallization was also observed to increase with addition of graphene. Overall, it can be concluded that significant increment in composite properties have been gained in the polyethylene/graphene nanocomposites by optimizing the synthesis method, filler content, surface functionalization, etc. The obtained nanocomposites represent advanced materials with a high potential of use in a wide variety of applications.

Acknowledgement

The authors are indebted to Dr. Ali Chaudhry for generating and characterizing the PE-EAA-G composites.

References

1. Mittal, V., and Matsko, N. (2014) Characterization of polyethylene-based multiphase systems containing zero-and two-dimensional nanoparticulate reinforcement materials by analytical electron and atomic force microscopy. *Journal of Thermoplastic Composite Materials*, **27**(6), 845-864.
2. Hu, K., Kulkarni, D. D., Choi, I., and Tsukruk, V. V. (2014) Graphene-polymer nanocomposites for structural and functional application. *Progress in Polymer Science*, **39**, 1934-1972.
3. Lee, C., Wei, X., and Kysar, J. W. (2008). Measurement of the elastic properties and intrinsic properties of monolayer graphene. *Science*, **321**, 385-388.
4. Kim, H., Abdala, A. A., and Macosko, C. W. (2010) Graphene/polymer nanocomposites. *Macromolecules*, **43**, 6515-6530.
5. Mittal, V. (2014) Functional polymer nanocomposites with graphene: a review. *Macromolecular Materials and Engineering*, **299**(8), 906-931.
6. Worsley, K. A., Ramesh, P., Mandal, S. K., Niyogi, S., Itkis, M. E., and Haddon, R. C. (2007) Soluble graphene derived from graphite fluoride. *Chemical Physics Letters*, **445**, 51-56.
7. Novoselov, K. S., Geim, A. K., Morozov, S. V., Jiang, D., Zhang, Y., Dubonos, S. V., Grigorieva, I. V., and Firsov, A. A. (2004) Electric field effect in atomically thin carbon films. *Science*, **306**, 666-669.
8. Behabtu, N., Lomeda, J. R., Green, M. J., Higginbotham, A. L., Sinitskii, A., Kosynkin, D. V., Tsentalovich, D., Para-Vasquez, A. N. G., Schmidt, J., Kesselman, E., Cohen, Y., Talmon, Y., Tour, J. M., and Pasquali, M. (2010) Spontaneous high-concentration dispersions and liquid crystals of graphene. *Nature Nanotechnology*, **5**, 406-411.
9. Liu, N., Luo, F., Wu, H., Liu, Y., Zhang, C., and Chen, J. (2008) One-step ionic-liquid-assisted electrochemical synthesis of ionic-liquid functionalized graphene sheets directly from graphite. *Advanced Functional Materials*, **18**, 1518-1525.
10. Hernandez, Y., Nicolosi, V., Lotya, M., Blighe, F. M., Sun, Z., De, S., McGovern, I. T., Holland, B., Byrne, M., Gun'Ko, Y. K., Boland, J. J., Niraj, P., Duesberg, G., Krishnamurthy, S., Goodhue, R., Hutchison, J., Scardaci, V., Ferrari, A. C., and Coleman, J. N. (2008). High-yield production of graphene by liquid-phase exfoliation of graphite. *Nature Nanotechnology*, **3**, 563-568.
11. Hummers, W. S., and Offeman, R. E. (1958) Preparation of graphitic oxide. *Journal of American Chemical Society*, **80**, 1339-1339.
12. Dervishi, E., Li, Z., Watanabe, F., Biswas, A., Xu, Y., Biris, A. R., Saini, V., and Biris, A. S. (2009) Large-scale graphene production by RF-cCVD method. *Chemical Communications*, 4061-4063.
13. Zhang, W., Cui, J., Tao, C., Wu, Y., Li, Z., Ma, L., Wen, Y., and Li, G.

(2009) A strategy for producing pure single-layer graphene sheets based on a confined self-assembly approach. *Angewandte Chemie International Edition*, **48**, 5864-5868.
14. Kim, C., Min, B., and Jung, W. (2009) Preparation of graphene sheets by the reduction of carbon monoxide. *Carbon*, **47**, 1605-1612.
15. Li, N., Wang, Z., Zhao, K., Shi, Z., Gu, Z., and Xu, S. (2010) Large scale synthesis of N-doped multi-layered graphene sheets by simple arc-discharge method. *Carbon*, **48**, 255-259.
16. Schaefer, D. W., and Justice, R. S. (2007). How nano are nanocomposites? *Macromolecules*, **40**, 8501-8517.
17. Potts, J. R., Dreyer, D. R., Bielawski, C. W., and Ruoff, R. S. (2011) Graphene-based polymer nanocomposites. *Polymer*, **52**, 5-25.
18. Kuilla, T., Bhadra, S., Yao, D., Kim, N. H., Bose, S., and Lee, J. H. (2010) Recent advances in graphene based polymer composites. *Progress in Polymer Science*, **35**, 1350-1375.
19. Yang, X., Mei, T., Yang, J., Zhang, C., Lv, M., and Wang, X. (2014) Synthesis and characterization of alkylamine-functionalized graphene for polyolefin-based nanocomposites. *Applied Surface Science*, **305**, 725-731.
20. Lewandowska, K. (2012) Surface studies of microcrystalline chitosan/poly (vinyl alcohol) mixtures. *Applied Surface Science*, **263**, 115-123.
21. Tan, Y., Fang, L., Xiao, J., Song, Y., and Zheng, Q. (2013) Grafting copolymers onto graphenes by miniemulsion polymerization for conductive polymer composites: improved electrical conductivity and compatibility induced by interfacial distribution of graphene. *Polymer Chemistry*, **4**, 2939-2944.
22. Liao, K. H., Park, Y. T., Abdala, A., and Macosko, C. (2013). Aqueous reduced graphene/thermoplastic polyurethane nanocomposites. *Polymer*, **54**, 4555-4559.
23. Stankovich, S., Piner, R. D., Nguyen, S. T., and Ruoff, R. S. (2006) Synthesis and exfoliation of isocyanate-treated graphene oxide nanoplatelets. *Carbon*, **44**, 3342-3347.
24. Yang, X. Y., Wang, X. B., Yang, J., Li, J., and Wan, L. (2013) Functionalization of graphene using trimethoxysilanes and its reinforcement of polypropylene nanocomposites. *Chemical Physics Letters*, **570**, 125-131.
25. Ebewele, R. O. (2000) *Polymer Science and Technology*, CRC press, USA.
26. Lahiri, D., Dua, R., Zhang, C., Socarraz-Novoa, I., Bhat, A., Ramaswamy, S., and Agarwal, A. (2012) Graphene nanoplatelet-induced strengthening of ultra-high molecular weight polyethylene and biocompatibility in vitro. *Applied Materials & Interfaces*, **4**, 2234-2241.
27. Suner, S., Joffe, R., Tipper, J. L., and Emami, N. (2015) Ultra-high mo-

lecular polyethylene/graphene nanocomposites: thermal, mechanical and wettability characterization. *Composites Part B*, **78**, 185-191.

28. Sturzel, M., Kempe, F., Thomann, Y., Mark, S., Enders, M., and Mulhaupt, R. (2012) Novel graphene UHMWPE nanocomposites prepared by polymerization filling using single-site catalysts supported on functionalized graphene nanosheet dispersions. *Macromolecules*, **45**, 6878-6887.
29. Chen, Y., Qi, Y., Tai, Z., Yan, X., Zhu, F., and Xue, Q. (2012) Preparation, mechanical properties and biocompatibility of graphene oxide/ultra-high molecular weight polyethylene composites. *European Polymer Journal*, **48**, 1026-1033.
30. Kuila, T., Bose, S., Mishra, A. K., Khanra, P., Kim, N. H., and Lee, J. H. (2012) Effect of functionalized graphene on the physical properties of linear low density polyethylene nanocomposites. *Polymer Testing*, **31**, 31-38.
31. Mittal, V., and Chaudhry, A. U. (2015) Effect of amphiphilic compatibilizers on the filler dispersion and properties of polyethylene – thermally reduced graphene nanocomposites. *Journal of Applied Polymer Science*, **132**(35), doi: 10.1002/app.42484.
32. Park, S., He, S., Wang, J., Stein, A., and Macosko, C. W. (2016). Graphene - polyethylene nanocomposites: Effect of graphene functionalization. *Polymer*, **104**, 1-9.
33. Kim, H., Kobayashi, S., AbdurRahim, M. A., Zhang, M. J., Khusainova, A., Hillmyer, M. A., Abdala, A. A., and Macosko, C. W. (2011) Graphene/polyethylene nanocomposites: Effect of polyethylene functionalization and blending methods. *Polymer*, **52**, 1837-1846.
34. Jiang, X., and Drzal, L. T. (2012) Multifunctional high-density polyethylene nanocomposites produced by incorporation of exfoliated graphene nanoplatelets 2: Crystallization, thermal and electrical properties. *Polymer Composites*, 33, 636-642.
35. Jiang, X., and Drzal, L. T. (2011) Improving electrical conductivity and mechanical properties of high density polyethylene through incorporation of paraffin wax coated exfoliated graphene nanoplatelets and multi-wall carbon nano-tubes. *Composites: Part A*, **42**, 1840-1849.
36. El Achaby, M., and Qaiss, A. (2013) Processing and properties of polyethylene reinforced by graphene nanosheets and carbon nanotubes. *Materials and Design*, **44**, 81-89.
37. Mittal, V., and Chaudhry, A. U. (2015) Polymer-graphene nanocomposites: Effect of polymer matrix and filler amount on properties. *Macromolecular Materials and Engineering*, **300**, 510-521.
38. Chaudhry, A. U., and Mittal, V. (2013) High-density polyethylene nanocomposites using masterbatches of chlorinated polyethylene/graphene oxide. *Polymer Engineering and Science*, **53**(1), 78-

88.

39. Mittal, V., and Matsko, N. (2013) PE-CPE blends and their graphene oxide nanocomposites with reduced low temperature brittleness, *Colloid and Polymer Science*, **291**(8), 1949-1961.
40. Fim, F. C., Basso, N. R. S., Greabin, A. P., Azambuja, D. S., and Galland, G. B. (2013) Thermal, electrical and mechanical properties of polyethylene-graphene nanocomposites obtained by in-situ polymerization. *Journal of Applied Polymer Science*, **128**, 2630-2637.
41. Zhang, H., Park, J., Ko, E., Moon, Y., Lee, Y., Hu, Y., Zhang, X., and Yoon, K. (2016) Comparison of the properties of graphene- and graphene oxide- based polyethylene nanocomposites prepared by an in-situ polymerization method. *RSC Advances*, **6**, 73013-73019.
42. Zhang, H., Ko, E., Park, J., Moon, Y., Zhang, X., and Yoon, K. (2016) Fabrication of polyethylene/graphene nanocomposites through in-situ polymerization with a spherical graphene/$MgCl_2$-supported Ziegler-Natta catalyst. *Composites Science and Technology*, **136**, 61-66.
43. Mittal, V., Kim, S., Neuhofer, S., Paulik, C. (2016) Polyethylene/graphene nanocomposites: effect of molecular weight on mechanical, thermal, rheological and morphological properties. *Colloid and Polymer Science*, **294**, 691-704.
44. Mittal, V., and Chaudhry, A. U. (2016) Polyethylene-thermally reduced graphene nanocomposites: comparison of masterbatch and direct melt mixing approaches on mechanical, thermal, rheological, and morphological properties. *Colloid and Polymer Science*, **294**, 1659-1670.
45. Wang, J., Xu, C., Hu, H., Wan, L., Chen, R., Zheng, H., Liu, F., Zhang, M., Shang, X., and Wang, X. (2011) Synthesis, mechanical, and barrier properties of LDPE/graphene nanocomposites using vinyl triethyoxysilane as a coupling agent. *Journal of Nanoparticle Research*, **13**, 869-878.
46. Huang, H., Zhou, S., Ren, P., Ji, X., and Li, Z. (2015) Improved mechanical and barrier properties of low-density polyethylene nanocomposite films by incorporating hydrophobic graphene oxide nanosheets. *RSC Advances*, **5**, 80739-80748.
47. Patwary, F., and Mittal, V. (2015) Degradable polyethylene nanocomposites with silica, silicate and thermally reduced graphene using oxo-degradable pro-oxidant. *Heliyon*, **1**(4), e00050.
48. Mittal, V., and Krauss, L. (2014) Compatibilized polyethylene-thermally reduced graphene nanocomposites: Interfacial interactions and hyperspectral mapping for component distribution. *Colloid and Polymer Science*, **292**(10), 2509-2518
49. Castelain, M., Martinez, G., Marco, C., Ellis, G., and Salavagione, H. J. (2013) Effect of click chemistry approaches for graphene modification on the electrical, thermal, and mechanical properties of poly-

ethylene/graphene nanocomposites. *Macromolecules*, **46**, 8980-8987.

50. Vasileiou, A. A., Kontopoulou, M., and Docoslis, A. (2014) A noncovalent compatibilization approach to improve the filler dispersion and properties of polyethylene/graphene composites. *ACS Applied Materials & Interfaces*, **6**, 1916-1925.
51. Tarani, E., Wurm, A., Schick, C., Bikiaris, D. N., Chrissafis, K., and Vourlias, G. (2016) Effect of graphene nanoplatelets diameter on non-isothermal crystallization kinetics and melting behavior of high density polyethylene nanocomposites. *Thermochimica Acta*, **643**, 94-103.
52. Cheng, S., Chen, X., Hsuan, G., and Li, C. Y. (2012) Reduced graphene oxide-induced polyethylene crystallization in solution and nanocomposites. *Macromolecules*, **45**, 993-1000.
53. Avrami, M. (1939) Kinetics of phase change. I General theory. *The Journal of Chemical Physics*, **7**, 1103-1112.
54. Hu, W., Zhan, J., Hong, N., Wang, B., Song, L., Stec, A. A., Hull, T. R., Wang, J., and Hu, Y. (2014) Effect of functionalized graphene oxide with hyper-branched flame retardant on flammability and thermal stability of cross-linked polyethylene. *Industrial & Engineering Chemistry Research*, **53**, 3073-3083.
55. Han, Z., Wang, Y., Dong, W., and Wang, P. (2014) Enhanced fire retardancy of polyethylene/alumina trihydrate composites by graphene nanoplatelets. *Materials Letters*, **128**, 275-278.
56. Ran, S., Guo, Z., Han, L., and Fang, Z. (2014) Effect of Friedel-Crafts reaction on the thermal stability and flammability of high-density polyethylene/brominated polystyrene/graphene nanoplatelet composites. *Polymer International*, **63**, 1835-1841.
57. Carotenuto, C., Nicola, S. D., Ausanio, G., Massarotti, D., Nicolais, L., and Pepe, G. P. (2014) Synthesis and characterization of electrically conductive polyethylene-supported graphene films. *Nanoscale Research Letters*, **9**, 475.
58. Wang, B., Li, H., Li, L., Chen, P., Wang, Z., and Gu, Q. (2013) Electrostatic adsorption method for preparing electrically conducting ultrahigh molecular weight polyethylene/graphene nanosheets composites with a segregated network. *Composites Science and Technology*, **89**, 180-185.
59. Du, J., Zhao, L., Zeng, Y., Zhang, L., Li, F., Liu, P., and Liu, C. (2011) Comparison of electrical properties between multi-walled carbon nanotube and graphene nanosheet/high density polyethylene composites with a segregated network structure. *Carbon*, **49**, 1094-1100.

9

Polyethylene Nanocomposites with Psyllium Husk, Graphene, Sepiolite, and Their Hybrids

9.1 Introduction

Polyolefins have one of the largest tonnage in the world with respect to production. It is estimated that more than 60% of produced polyolefins (mainly polyethylene (PE) and polypropylene (PP)) is introduced to the market as compounds, while only about 23% of the volume of other thermoplastics is used for compounding. The prominent properties of the polyolefins include low cost, recyclability, good processability, non-toxicity, and biocompatibility, thus, leading to a wide range of applications including orthopedic implants, automobile parts, consumer goods, durable equipment and industrial machinery [1]. Especially in the case of polyethylene, there is a need to further improve the stiffness, tensile strength, wear resistance, gas barrier and rigidity so as to further expand its wide spectrum of applications [2,3].

The combination of polymers and reinforcing fillers is a fast and practical technique for the development of new materials having improved properties. Significant attempts have been made to enhance the properties of high density polyethylene (HDPE), by the incorporation of different varieties of fillers. For instance, the effect of carbon nanotubes (CNTs) on the mechanical properties such as Young's modulus, ultimate stress, and toughness has been studied in nanocomposites of HDPE and CNT [4]. It was observed that with increase in percentage of CNTs, the composite properties also exhibited improvement owing to good load transfer effect and interface linking between polymer matrix and CNTs [4]. In another study, solution and melt mixing using co-counter- and modified co-rotating screw system were employed to achieve uniform dispersion of graphene in linear low density polyethylene (LLDPE) and the effect on the mechanical

Vikas Mittal[a], Ali U. Chaudhry[a,], M. Cole[a,**] and Nadejda Matsko[b]*
[a]Khalifa University of Science and Technology, Abu Dhabi, UAE; [b]Graz Centre for Electron Microscopy, Graz, Austria
**Current address: Texas A&M University, Qatar; **Current address: University of Waterloo, Canada*

properties was reported [5]. It was observed that solution mixing resulted in better reinforcement of the polymer [5]. In another study reporting the use of graphene, solution mixing of the graphene oxide filler with chlorinated polyethylene compatibilizers followed by melt mixing of the solution mixed masterbatches with high density polyethylene was reported [6]. The mechanical performance of the nanocomposites at ambient conditions was observed to be affected by the factors like filler delamination, plasticization of the matrix, compatibilizer content as well as their chlorination level and amorphous/semi-crystalline nature. Thus, the use of both 1-D and 2-D fillers has become predominant for the enhancement of polyolefin properties.

Graphene has specifically evolved as the filler of interest in recent years. It consists of one atomic thick sheets of covalently sp^2-bonded carbon atoms in a hexagonal arrangement [7]. Its choice as a filler is due to its excellent electrical and mechanical properties, which are significantly better than other inorganic filler materials. Among different 1-D filler materials, sepiolite is a non-swelling natural, abundant and low-cost clay having theoretical unit cell formula $(Si_{12}Mg_8O_{30})(OH)_4(OH_2)_4\cdot 8H_2O$. It belongs to the class of layered silicates comprising of hydrated magnesium silicate having micro-fibrous needle like morphology. As compared to phyllosilicates (BET surface area 82±1 m^2/g), it has high surface area (BET 374±7 m^2/g) and contains large amount of silanol groups present at the surface [8]. The silanol groups have been proven to be effective as sepiolite acts as reinforcing and compatibilizing agent for different polymer matrices owing to coupling reactions. In a related study, Chen *et al.* [9] prepared well dispersed sepiolite/polyurethane nanocomposite using in-situ polymerization technique. Sepiolite was observed to improve the tensile strength and modulus of polyurethane due to the strong interfacial interaction between the organic and the inorganic phases [9].

Many biofillers have also been developed for enhancement of polymer properties as well as to reduce cost. For example, Ghazanfari *et al.* [10] reported the incorporation of date seed flour in HDPE. The melt flow index of the composites was decreased, whereas the thermal conductivity of the composites increased. The advantages of using biofillers over conventional fillers are their low cost, recyclability, biodegradability, adequate physical and mechanical properties, etc. Another example of such composites reporting the use of biofillers include LDPE incorporated with oyster shell hybrids, which exhibited effective compatibility and good oxidative properties [11]. Similarly, HDPE/rice husk composites have been fabricated in order to add biodegradability to HDPE and also exhibited improved tensile and flexural properties [12]. Mechanically removed natural polysaccharides such as psyllium husk seeds of *Plantago ovata Forsk* are also known for use in many applications as binding, thickening, emulsifying, gelling and controlled release agent [13,14]. These are also known for usefulness in numerous medicinal applications. The composites of psyllium husk with polymers have also been reported in literature, however, only a few studies are available, thus, requiring further development and characterization.

In recent years, the use of hybrid fillers for generating composites has also gained considerable research attention. Hybrid filler systems contain two or more different types of incorporated fillers in order to achieve synergistic effect on the properties associated with each phase. In a study reporting the use of hybrid fillers, it was observed that the incorporation of hybrid filler in PE, using different combination of aluminum nitride (AlN), wollastonite, silicon carbide whisker (SiC) and boron nitride (BN), enhanced the thermal conductivity of the composite system [15]. The enhancement of thermal conductivity was owing to the presence of fibrous filler acting as an interactive agent between the spherical shaped filler, hence facilitating conduction [15]. Similarly, using the same concept, systems of carbon nanotube/graphite nano-platelet/epoxy [16] and carbon nanotube/silica/carbon black/EPDM rubber [17] enhanced the thermal and electrical conductivity of filled composites. Further, the use of biofiller additionally within the conventional polymer composite systems can also enhance the biodegradability of the system without sacrificing its mechanical or thermal properties. For instance, Gwon *et al.* [18] reported the hybridization of mineral fillers with wood fibers in polyolefin matrices.

In the current study, further efforts have been made to gain insights into the various aspects related to the composites of polyethylene with a variety of 0-D, 1-D and 2-D fillers as well as their hybrids. Firstly, the composites of polyethylene with increasing fraction of psyllium husk were analyzed for the impact of biofiller on the composite properties. This was followed by a comparison in the performance of psyllium husk (0-D), sepiolite (1-D) and thermally reduced graphene (2-D) containing composites. Finally, the physically mixed hybrid fillers in the combinations of graphene/sepiolite, husk/graphene, husk/sepiolite as well as husk/graphene/sepiolite were also incorporated in polyethylene and composite performance was characterized. Figure 9.1 depicts the unit cell/structural representation of sepiolite, thermally reduced graphene and psyllium husk. The complex structure of psyllium husk was described by Fischer *et al.*, Edwards *et al.*, and Marlett & Fischer, among other authors [13,19-22]. The main part is a complex heteroxylan having monosaccharides, arabinose and xylose, collectively called arabinoxylan. It is highly branched non-starch polysaccharide composed of b-(1,4)-linked xylopyranose residues, arabinofuranose and xylopyranose residues, rhamnose and galacturonic acid [21].

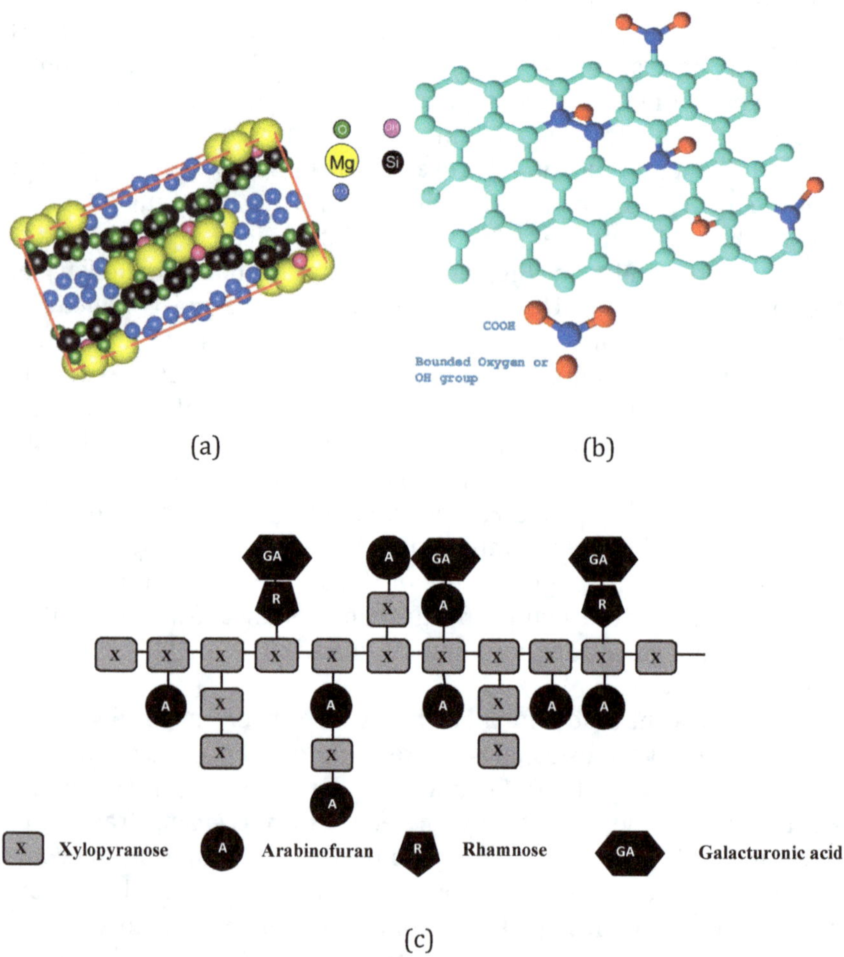

Figure 9.1 Structural representation of (a) sepiolite, (b) graphene and (c) psyllium seed husk.

9.2 Experimental

9.2.1 Materials

HDPE BB2581 was supplied by Abu Dhabi Polymers Company Limited (Borouge), UAE. It was in the form of white pellets with specific gravity of 0.958. The polymer was used as obtained. Thermally reduced graphene was prepared through thermal exfoliation of precursor graphite oxide [23] using modified Hummer's method [24], as reported earlier [6]. A short description for the preparation of graphene is as follows: 5 g of graphite powder was added with 125 ml concentrated H_2SO_4, to which 2.5 g of $NaNO_3$ was subsequently added. After stirring the mixture for 30 min while keeping in an ice-bath (5 °C), 15 g of $KMnO_4$ was added and the temperature was raised gradually to 35 °C. After stirring for 2 h, deionized water was added and the temperature was raised to 100 °C. Subsequently, the mixture was quenched and diluted by pouring it into 1.5 l deionized water. This was followed by addition of 30% H_2O_2 until the evolution of bubbles of hydrogen stopped. The solution was then filtered and the residues were dispersed in 2 l deionized water and were added with dilute HCl (6%) (2 l) to remove the SO_4^{2-} ions. The dispersion was filtered and the filtrate was analyzed for SO_4^{2-} and Cl^- ions using $BaSO_4$ and $AgNO_3$, respectively. The washed graphite oxide was dried under vacuum at 60 °C for 24 h. Graphene was generated via thermal exfoliation of graphite oxide by placing 1 g in a long quartz tube with 25 mm internal diameter and sealed at one end. The sample was flushed with nitrogen, followed by insertion of the tube in a tube furnace preheated to 1050 °C. The tube was held in the furnace for 30 s. The confirmation of graphene formation was achieved through Raman spectroscopy, microscopy and X-ray diffraction. Sepiolite was procured from Sigma Aldrich, whereas psyllium seed husk was procured locally.

9.2.2 Generation of Composites

To form composites, the fillers and polymer pellets were physically mixed before adding to the compounder. One step melt mixing of polymer and fillers was carried out at 190 °C using a mini twin screw extruder (MiniLab HAAKE Rheomex CTW5, Germany). The screw length and screw diameter were 109.5 mm and 5/14 mm conical respectively. Batch size of 5 g was used in all cases and shear mixing was performed for seven minutes at 80 rpm. Pure HDPE was also processed by subjecting it to similar shear and thermal conditions. Table 9.1 also describes the compositions of different composites. Disc- and rectangular bar-shaped test specimens were injection molded using a mini injection molding machine (HAAKE MiniJet, Germany) at a processing temperature of 190 °C. The injection pressure was 700 bar for 6 s, whereas holding pressure was 400 bar for 3 s. The temperature of the mold was kept at 50°C.

Table 9.1 Composition of the composites (in wt%)

Notation	PE content (%)	Husk content (%)	Graphene content (%)	Sepiolite content (%)
PE	100	-	-	-
PE+2.5H	97.5	2.5	-	-
PE+5H	95	5	-	-
PE+10H	90	10	-	-
PE+5G	95	-	5	-
PE+5S	95	-	-	5
PE+2.5G+2.5S	95	-	2.5	2.5
PE+2.5H+5G	92.5	2.5	5	-
PE+2.5H+5S	92.5	2.5	-	5
PE+2.5H+2.5G+2.5S	92.5	2.5	2.5	2.5

9.2.3 Material Characterization

Calorimetric properties of composites were recorded on a Perkin-Elmer Pyris-1 differential scanning calorimeter under nitrogen atmosphere. The scans were obtained from 30 to 190 °C at a heating rate of 10 °C/min (second heating runs). The heat enthalpies were measured with an error of ±0.1% and were confirmed by repeating the runs. Thermal degradation properties of the composites were recorded using Netzsch thermogravimetric analyzer (TGA). Nitrogen was used as a carrier gas and the scans were obtained from 50 to 700 °C at a heating rate of 20 °C/min.

Rheological properties such as storage modulus (G'), loss modulus (G''), viscosity and elasticity of the composites were measured using AR 2000 Rheometer from TA Instruments. The measuring temperature and gap opening were 185 °C and 1.6 mm respectively. Disc-shaped samples of diameter 25 mm and thickness 2 mm were used.

Strain sweeps were recorded at ω 1 rad/s from 0.1 to 100% strain. The shear stability of the samples was observed up to 10% strain. Hence, as a safe approach, frequency sweeps (dynamic testing) were recorded at 4% strain from ω 0.1-100 rad/s. DMA (dynamic mechanical analyzer) RSA3 from TA Instruments was used to study the change in modulus with temperature (25-100 °C). Dynamic temperature ramp test was used at a frequency of 5 rad/s, ramp rate 3 °C/min, and 1% strain for all samples.

Ultrathin sections (30-90 nm) as well as block faces of the composite samples were obtained using a Leica Ultracut E microtome (Leica, Austria) equipped with a diamond knife (Diatome, Switzerland) at -120 °C. The so-obtained sections were collected on 400 mesh formvar electron microscopy grids (coated with holey carbon film), and were examined in a Philips CM 20 (Philips/FEI, Germany) electron microscope at 200 kV at room temperature without additional staining. TEM image processing was performed using DigitalMicrograph software (Gatan, USA). All energy-dispersive X-ray spectroscopy (EDXS) experiments on the composites were performed in STEM mode with a probe current of 20 nA and a beam diameter of 100 nm. EDXS spectra were collected using a HPGe detector (solid angle 0.13 sr). For the calculation of elemental distribution images, jump ratio method was used and EDXS spectra have been visualized using DigitalMicrograph software (Gatan, USA). The block faces of specimens after cryo-ultramicrotomy were also investigated at ambient conditions using ZEISS Axioplan light microscope (Zeiss, Germany) equipped with Zeiss Axio Cam ICc 1 CCD camera in reflected polarized light.

9.3 Results and Discussion

The current study aims to achieve multiple objectives. The first objective is to demonstrate the use of psyllium husk as a useful filler for polyethylene to enhance its properties as well as to reduce cost. Secondly, the performance of the different 0-D, 1-D and 2-D fillers has been compared in achieving enhancement of mechanical, thermal and rheological performance of polyethylene. Finally, the binary and ternary hybrids of 0-D, 1-D and 2-D fillers have been incorporated in polyethylene in order to gain insights into the effect of these combinations on the composite behavior. Along with the conventionally used filler systems, the use of hybrid fillers (both physically mixed and chemically interacting hybrid fillers) has gained recent research interest. Specifically, the use of fillers of different dimensions can lead to synergistic effects on composite properties. In addition, the hybrid fillers are advantageous as one of the components of the hybrid filler mixture may have the main role of reducing the material's cost, while still retaining (or enhancing) the composite properties achieved through the other filler.

Figure 9.2 demonstrates the TGA thermograms of PE composites as a function of psyllium husk content. The husk exhibited degradation around 325 °C and the magnitude of the degradation peak increased with increasing husk content in the composites. It indicated that the filler resulted in lowering of thermal stability of PE, however, the degradation of the filler temperatures >300 °C indicated that the composites were still useful for high temperature applications as well as had stability towards high temperature processing. It was also confirmed by carrying out the dynamic TGA analysis, where the sample weight was recorded as a function of time at the compounding temperature. No significant weight loss was observed for the test duration equivalent to the melt compounding time, thus, confirming the thermal stability of the composites for processing. The thermal stability was also retained even after subjecting the composites to few processing cycles.

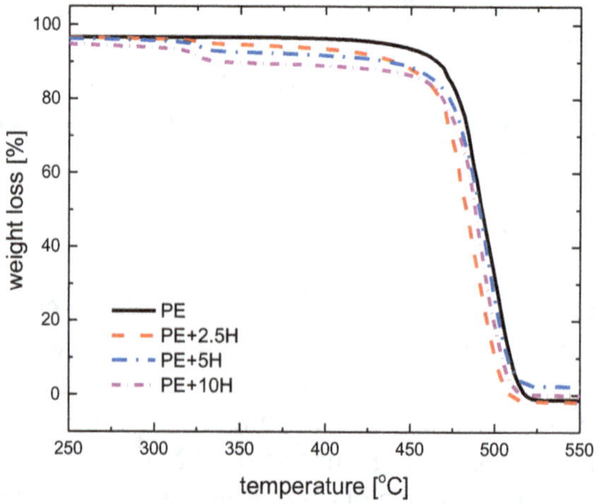

Figure 9.2 TGA thermograms of the PE composites as a function of psyllium husk content.

Table 9.2 and Figure 9.3 show the calorimetric properties of polyethylene composites as a function of husk content. Pure polyethylene

Table 9.2 Calorimetric properties of polyethylene composites with husk

Polymer/Composite	Peak melting temperature, °C	Melt enthalpy, J/g	Peak crystallization temperature, °C
PE	135.0	154	117.0
PE+2.5H	135.4	162	115.2
PE+5H	137.4	161	115.6
PE+10H	137.6	160	115.1

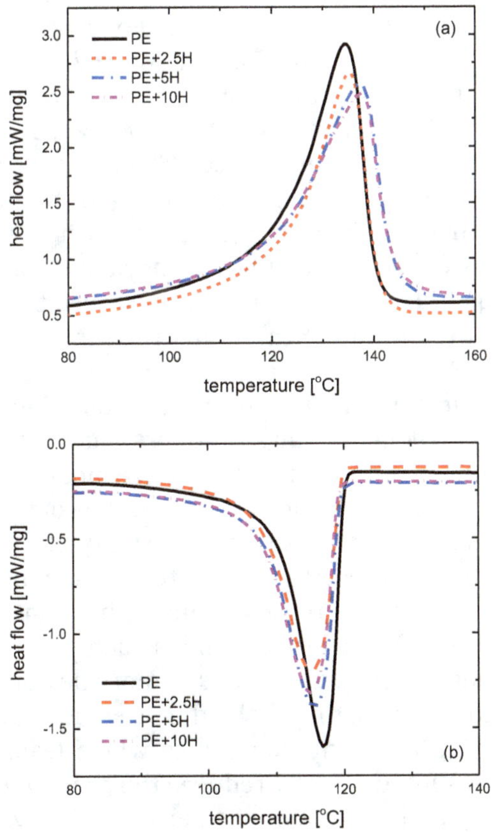

Figure 9.3 DSC thermograms of the composites as a function of psyllium husk content.

exhibited the peak melting point of 135 °C, which was increased slightly on the addition of filler. For instance, the composite with 10% husk had the peak melting temperature of 137.6 °C. It indicated that the crystallization of the polymer was affected by the incorporation of the filler. The melt enthalpy of the polymer was also observed to increase in the range of 4-5% in the composites, thus, confirming the effect of filler on the polymer crystallization and degree of crystallinity. However, the filler did not exhibit any nucleation effect, as the peak crystallization temperature (T_c) was slightly reduced on the addition of filler. Also, the filler content did not have any impact on T_c.

Figure 9.4 presents the rheological analysis of the HDPE composites as a function of psyllium husk content. As evident in Figure 9.4a, the addition of filler exhibited sharp increase in the storage modulus at lower angular frequency. The composite with 10% filler content had the highest storage modulus values. For instance, the storage modulus of the composite with 10% husk was 62940 Pa, in comparison with 39775 Pa for the composites with 2.5% husk. The rate of increment of modulus was observed to slow down at higher angular frequency due to shear thinning effect. The difference between the different composites was retained even at higher frequency, indicating frequency dependent behavior of the composites. The loss modulus of the composites indicated only a slight change till a frequency of 1 rad/s. Subsequently, the modulus was observed to increase exponentially. Similar to the storage modulus, the curves for the different composites did not merge with each other even at high angular frequency, thus, indicating the frequency dependent behavior. For the complete angular frequency range, the loss modulus of the composites was lower in magnitude than the storage modulus, thus, indicating the dominance of the elastic behavior in the composites. Figures 9.4c and d also show the viscosity and elasticity of the samples as a function of angular frequency. Contrary to the modulus, these quantities were observed to decrease on increasing angular frequency. Due to the increase in frequency, polymer structure shows temporary network of entanglements which leads to more flexibility (lower viscosity). In the composites, this allowed more deformation energy to be stored resulting in elastic dominance which leads to increased modulus. Thus, it simultaneously reduced the contribution of lost deformation energy and hence viscous behavior. The viscosity of the composites was almost same at lower frequency whereas higher frequency exhibited some differences depending on the filler content. Also, the slope of decrease in viscosity decreased after a frequency of ~10 rad/s. On the contrary, the elasticity of the composites was a function of filler content at lower frequency, but did not show any difference on increasing the frequency. Similar to modulus, the elasticity of the composites was higher in magnitude than the viscosity for the whole frequency range. It indicated that the psyllium husk was not a passive filler added with an aim to reduce the cost, and contributed significantly to the enhancement of polyethylene properties.

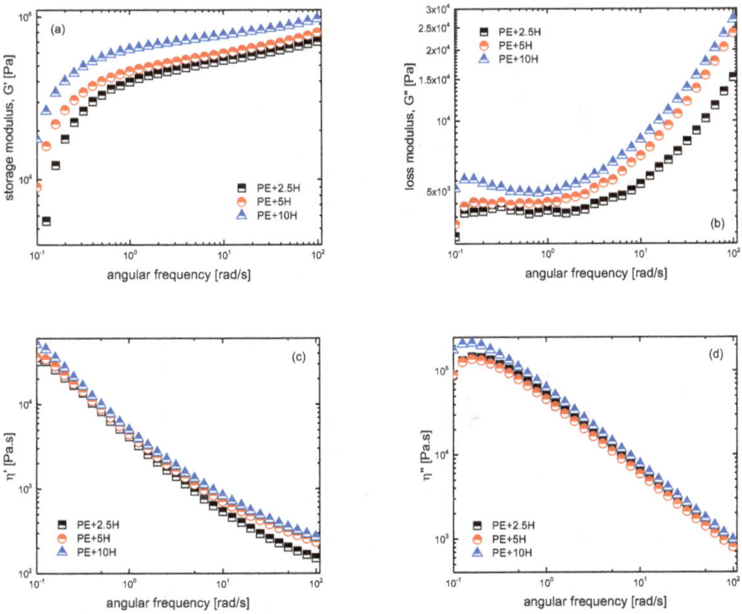

Figure 9.4 Rheological analysis of the composites as a function of psyllium husk content: (a) storage modulus, (b) loss modulus, (c) viscosity, η' and (d) elasticity, η". The angular frequency range from 0.1 to 100 rad/s was used.

Figure 9.5 also presents the dynamic mechanical properties of the composites as a function of filler content as well as temperature. Similar to the modulus in shear rheology, the storage modulus was observed to increase with filler fraction. There was a gradual drop in modulus with temperature, but the composites exhibited better response than the pure polymer. For instance, the modulus of the 10% husk composite at 50 °C and 70 °C was 5.67E8 Pa and 3.17E8 Pa, in comparison with 3.28E8 Pa and 1.89E8 Pa for pure polymer, respectively. This indicated that the addition of filler enhanced the high temperature performance of polyethylene. Though the incorporation of psyllium husk powder was observed to reduce the thermal degradation temperature of polyethylene, but useful performance at service temperatures was, nonetheless, enhanced. It also confirmed the role of psyllium husk as a reinforcing filler. The loss modulus of the composites was also observed to increase with filler fraction, but was lower than the storage modulus for the complete temperature range. In addition, the drop of loss modulus with temperature also exhibited two steps, first a lower rate of decrease till ~45 °C, followed by faster rate of modulus reduction.

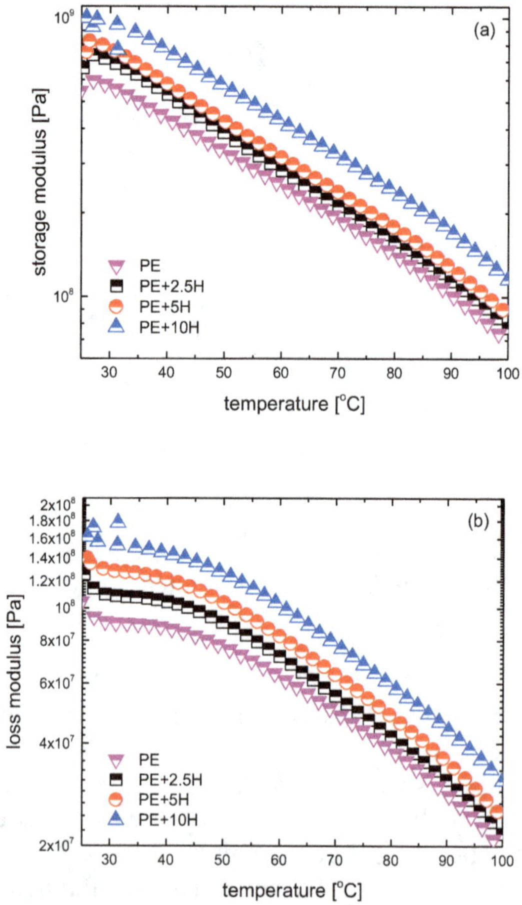

Figure 9.5 Dynamic mechanical analysis of the composites as a function of psyllium husk content and temperature: (a) storage modulus and (b) loss modulus.

The mechanical and rheological performance of the PE composites as a function of husk content confirmed significant reinforcement in the performance on the addition of filler. Further, the impact of husk on the properties of polyethylene was compared with other high performance fillers like 2-D graphene platelets and 1-D sepiolite nanofibers. The extent of filler in the composites was fixed to 5 wt%. In addition, the effect of 1:1 graphene/sepiolite hybrid filler on the composite properties was also studied. All the composites were processed using the same melt mixing parameters so as to relate any property differences to the effect of filler. Figure 9.6 shows the TGA thermograms of the composites. As mentioned earlier, the addition of 5% husk resulted in the thermal degradation peak around 325 °C. Sepiolite incorporation did not impact the onset degradation temperature of the polymer, but the peak degradation temperature was slightly decreased. Addition of graphene resulted in an increase in both onset and peak degradation temperature of polyethylene, indicating better thermal resistance provided by 2-D graphene platelets due to effective heat transfer from the polymer chains to the platelets. Higher surface area and high aspect ratio of graphene platelets would have resulted in enhanced interfacial contact between the polymer chains and the filler particles, thus, leading to enhanced thermal performance. The composite containing the hybrid of graphene and sepiolite exhibited similar degradation pattern as polyethylene, which was intermediate between the behavior of graphene and sepiolite, when added in the composite separately. Similar to the findings from the dynamic TGA analysis mentioned earlier, the weight loss with time at the compounding temperature of 190 °C was not significant, thus, confirming the stability of the composites for high temperature processing as well as applications.

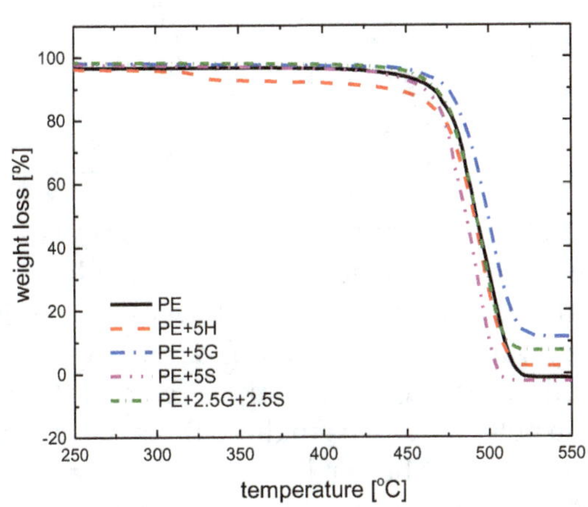

Figure 9.6 TGA thermograms of the composites with 5% filler content.

Table 9.3 and Figure 9.7 demonstrate the calorimetric behavior of the composites with 5% content of husk, graphene, sepiolite and graphene/sepiolite hybrid. As reported earlier, the incorporation of husk resulted in enhanced peak melting temperature as well as melt enthalpy as compared to pure PE, though the filler did not exhibit any nucleation effect. Addition of sepiolite to PE also enhanced the peak melting temperature, however, resulted in slight decrease in the enthalpy indicating the hindrance to crystallization in the presence of filler. Graphene also exhibited decrease in the melt enthalpy, along with peak melting temperature, further confirming the constraint to the polymer chain movement and crystallization due to the 2-D platelets. The hybrid filler exhibited even more pronounced decrease in the degree of crystallization as it was observed to decrease to 139 J/g (normalized to polymer amount in the composite), in comparison with 154 J/g for pure PE. It indicated that the hybrid filler caused more hindrance to polymer chains probably due to specific distribution of the filler particles in the matrix. Presence of both 1-D and 2-D filler particles can also be expected to result in higher degree of anisotropy in the composites. The peak melting point was also deceased the maximum in the composite with hybrid filler.

Table 9.3 Calorimetric properties of the composites with 5% filler content

Polymer/ Composite	Peak melting temperature, °C	Melt enthalpy, J/g	Peak crystallization temperature, °C
PE	135.0	154	117.0
PE+5H	137.4	161	115.6
PE+5G	135.3	149	115.5
PE+5S	137.3	150	115.2
PE+2.5G+2.5S	133.7	139	115.1

The rheological performance of the composites is also compared in Figure 9.8. As observed earlier, the storage modulus enhanced exponentially at the lower frequency, followed by leveling off at the higher frequency values. The storage modulus was observed to be maximum in the case of graphene composite, indicating better stress transfer from the polymer chains to the 2-D platelets, also supporting the earlier observation of better heat transfer due to the dispersed graphene platelets. The storage modulus in the sepiolite containing composite was marginally better than husk as filler, which indicated

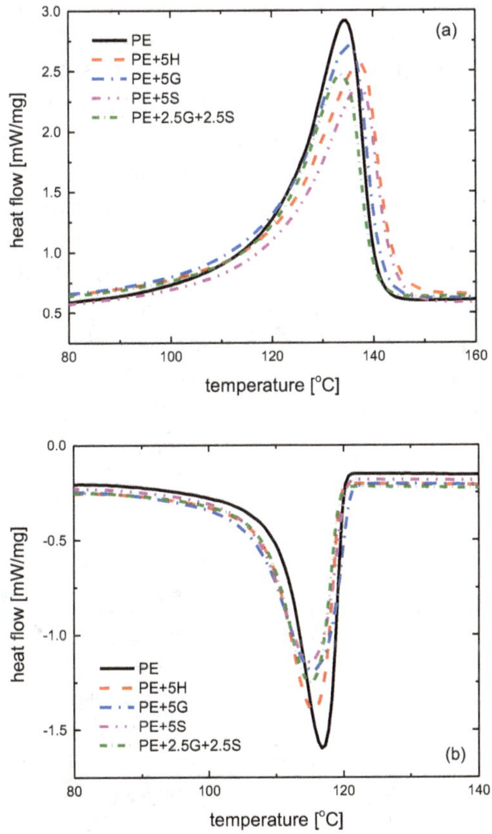

Figure 9.7 DSC thermograms of the composites with 5% filler content.

lower degree of reinforcement of 1-D sepiolite needles as compared to 2-D graphene platelets. Only at higher angular frequency, the resistance of the sepiolite composite was observed to be better than husk based composite. The hybrid filler containing composites had similar modulus as the graphene composite, except at very low and high angular frequency. It demonstrated that higher degree of performance could be achieved in this case as compared to sepiolite alone. The loss modulus of the hybrid filler composite was also observed in between the moduli of graphene and sepiolite containing composites. In addition, the loss modulus of the composites was lower in magnitude than the storage modulus due to dominant elastic character. As earlier, the loss modulus also exhibited slower increase at lower frequencies, followed by higher rate of increment at higher angular frequency values. The viscosity of the composites (as shown in Figure

9.8c) also exhibited similar behavior as loss modulus, with graphene composites demonstrating maximum enhancement in viscosity followed by the composites with hybrid filler and sepiolite. However, it should be noted that even though significant increment in the viscosity was observed in the composites, these were still processable the same way as the pure polymer. The elasticity of the composites was higher in magnitude than the viscosity, similar to the observation made earlier for storage and loss modulus. These findings, thus, indicate that the incorporation of graphene and graphene/sepiolite hybrid filler resulted in highest extent of property increase, however, the organic filler also resulted in considerable improvement and should be further explored as a component of the hybrid filler system.

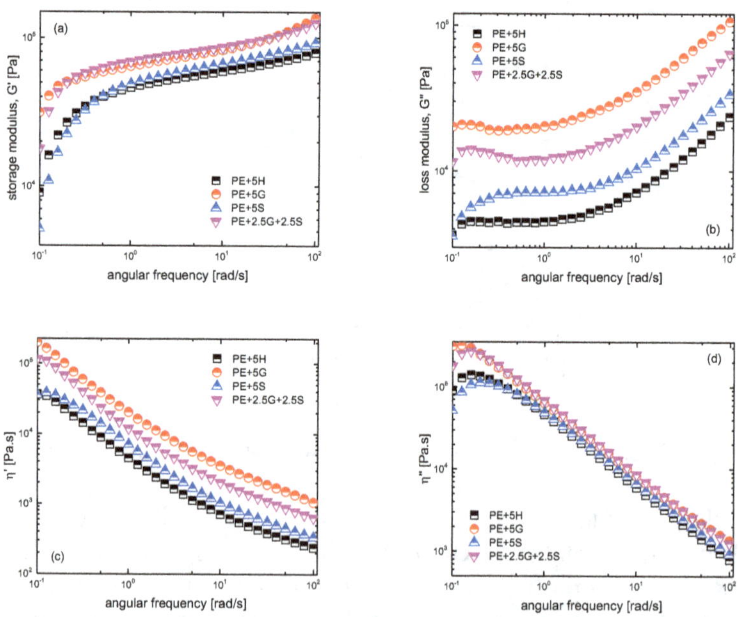

Figure 9.8 Rheological analysis of the 5% filler composites as a function of angular frequency: (a) storage modulus, (b) loss modulus, (c) viscosity, η' and (d) elasticity, η". The angular frequency range was 0.1 to 100 rad/s.

Figure 9.9 also demonstrates the dynamic mechanical analysis of the composite with 5% husk, graphene, sepiolite and graphene/sepiolite hybrid. Similar trends were observed for storage modulus, as observed earlier in rheological analysis. The graphene composite exhibited higher modulus over the whole temperature range due to effective stress transfer. The other composites though exhibited lower modulus values, however, the difference was not significant. Overall, though the modulus decreased by an order of magnitude at 100 °C, however, the composites still exhibited higher modulus, thus, exhibiting improved property profiles. The loss modulus of the composites exhibited similar trend, especially at higher temperatures. The elastic nature of the composites was also confirmed as the loss modulus of these composites was almost an order of magnitude lower than the storage modulus.

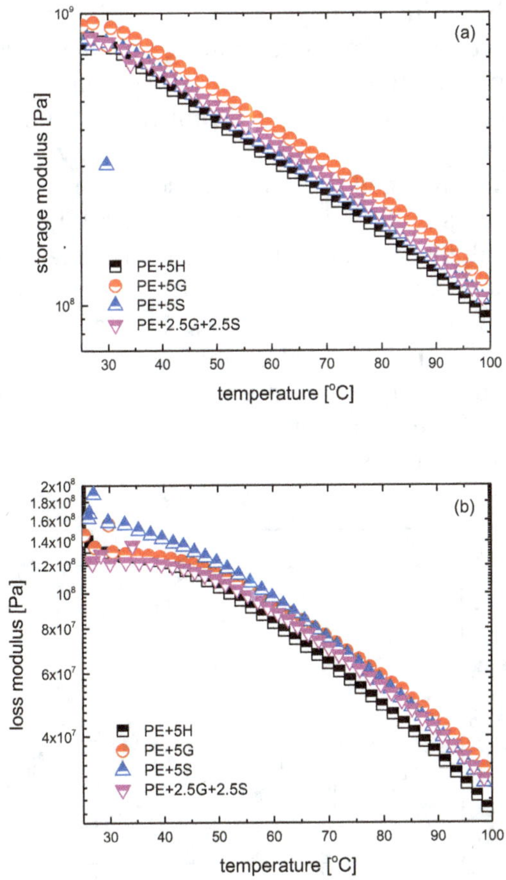

Figure 9.9 Dynamic mechanical analysis of the 5% filler (husk, graphene, sepiolite, graphene/sepiolite hybrid) composites as a function of temperature.

Finally, the effect of hybrid fillers was explored further by incorporating hybrids of husk/graphene, husk/sepiolite as well as husk/graphene/sepiolite. For comparison, the extent of filler was fixed at 7.5 wt% and same processing protocol was used to generate the composites so as to relate the observed morphology and properties solely to the filler phase in the polymer matrix. Figure 9.10 presents the TGA thermograms of the composites in comparison with pure polymer. As observed earlier, the composite with husk/sepiolite hybrid exhibited a degradation peak around 325 °C which was attributed to the degradation of husk. In addition, the composite had reduction in onset and peak degradation temperature of PE phase, which was attributed to the degradation of sepiolite. Due to better thermal resistance, the hybrids containing graphene did not show any negative impact on the onset and peak degradation temperature of PE. In addition, the degradation owing to husk phase was also largely improved in the case of husk/graphene hybrid composite. These findings re-confirmed the earlier observations of slight reduction in the thermal stability of the polymer with husk and sepiolite fillers, whereas enhancement in the thermal performance in the presence of graphene. However, the composites were still suitable for high temperature applications and processing cycles. The thermal stability was maintained even after multiple processing cycles at compounding/processing temperature. As mentioned earlier, the thermal stability of the composites was also confirmed by carrying out the dynamic TGA analysis.

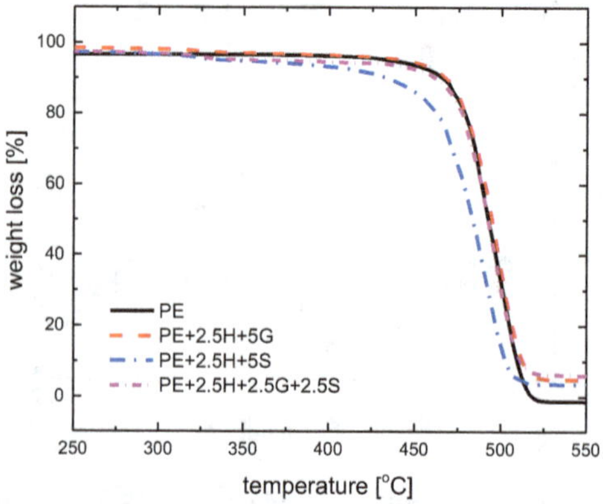

Figure 9.10 TGA thermograms of the composites with 7.5 wt% of hybrid fillers.

The calorimetric response of the composites with hybrid fillers is demonstrated in Table 9.4 and Figure 9.11. The composite with husk/graphene hybrid exhibited a significant reduction in the melt enthalpy to 138 J/g, as compared to 154 J/g for pure PE. Similar reduction was also observed for the composite with hybrid of graphene/sepiolite, indicating hindrance to the polymer chain movement in the presence of graphene. Relatively lower reduction in melt enthalpy was observed for the composite with husk/sepiolite composite, as also observed earlier for sepiolite composite in comparison with graphene. It should also be noted that the enthalpy of the graphene and sepiolite composites was lower when incorporated with husk, as compared to the composites with graphene and sepiolite alone. Though incorporation of husk alone was also observed earlier to enhance the melt enthalpy, but in the hybrid form with graphene and sepiolite, this effect was observed to be opposite, thus indicating a specific morphology of these hybrid fillers in the polymer matrix, as suggested earlier. The composite with hybrid of husk/graphene/sepiolite also exhibited significantly reduced melt enthalpy predominantly due to the presence of graphene. Thus, it indicated that the effect of the hybrid fillers on the crystallization of PE was not directly correlated with the effect of the individual fillers. Specific changes to the composite morphology in the presence of hybrid fillers would have resulted in such effect, even though the fillers were physically mixed with each other and were not expected to have any chemical interaction with each other. Presence of particles of different dimensions are generally envisaged to generate a more stabilized dispersion and intermixing with the polymer. The similar enthalpy was observed in the samples processed for different cycles indicating that the composite morphology was stable.

Similar to the earlier case, the rheological characterization of the composites with hybrid filler systems (husk/graphene, husk/sepiolite and husk/graphene/sepiolite) was also carried out, as shown in Figure 9.12. The composite with husk/sepiolite hybrid exhibited the

Table 9.4 Calorimetric properties of the composites with 7.5 wt% of hybrid fillers

Polymer/Composite	Peak melting temperature, °C	Melt enthalpy, J/g	Peak crystallization temperature, °C
PE	135.0	154	117.0
PE+2.5H+5G	135.5	138	115.2
PE+2.5H+5S	136.9	146	115.0
PE+2.5H+2.5G+2.5S	135.8	137	115.4

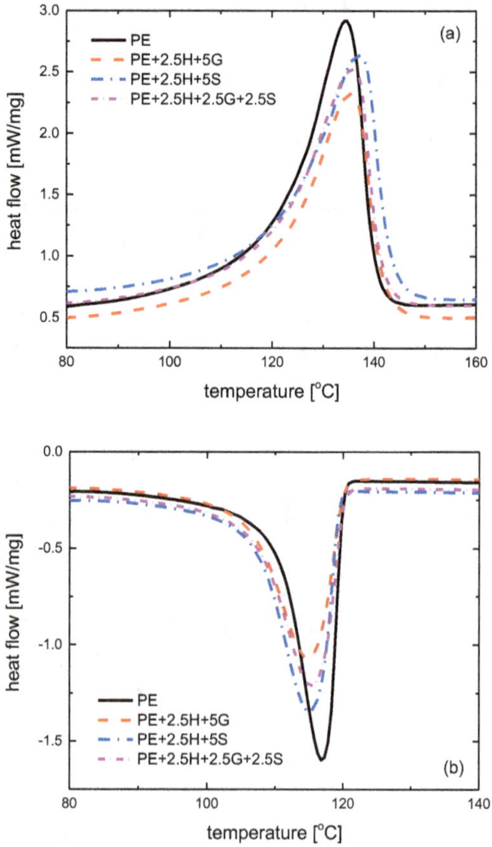

Figure 9.11 DSC thermograms of the composites with 7.5 wt% of hybrid fillers.

lowest modulus as compared to the composites with husk/graphene and husk/graphene/sepiolite hybrid fillers. Specifically, the composite with husk/graphene/sepiolite hybrid exhibited a synergistic enhancement in the modulus as it had higher modulus than both husk/graphene and husk/sepiolite containing composites. Moreover, the incorporation of husk in addition to graphene in the composites led to the increment in the modulus. For instance, the storage modulus of the PE+5G composite at 1 rad/s was 64580 Pa, which was increased to 83630 Pa for the PE+2.5H+5G composite. It indicated the addition of husk did not deteriorate mechanical performance of graphene, and, in fact, enhanced it. The modulus of the sepiolite composite was observed to remain unchanged with the addition of husk, thus, reconfirming that husk did not negatively impact the properties

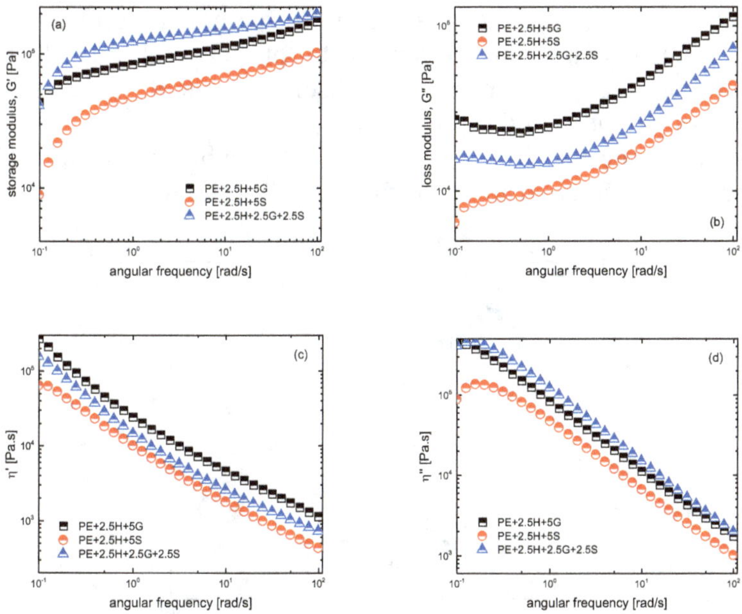

Figure 9.12 Rheological analysis of the 7.5 wt% hybrid filler composites as a function of angular frequency: (a) storage modulus, (b) loss modulus, (c) viscosity, η' and (d) elasticity, η''. The angular frequency range was 0.1 to 100 rad/s.

and can be useful in this case to reduce the material cost. Similarly, the modulus of the husk/graphene/sepiolite ternary hybrid filler composite was much higher than the graphene/sepiolite binary hybrid filler composite. For all composites, the modulus had much faster rate of enhancement at the lower frequency, followed by slower rate of modulus increase, as also observed earlier. The loss modulus of the ternary hybrid filler composites was observed to in between the husk/graphene and husk/sepiolite containing composites. As also observed earlier, the loss modulus exhibited faster rate of increment only after a frequency of 1 rad/s. In addition, the loss modulus of the composites was smaller in magnitude than the storage modulus over the whole angular frequency range due to the elastic nature of the composites. The viscosity of the composites also exhibited similar behavior as the loss modulus. Similar to the other composites, the elasticity of the composites was higher in magnitude than the viscosity, thus, confirming the elastic nature of the composites. Similar to storage modulus, the elasticity of the ternary filler composite was the highest among the composites with hybrid fillers as well as the composites with 5% graphene or sepiolite. These findings confirmed the positive impact of husk on the composites properties as no reduction in the rheological performance was observed when husk was added along with graphene and sepiolite.

The dynamic mechanical properties of the hybrid filler containing composites is demonstrated in Figure 9.13. The addition of husk with

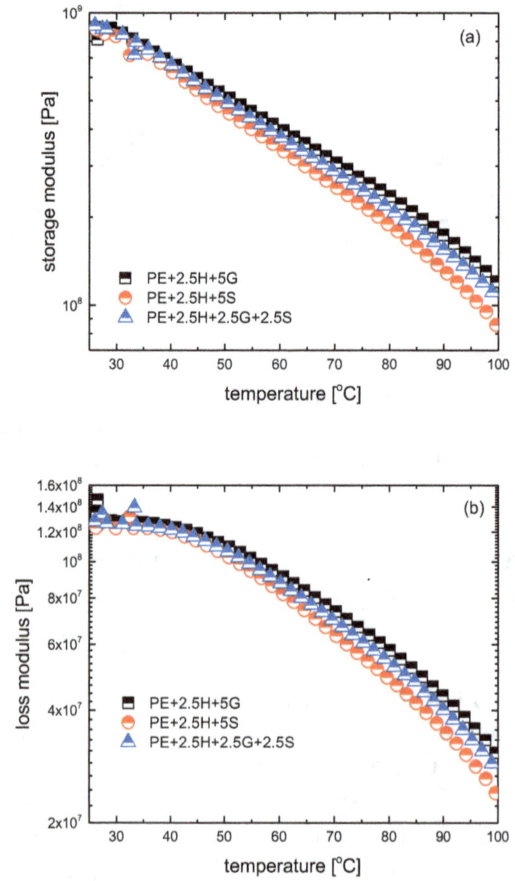

Figure 9.13 Dynamic mechanical analysis of the 7.5 wt% hybrid filler composites as a function of temperature.

other fillers was observed to retain the modulus to the similar values as observed in Figure 9.9 and no negative impact on the mechanical performance was observed over the whole temperature range. Thus, even though the modulus did not enhance the way as observed in the

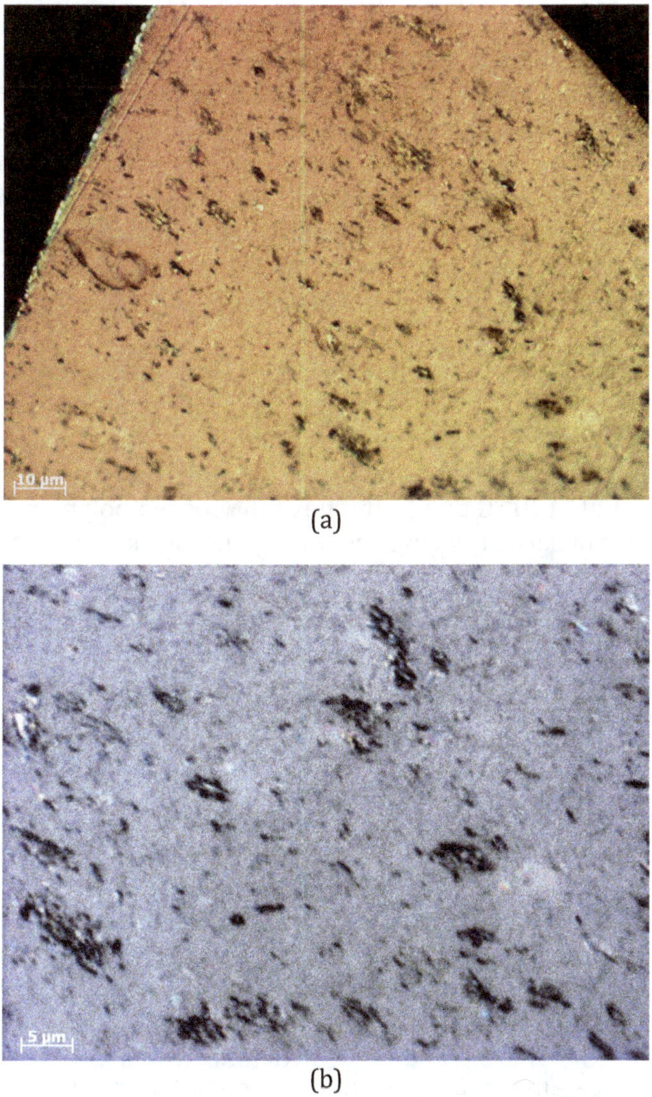

Figure 9.14 Optical micrographs of PE+2.5H+2.5G+2.5S composite.

rheological analysis, however, the overall observed behavior of the husk containing composites confirmed that the properties of the composites with graphene or sepiolite could be retained on the addition of husk, which would, thus, provide cost efficiency due to much lower price of husk. Overall, the composite with husk/graphene exhibited the highest modulus followed by the ternary hybrid filler and husk/sepiolite composites. The modulus of composites decreased gradually with temperature and had an order of magnitude reduction in the storage modulus at 100 °C, but the composites still had appreciable mechanical resistance. Similar to the storage modulus, the loss modulus of the hybrid composites also exhibited similar values as observed earlier in Figure 9.9. As is evident, the husk/graphene reinforced composite exhibited the highest magnitude. The loss modulus of the composites was also significantly lower in magnitude than the storage modulus due to the dominant elastic nature of the composites, which has been observed for all the composites generated in the study.

Further insights in the morphology of the ternary hybrid filler composite were also gained through microscopy analysis. Figure 9.14 shows the optical micrographs of the composite at two different magnifications. Block face preparation for the microscopy analysis was carried out at -120 °C due to the flexibility of the polymer. No filler pullout was observed during sectioning and block face preparation indicating good degree of filler adhesion with the polymer. The filler distribution was observed to be uniform with no presence of large filler aggregates. It has to be noted that the composites were generated without the addition of commonly used compatibilizers, thus, the good degree of polymer-filler intermixing indicated that the processing parameters used to generate the composites were appropriate.

Figure 9.15 demonstrates the TEM analysis of the PE+2.5H+2.5G+2.5S composite. As evident in Figure 9.15a and c, uniform filler distribution was confirmed. As the fillers were physically mixed with each other during compounding and no thermodynamic interaction among these was expected, therefore, the filler particles were observed to be present as separate phases in the composite morphology. Large aspect ratio platelets of graphene were observed in Figure 9.15c, whereas Figure 9.15b demonstrates the fibrillar morphology of sepiolite. The Si ratio image in Figure 9.15d also confirmed the presence of sepiolite (bright signal). The husk particles could not be observed in the composites probably due to the difficulty in generating phase contrast with the organic PE phase.

EDXS analysis was also carried out on the PE+2.5H+2.5G+2.5S composite, as shown in Figure 9.16. Different locations were chosen on the TEM micrograph and were subsequently analyzed for the elemental analysis. The area marked as 1 in the TEM micrograph indicated the presence of Si and Mg signals, thus, further confirming the presence of sepiolite in the composite. Similarly, the area marked as

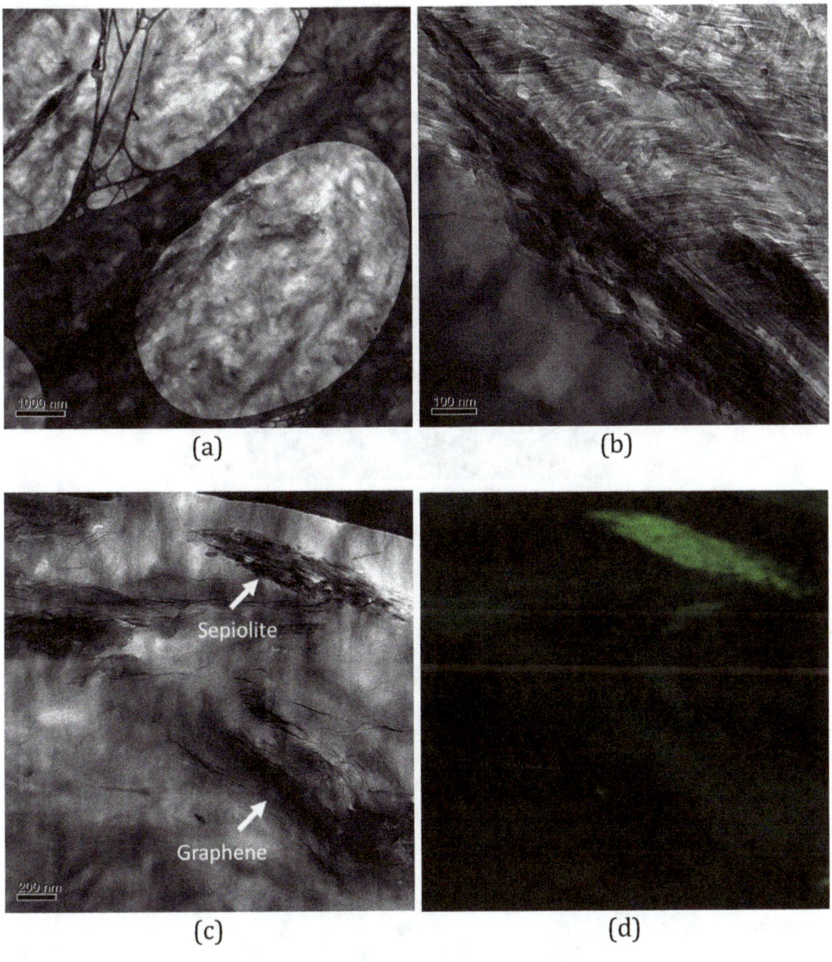

Figure 9.15 Electron microscopy analysis of PE+2.5H+2.5G+2.5S composite: (a) low magnification image, (b) sepiolite distribution, (c) high magnification image depicting the distribution of graphene and sepiolite and (d) corresponding Si ratio image to confirm the presence of sepiolite fibers.

3 on the TEM micrograph exhibited the C signal, due to the presence of graphene. The area marked as 2 on the micrograph exhibited diffuse signals of sepiolite along with C, probably due to the joint presence of sepiolite and graphene due to occasional intermixing. Thus, the hybrid composite exhibited uniform morphology where the three fillers co-existed with each other without leading to any antagonistic effect on the composite properties. These findings also confirmed the

potential of husk as a functional filler which can either enhance or retain the composite properties, along with providing cost efficiency.

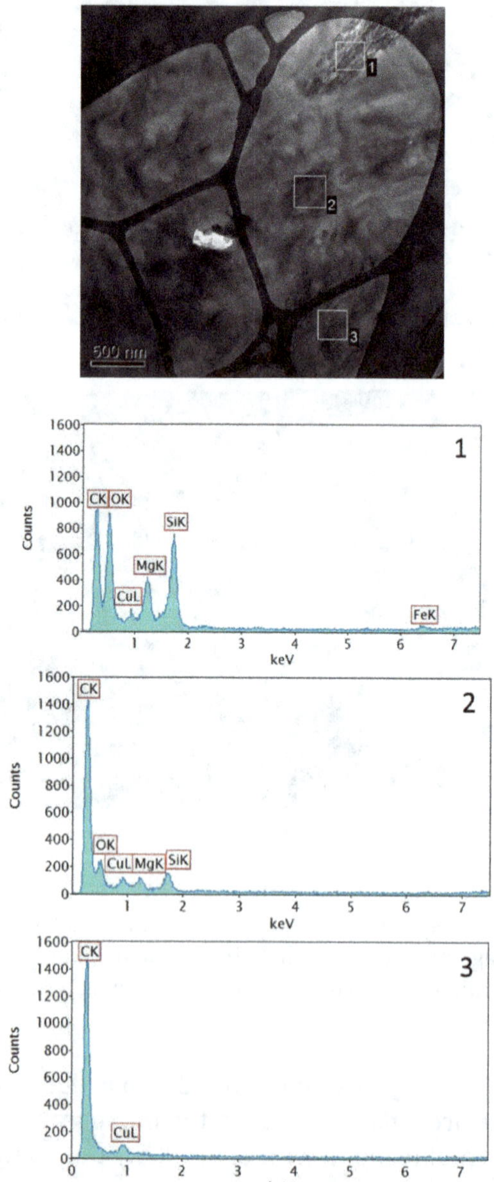

Figure 9.16 (a) TEM image of the PE+2.5H+2.5G+2.5S composite; (c-d) EDXS analysis at the three locations in the composite depicting sepiolite, hybrid filler and graphene (marked at 1, 2 and 3).

9.4 Conclusions

In the current study, the composites of polyethylene were prepared with a variety of 0-D, 1-D and 2-D fillers, along with their hybrid systems in order to analyze their impact on thermal, mechanical and rheological properties. Incorporation of varying amounts of 0-D psyllium husk particles to HDPE led to slight reduction in the thermal stability due to the degradation of husk at 325 °C, however, the composites were still suitable for high temperature applications. The husk particles also enhanced the peak melting temperature and melt enthalpy of polyethylene with increasing filler content, however, the crystallization temperature did not change in the composites. The storage modulus from shear rheology was observed to increase as a function of husk content and was also higher than loss modulus due to dominant elastic nature of the composites. Similarly, the elasticity of the composites was higher than the viscosity due to the elastic nature. The dynamic mechanical analysis also exhibited increase in modulus with increasing filler content, thus, confirming the positive impact of husk on the composite properties. Though the modulus of the composites was observed to decrease by an order of magnitude by 100 °C, however, it was still significant, thus, confirming the enhancement of high temperature performance. On comparing the composites with husk, sepiolite, graphene, and graphene/sepiolite hybrid as fillers, the graphene composite was observed to have the best thermal performance due to efficient heat transfer from the polymer chains to the high aspect ratio 2-D graphene platelets. However, the other filler systems also exhibited composite stability against high temperature processing. The use of graphene also resulted in decreased melt enthalpy and peak melting point of PE due to hindrance to chain movement and crystallization. The storage modulus of both graphene and graphene/sepiolite based composites was observed to be higher than other composites due to effective stress transfer from the polymer chains to the graphene platelets owing to higher degree of interfacial contacts between the polymer and filler phases. The addition of husk particles to graphene, sepiolite and graphene/sepiolite led to further reduction in the melt enthalpy, thus, indicating combined effect of fillers, though the addition of husk alone was observed to increase the melt enthalpy of the polymer. The addition of husk also resulted in significant enhancement of the storage modulus (from shear rheology) in both graphene and graphene/sepiolite composites, whereas the modulus was retained in the case of husk/sepiolite composite.

The dynamic mechanical analysis also exhibited that the incorporation of the husk particles to graphene, sepiolite and graphene/sepiolite retained the storage modulus. Overall, the addition of husk either improved or retained the composite properties, which resulted in the cost-efficiency of the composites due to the lower cost of husk particles. The morphology of the composites with hybrid filler systems was observed to be uniform with filler phases distributed effectively in the polymer matrix, even without the use of externally added compatibilizer. This also indicated that the processing parameters used for the generation of the composites were appropriate. Thus, these findings confirmed the usefulness of employing filler hybrids to achieve synergistic improvement in the composite properties. The obtained composites are effective alternatives to many polymeric materials in terms of properties as well as cost.

References

1. Mittal, V. (2010) *Advances in Polyolefin Nanocomposites*, CRC Press, USA.
2. Jancar, J. (1999) *Mineral Fillers in Thermoplastics I: Raw Materials and Processing*, Springer-Verlag, Germany.
3. Nwabunma, D., and Kyu, T. (2008) *Polyolefin Composites*, Wiley, USA.
4. Mittal, V. (2010) *Polymer Nanotube Nanocomposites: Synthesis, Properties, and Applications*, John Wiley & Sons, USA.
5. Kim, S., Do, I., and Drzal, L. T. (2009) Multifunctional xGnP/LLDPE nanocomposites prepared by solution compounding using various screw rotating systems. *Macromolecular Materials and Engineering*, **294**(3), 196-205.
6. Chaudhry, A. U., and Mittal, V. (2013) High-density polyethylene nanocomposites using masterbatches of chlorinated polyethylene/graphene oxide. *Polymer Engineering and Science*, **53**(1), 78-88.
7. Mukhopadhyay, P., and Gupta, R. K. (2011) Trends and frontiers in graphene-based polymer nanocomposites. *Plastics Engineering*, **67**, 32-42 (2011).
8. Tartaglione, G., Tabuani, D., Camino, G., and Moisio, M. (2008) PP and PBT composites filled with sepiolite: Morphology and thermal behaviour. *Composites Science and Technology*, **68**(2), 451-460.
9. Chen, H., Zheng, M., Sun, H., and Jia, Q. (2007) Characterization and properties of sepiolite/polyurethane nanocomposites. *Materials Science and Engineering: A*, **445-446**(0), 725-730.

10. Ghazanfari, A., Emami, S., Panigrahi, S., and Tabil, L. G. (2008) Themal and mechanical properties of blends and composites from HDPE and date pits particles. *Journal of Composite Materials*, **42**, 77-89.
11. Shnawa, H. A., Abdulah, N. A., and Mohamad, F. J. (2011) Thermal properties of low density polyethylene with oyster shell composite: DSC study. *World Applied Sciences Journal*, **14**(11), 1730-1733.
12. Tong, J. Y., Royan, N. R. R., Ng, Y. C., Ghani, M. H. A., and Ahmad, S. (2014) Study of the mechanical and morphology properties of recycled HDPE composite using rice husk filler. *Advances in Materials Science and Engineering*, Article ID 938961.
13. Singh, B., and Kumar, S. (2008) Synthesis and characterization of psyllium-NVP based drug delivery system through radiation crosslinking polymerization. *Nuclear Instruments and Methods in Physics Research Section B: Beam Interactions with Materials and Atoms*, **266**(15), 3417-3430.
14. Bhatia, M., and Ahuja, M. (2014) Synthesis and evaluation of psyllium-g-poly(N-vinyl-2-pyrrolidone) as mucoadhesive gel. *Der Pharmacia Lettre*, **6**(2), 127-131.
15. Lee, G.-W., Park, M., Kim, J., Lee, J. I., and Yoon, H. G. (2006) Enhanced thermal conductivity of polymer composites filled with hybrid filler. *Composites Part A: Applied Science and Manufacturing*, **37**(5), 727-734.
16. Yu, A., Ramesh, P., Sun, X., Bekyarova, E., Itkis, M. E., and Haddon, R. C. (2008) Enhanced thermal conductivity in a hybrid graphite nanoplatelet - carbon nanotube filler for epoxy composites. *Advanced Materials*, **20**(24), 4740-4744.
17. Lorenz, H., Fritzsche, J., Das, A., Stockelhuber, K. W., Jurk, R., Heinrich, G., and Kluppel, M. (2009) Advanced elastomer nanocomposites based on CNT-hybrid filler systems. *Composites Science and Technology*, **69**(13), 2135-2143.
18. Gwon, J. G., Lee, S. Y., Chun, S. J., Doh, G. H., and Kim, J. H. (2010) Effects of chemical treatments of hybrid fillers on the physical and thermal properties of wood plastic composites. *Composites Part A: Applied Science and Manufacturing*, **41**(10), 1491-1497.
19. Marlett, J. A., and Fischer, M. H. (2002) A poorly fermented gel from psyllium seed husk increases excreta moisture and bile acid excretion in rats. *The Journal of Nutrition*, **132**(9), 2638-2643.
20. Fischer, M. H., Yu, N., Gray, G. R., Ralph, J., Anderson, L., and Marlett, J. A. (2004) The gel forming polysaccharide of psyllium husk (Plantago ovata Forsk). *Carbohydrate Research*, **339**(11), 2009-2017.
21. van Craeyveld, V., Delcour, J. A., and Courtin, C. M. (2009) Extractability and chemical and enzymic degradation of psyllium (Plantago ovata Forsk) seed husk arabinoxylans. *Food Chemistry*, **112**(4), 812-819.

22. Edwards, S., Chaplin, M. F., Blackwood, A. D., and Dettmar, P. W. (2003) Primary structure of arabinoxylans of ispaghula husk and wheat bran. *Proceedings of the Nutrition Society*, **62**(1), 217-222.
23. McAllister, M. J., Li, J. L., Adamson, D. H., Schniepp, H. C., Abdala, A. A., Liu, J., Herrera-Alonso, M., Milius, D. L., Car, R., Prud'homme, R. K., and Aksay, I. A. (2007) Single sheet functionalized graphene by oxidation and thermal expansion of graphite. *Chemistry of Materials*, **19**, 4396-4404.
24. Hummers, W. S., and Offeman, R. E. (1958) Preparation of graphitic oxide. *Journal of American Chemical Society*, **80**, 1339.

10

Polyethylene-Thermally Reduced Graphene Nanocomposites: Comparison of Masterbatch and Direct Melt Mixing Approaches on Mechanical, Thermal, Rheological and Morphological Properties

10.1 Introduction

Nanocomposites represent a new class of materials having superior mechanical, electrical, thermal and gas barrier properties along with lightweight and optical transparency [1-7]. Various 1-D and 2-D nanofillers (e.g., fibrous and platy fillers) due to their high aspect ratio and geometrical shape have the potential to influence the polymer properties at very low fractions. However, to achieve superior performance especially at lower filler fractions, dispersion of the reinforcement phase at nanoscale is of utmost importance [1,6]. Thermodynamic factors such as interfacial compatibility of polymer and filler phases as well as kinetic factors such as filler shape and size, dispersion techniques and equipment, time of mixing and applied shear lead to final morphology in the polymer nanocomposites [8-13]. Absence of any interfacial interactions between the polymer and filler phases and mismatch of polarity between the low surface energy polymers, especially polyolefins, and high surface energy reinforcements often leads to the non-optimum filler dispersion, thus, resulting in performance similar to conventional macro-composites.

The development of free-standing single-layer graphene from inexpensive sources (graphite) has opened the gateway for novel functional polymer nanocomposites with much superior properties as compared to property increments usually achieved from other conventional fillers [14-20]. Graphene sheets are one-atom-thick two-dimensional layers of sp^2-bonded carbon and can effectively enhance electrical and thermal conductivity, gas impermeability and mechanical properties of polymers [14-20]. For instance, a single defect-free graphene layer has been estimated to have Young's

Ali U. Chaudhry and Vikas Mittal*
Khalifa University of Science and Technology, Abu Dhabi, UAE
**Current address: Texas A&M University, Qatar*

modulus of ≈1.0 TPa, intrinsic strength ≈42 N/m, thermal conductivity ≈4840-5300 W/(m.K), electron mobility exceeding 25,000 $cm^2/V.s$, excellent gas impermeability and specific surface area of ≈2630 m^2/g [21]. Thus, even with some degree of defects in the structure, graphene platelets represent functional reinforcement material to achieve nanocomposites with exceptional properties. It is also confirmed by a large number of recent research studies reported on various polymer-graphene nanocomposites systems [21-26].

Significant research efforts have been dedicated to enhance the dispersion of nanofillers by using different mixing techniques, modification of polymer backbone or filler surfaces, use of compatibilizers (functional polymers) and coupling agents, etc. [27]. Thermally reduced graphene as compared to graphene oxide (GO) lacks the presence of functional groups and has strong interlayer cohesive energy and surface inertia, which make it less compatible with polymers for generating nanocomposites [28]. In order to improve the dispersion of thermally reduced graphene in polymer matrices, different routes have been adopted, such as surface modification by phenyl isocyanate and porphyrin [29] as well as use of surfactants and compatibilizers for exfoliation and dispersion [30,31]. Fang *et al.* [29] covalently linked graphene with polystyrene using atom transfer radical polymerization and showed improvement in thermal and mechanical properties. Castelain *et al.* [32] reported different chemical routes to functionalize graphene with short-chain polyethylene which influenced the mechanical properties of graphene-based high density polyethylene (HDPE) nanocomposites. Vasileiou *et al.* [33] reported the dispersion and properties of polyethylene-graphene composite using maleated linear low-density polyethylene (LLDPE) derivatives and thermally reduced graphene oxide through a non-covalent compatibilization approach. Yu *et al.* [34] prepared a flame retardant functionalized GO (FRs-FGO) using amine functional GO and phosphoramide oligomer, which was incorporated into PP along with PP-grafted maleic anhydride (PP-g-MA). In an interesting study, Seo *et al.* [35] also reported the compatibility of functionalized graphene with polyethylene as well as its copolymers as a function of molecular weight and polarity.

Many studies have also focused on different methods of nanocomposite generation to achieve efficient filler dispersion. Wang *et al.* [36] prepared low density polyethylene (LDPE)/graphene nanocomposites using vinyl functionalized graphene sheets and LDPE through solution blending method. In a recent study, different methods of dispersion for graphene in LLDPE were adopted, and their effect on mechanical properties was reported. The methods included solution mixing and melt mixing by co-counter-rotating and modified co-rotating screw systems [37]. It was observed that solution mixing resulted in better properties than co-counter-rotating and modified co-rotating methods [37]. Giannelis *et al.* [38] incorporated functionalized graphene sheets in poly(vinylidene fluoride) matrix by solution mixing and melt blending which resulted in percolation at a much lower filler amount. As the high molecular weight polymers, especially polyolefins, are not soluble in common solvents, solution mixing method is, thus, less effective. As an alternative, solvent based dispersion of fillers in compatibilizers or functional polymers (solution mixing) is a simple and effective processing route for the fabrication of hybrid masterbatches. In this method, the filler and the functional polymer interact through relatively weak dispersive forces [31,39]. Subsequently, the incorporation of masterbatch by melt mixing with high molecular weight polymer matrices can lead to a better state of filler dispersion than directly melt mixed polymer, compatibilizer and filler. In a recent study, HDPE nanocomposites were generated by melt mixing with chlorinated polyethylene-graphene masterbatch [40], however, more efforts are needed to establish direct comparison between the masterbatch and direct melt mixing methods.

The goal of the current study was to explore the effectiveness of the masterbatch approach over direct melt mixing method in improving the filler dispersion and resulting composite properties. Compatibilizers of different chemical architectures and polarities were chosen. The amount of the composite constituents was held constant in order to correlate the observed property changes to the method of nanocomposite generation and specific compatibilizer.

10.2 Experimental

10.2.1 Materials

High density polyethylene FB1460 was supplied by Abu Dhabi Polymers Company Limited, UAE. It was in the form of white pellets with specific gravity 0.946 and peak melting point ~131 °C (supplier data). The melt flow index (MFI) for the polymer at 2.16 kg and 190 °C was reported by the supplier as <0.1 g/10 min, indicating high molecular weight. Three copolymer based compatibilizers (ethylene acrylic acid copolymer (EAA), ethylene vinyl acetate-*graft*-maleic anhydride (EVA-MA) and ethylene methacrylic acid copolymer with zinc ion (EMAZ)) of different specifications and functionalities were procured from suppliers as mentioned in Table 10.1. The polymer materials were used as obtained. Thermally reduced graphene (G) was prepared through thermal exfoliation of precursor graphite oxide [41] using modified Hummer's method [42], as reported earlier [40,43]. The confirmation of graphene formation was achieved through Raman spectroscopy, microscopy and X-ray diffraction. The average size of the graphene platelets was 2 μm, whereas the average thickness was 10 nm (few multilayers).

Table 10.1 Details of the polymer, compatibilizers, and filler used for the study

Name	Repeat Unit	Specifications
High density polyethylene (HDPE)	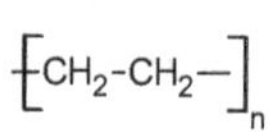	Trade name: Borouge FB1460 MFI (2.16 kg, 190 °C): <0.1 g/10 min Density: 0.946 g/cm³ Peak melting temperature: 131 °C
Ethylene acrylic acid copolymer (EAA)	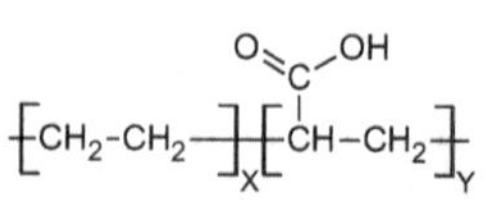	Trade name: Exxon Mobil ESCOR 5050 Co-monomer content: 9% AA MFI (2.16 kg, 190 °C): 8.4 g/10 min Density: 0.936 g/cm³ Peak melting temperature: 97 °C
Ethylene vinyl acetate-g-maleic anhydride (EVA-MA)	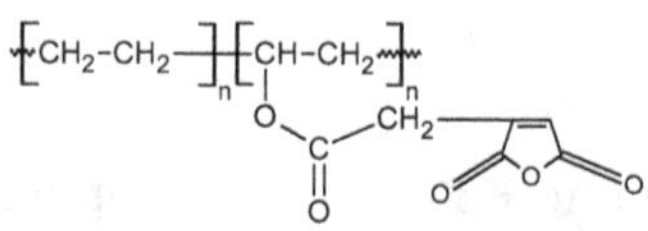	Trade name: Dupont Fusabond C190 Co-monomer content: 12-15% MA MFI (2.16 kg, 190 °C): 16 g/10 min Density: 0.950 g/cm³ Peak melting temperature: 71 °C

Ethylene methacrylic acid copolymer with zinc ion (EMAZ)	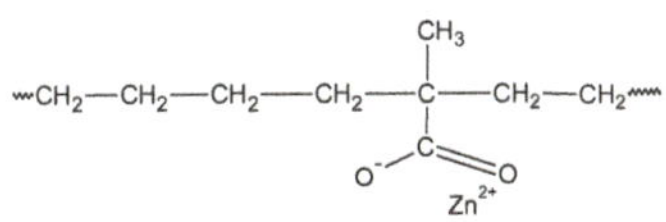	Trade name: Dupont Surlyn 9020 Co-monomer content: 14% MFI (2.16 kg, 190 °C): 1 g/10 min Density: 0.960 g/cm^3 Peak melting temperature: 85 °C
Thermally reduced graphene (G)	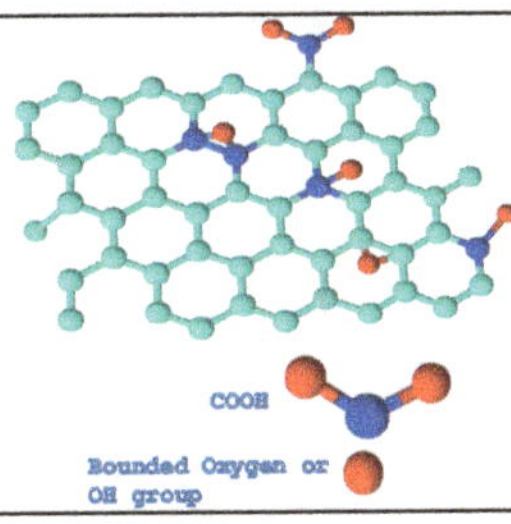	Obtained using modified Hummer's method C and O fractions: 77.5% and 18.4% respectively Density: 0.0161 g/cm^3

10.2.2 Generation of Nanocomposites

The nanocomposites were generated either by masterbatch approach (solution mixing followed by melt mixing) or direct melt mixing method, as shown in Scheme 10.1. For the masterbatch method, solution of compatibilizer (functional polymer (FP)) in *p*-xylene was generated by stirring at 100 °C for 2 h (3% solid content). Suspension of graphene in *p*-xylene was obtained by stirring at room temperature for 1 h followed by sonication for 10 min. The two components were then mixed at room temperature, sonicated and gradually mixed at 100 °C for further 1 h. The blend was stirred and gradually brought to room temperature. The solution was kept overnight under vacuum at room temperature followed by vacuum drying at 40 °C. Two masterbatches with graphene concentration of 9% and 17% were obtained. The so-obtained masterbatches of compatibilizer with graphene were subsequently melt mixed in the extruder with appropriate amount of HDPE to generate nanocomposite with 1% filler content and compatibilizer content 5% (using masterbatch with 17% graphene concentration) or 10% ((using masterbatch with 9% graphene concentration). In the direct melt mixing method, all three components of the composite system were directly mixed in the extruder to generate nanocomposites. HDPE/G composite was also generated similarly by melt mixing. For notation

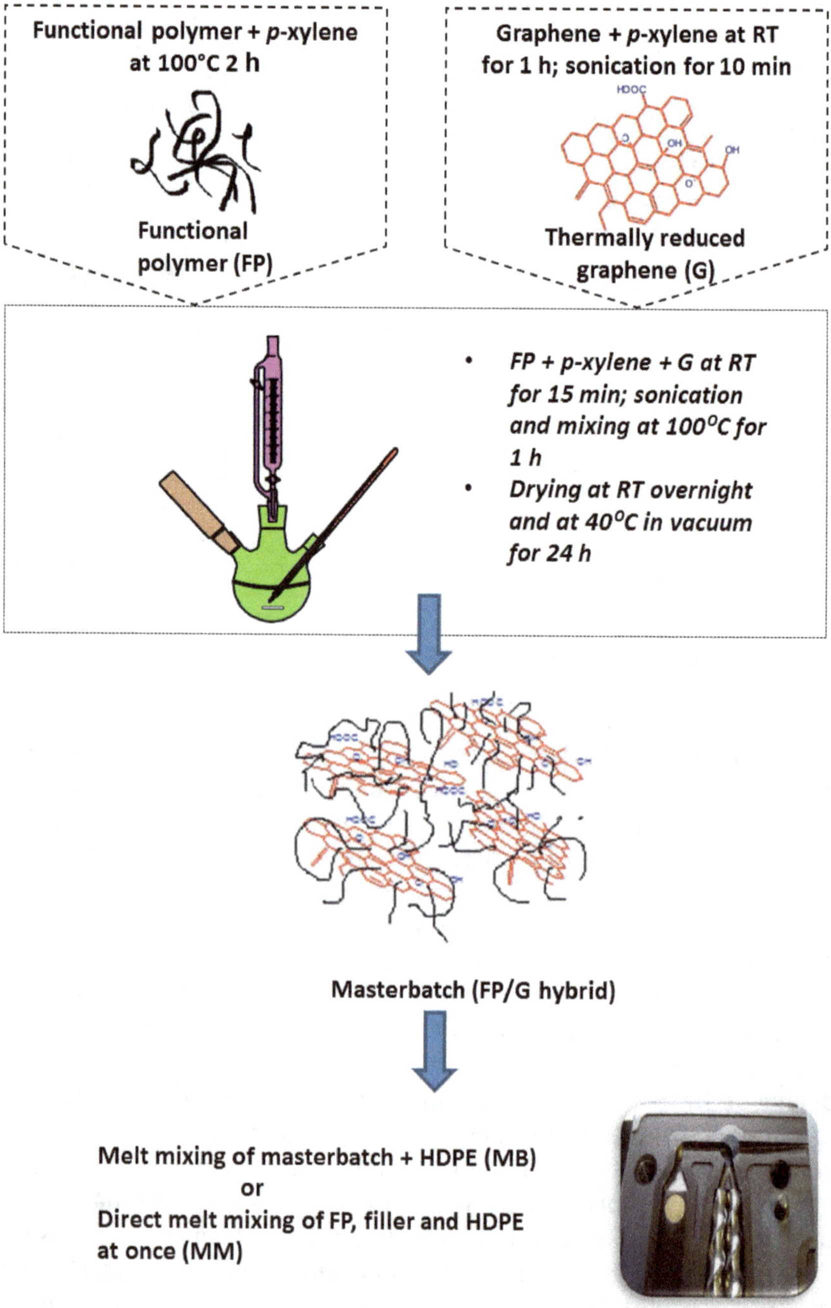

Scheme 10.1 Schematic representation of the generation of nanocomposites using masterbatch and direct melt mixing approaches.

purposes, the technique using masterbatch approach was denoted as 'MB', whereas the direct melt mixing was named as 'MM'. Melt mixing was carried out using a mini twin screw extruder (MiniLab HAAKE Rheomex CTW5, Germany), which had a screw length and screw diameter of 109.5 mm and 5/14 mm conical respectively. Batch size of 4.5 g was used and the shear mixing was performed for 6 min at 60 rpm. The compounding temperature used for nanocomposite generation was 190 °C. Pure polymer was also processed similarly by subjecting it to similar shear and thermal conditions. Disc and dumbbell shaped test specimens were injection molded using a mini injection molding machine (HAAKE MiniJet, Germany) using same processing temperature as used during compounding. The injection pressure was 700 bars for 6 s, whereas holding pressure was 400 bar for 3 s. The temperature of the mold was kept at 50 °C.

10.2.3 Characterization of Nanocomposites

For the characterization of thermally reduced graphene, XRD analysis was conducted on a Panalytical Powder Diffractometer (X'Pert PRO) using Cu-Kα radiation (λ=1.5406 Å, 45 kV, 40 mA). Diffraction pattern was collected in the range of 5-40° 2Θ scale, with a step size of 0.0170° s^{-1}. Raman spectrum was recorded on a Jobin Yvon Horiba LabRAM spectrometer with back-scattered confocal configuration using a HeNe laser (633 nm). A long working distance objective with magnification 50× was used both to collect the scattered light and to focus the laser beam on the sample surface. CCD (charge-coupled device) detector exposure time was 10 s and an average of 2 cycles was used to increase S/N ratio. Raman shifts were calibrated using Si wafer signal at 520.7 cm^{-1}.

Tensile testing of nanocomposites was performed on universal testing machine (Testometric, UK). The sample dimensions for tensile test were: sample length 73 mm, gauge length 30 mm, width 4 mm and thickness 2 mm (conforming to standard ASTM-638-V). A loading rate of 4 mm/min was used and the tests were carried out at room temperature. Tensile modulus and peak stress were calculated using built-in software Win Test Analysis. An average of six values is reported. Dynamic mechanical analysis (DMA) of the polymer materials was performed on RSA3 of TA Instruments to study the change in modulus with temperature (25-100 °C). Dynamic temperature ramp test was used at a frequency of 5 rad/s, ramp rate 5 °C/min and 0.1 % strain for all samples.

Thermal degradation properties of the nanocomposites were recorded using Netzsch thermogravimetric analyzer (TGA). Nitrogen was used as a carrier gas and the scans were obtained from 50 to 700 °C at a heating rate of 10 °C/min. Perkin-Elmer Pyris-1 differential scanning calorimeter under nitrogen atmosphere was used to study the calorimetric properties of nanocomposites. The scans were obtained from 50-200-50 °C (second heating runs) using heating and cooling rates of 5 °C/min respectively. The heat enthalpies were measured with an error of ±0.1 % and were confirmed by repeating the runs.

PRS-801 Resistance system was used to measure the resistance of the samples. It is capable of getting precision resistance measurements from 0.1 up to 2.0e14 Ω with an overall measurement tolerance of ±5 %. Geometric averages of the resistance measured from 3 to 6 different spots from each side of the samples were recorded. Samples were in the shape of circular discs of 2 cm diameter and 1.5 mm thickness. Test voltage was 100 V and test time of 8 sec was used for all samples.

AR 2000 rheometer from TA Instruments was used to measure the rheological performance of the nanocomposites as a function of angular frequency. The measurements were made at 190 °C using a gap opening of 1.6 mm. Strain sweeps were recorded at $\omega = 1$ rad/s from 0.1 to 100% strain and the specimens were observed to be stable up to 10% strain. Frequency sweeps (dynamic testing) were, thus, recorded at 4% strain from $\omega = 0.1$ to 100 rad/s [44].

The morphology of the graphene platelets and nanocomposites was analyzed using transmission electron microscope (EM 912 Omega (Zeiss, Oberkochen BRD)) at 120 kV and 200 kV accelerating voltages. Thin sections of 70-90 nm thickness were microtomed from the block of the nanocomposite specimens. The thin sections or graphene platelets were subsequently supported on 100 mesh grids sputter coated with a 3 nm thick carbon layer.

10.3 Results and Discussion

In the current study, a comparison between the masterbatch (solution mixing followed by melt mixing) and direct melt mixing techniques was made in terms of filler dispersion and properties of the polyethylene-thermally reduced graphene nanocomposites, generated in the presence of compatibilizers. The compounding conditions as well as amounts of composite constituents were held constant during the nanocomposite generation in order to attribute the observed properties solely to compatibilizers and the composite generation technique. The copolymer based compatibilizers used in the study were chosen depending on the functional groups and their amount in the copolymer structure. In order to attain compatibility with the matrix polymer, the non-polar component of the compatibilizers was based on polyethylene. As mentioned in Table 10.1, EAA had 9% acrylic acid content in its structure, whereas the rest was PE. Similarly, EVA-MA was expected to lead to thermodynamic interactions of maleic anhydride group with the filler surface. Ionomer EMAZ was used to study the effect of ionic species in generating interactions with the filler surface. The melt flow indices of the compatibilizers varied in the range of 1-16 g/10 min, thus, indicating a wide variation of their molecular weight characteristics.

The diffraction peak for pure graphene was observed at 26.4° 2Θ, confirming its successful generation. Figure 10.1(a) shows the TEM micrograph of the graphene platelets, whereas Figure 10.1(b) demonstrates the Raman spectrum of these platelets. The Raman spectra of the graphitic materials shows features such as the so-called G band appearing at 1582 cm^{-1} (graphite), the D band at about 1350 cm^{-1}, the D' band at about 1620 cm^{-1} and the G' band at about 2700 cm^{-1}. The evolution of disorder in the generated graphene was quantified using the well-known Tuinstra-Koenig relation $I_D/I_G = C(\lambda)/La$, where I_D/I_G accounts for the intensity ratio between the disorder-induced D band and the Raman-allowed first-order G band. I_D/I_G of the synthesized graphene for this study was observed to be 0.864 (raw peaks without deconvolution) and 0.815 (with deconvolution). In addition, EDX analysis of the graphene surface revealed the C and O fractions to be 77.5% and 18.4%. This indicated the presence of oxygen containing polar surface groups (hydroxyl, epoxide, carboxyl, etc.) in the graphene platelets. Thus, these functional groups present on the filler surface can be expected to interact with the polar compatibilizers used in the study leading to filler delamination in non-polar PE.

Table 10.2 describes the tensile properties of the nanocomposites. Addition of 1% graphene in the absence of any compatibilizers enhanced the tensile modulus by 13%. Addition of 5% EAA to the composite generated by direct melt mixing resulted in 20% increment in modulus, which was further enhanced to 25% for 10% EAA content in the composite. In the case of 5% EAA composite generated by masterbatch approach, much more significant increase of 41%

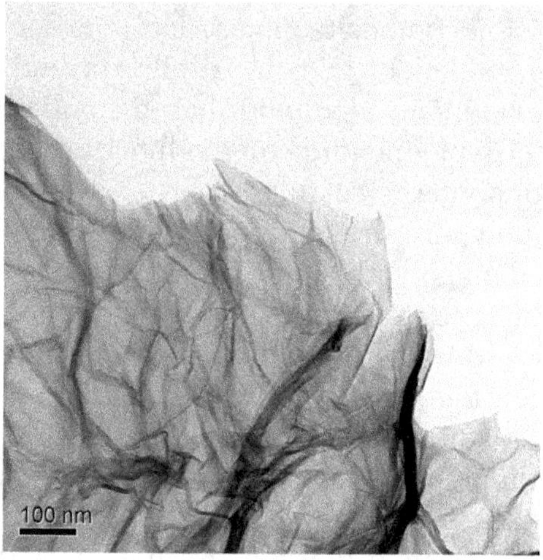

(a)

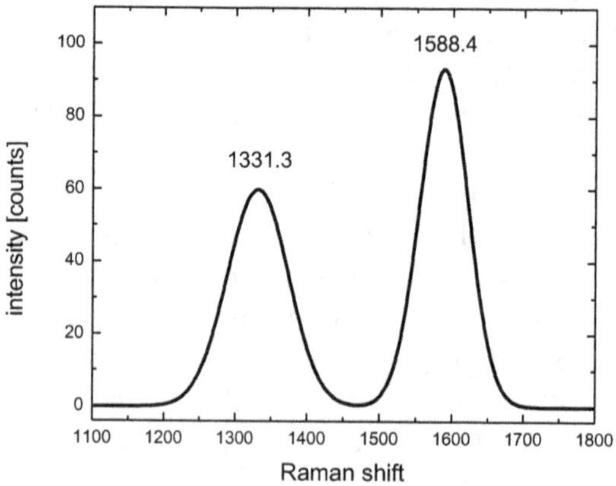

(b)

Figure 10.1 (a) TEM micrograph of the synthesized graphene and (b) Raman spectrum of the graphene sample.

in modulus was observed, indicating this method was able to attain higher degree of filler dispersion. Also, the gain in the modulus did not change with EAA content increasing to 10%, indicating that an efficient mixing of filler with the compatibilizer was obtained at lower compatibilizer content. Peak stress of the MB composites was also marginally better than the corresponding MM composites, though the values were lower than the pure HDPE probably due to

Table 10.2 Tensile properties of nanocomposites (graphene content: 1 wt%, compatibilizer content: 5 and 10 wt%)

Composite	Tensile modulus[a], MPa	Peak stress[b], MPa	Peak strain[c], %	Total elongation[d], mm
HDPE	870	24	7.2	3.4
HDPE/G	982	19	2.7	1.5
HDPE/G/5 EAA MM	1040	22	5.9	3.1
HDPE/G/10 EAA MM	1085	20	3.8	1.1
HDPE/G/5 EMAZ MM	1018	22	6.6	1.3
HDPE/G/10 EMAZ MM	976	20	4.8	1.2
HDPE/G/5 EVA-MA MM	945	22	5.4	1.8
HDPE/G/10 EVA-MA MM	841	20	4.8	1.2
HDPE/G/5 EAA MB	1225	23	4.6	1.3
HDPE/G/10 EAA MB	1222	22	4.4	1.2
HDPE/G/5 EMAZ MB	1207	23	3.5	1.5
HDPE/G/10 EMAZ MB	1011	22	3.1	1.0
HDPE/G/5 EVA-MA MB	1178	22	4.8	1.6
HDPE/G/10 EVA-MA MB	998	21	4.1	1.1

[a]Maximum relative probable error 5%
[b]Maximum relative probable error 15%
[c]Maximum relative probable error 5%
[d]Maximum relative probable error 15%

strain hardening. Reduction in peak strain was also observed in the composites as compared to pure HDPE due to the entrapment of the polymer chains in the filler interlayers and due to presence of some filler aggregates. It also confirmed that the matrix plasticization did not take place on addition of even 10% EAA to the composites. The masterbatch method also led to higher level of increment in the EMAZ composites, however, due to specific nature of the compatibilizer, its 10% content resulted in dominant plasticization effect. EVA-MA composites exhibited the least increment in the modulus. HDPE/G/5 EVA-MA/MM showed an increase of 9% in the modulus as compared to pure polymer. On the other hand, HDPE/G/10 EVA-MA/MM exhibited a slight decrease in the modulus as compared to pure polymer, which signaled poor interactions of the compatibilizer with the filler as well as matrix plasticization. The composites generated by masterbatch approach had 35% and 15% increase in the modulus for 5% and 10% compatibilizer content, which again confirmed the superiority of the method for achieving better filler mixing and property evolution. Thus, though the specific interactions of various compatibilizers with the filler surface resulted in different extents of enhancement in the mechanical performance, the method of nanocomposite generation also played a significant role in enhancing these interactions.

Similar findings were also observed from the DMA analysis shown in Figure 10.2. For instance, at 25 °C, the storage modulus of the composites HDPE/G/5 EAA/MB, HDPE/G/5 EMAZ/MB and HDPE/G/5 EVA-MA/MB respectively was 1.5E9, 1.45E9 and 1.2E9 Pa. The composites also exhibited much higher modulus at higher temperature (>60 °C) as compared to pure polymer as well as HDPE/G composite. This confirmed that the high temperature dimensional stability of the composites was enhanced significantly, though at high temperatures, no significant differences could be observed between the composites with varying compatibilizer contents as well as composites generated by two different methods. Also, some plasticization effects were observed in the composites with 10% compatibilizer content leading to reduction in tensile modulus at room temperatures, however, the higher temperature modulus was still superior than that of pure polymer and HDPE/G composite. This confirmed that the incorporation of the compatibilizers was necessary to enhance the dynamic mechanical response of the nanocomposites, which otherwise would not evolve due to poor dispersion of graphene in HDPE matrix, as observed in HDPE/G composite. Figure 10.3 also represents the loss modulus curves of the composites. In general, the loss modulus of masterbatch composites was higher than the melt mixed composites, due to an increase in internal friction, thus, promoting energy dissipation. The loss peaks indicated the efficiency of the composites to dissipate the mechanical energy. The peaks for all materials exhibited relaxation peak around 40 °C. The magnitude of the peaks for the masterbatch composites

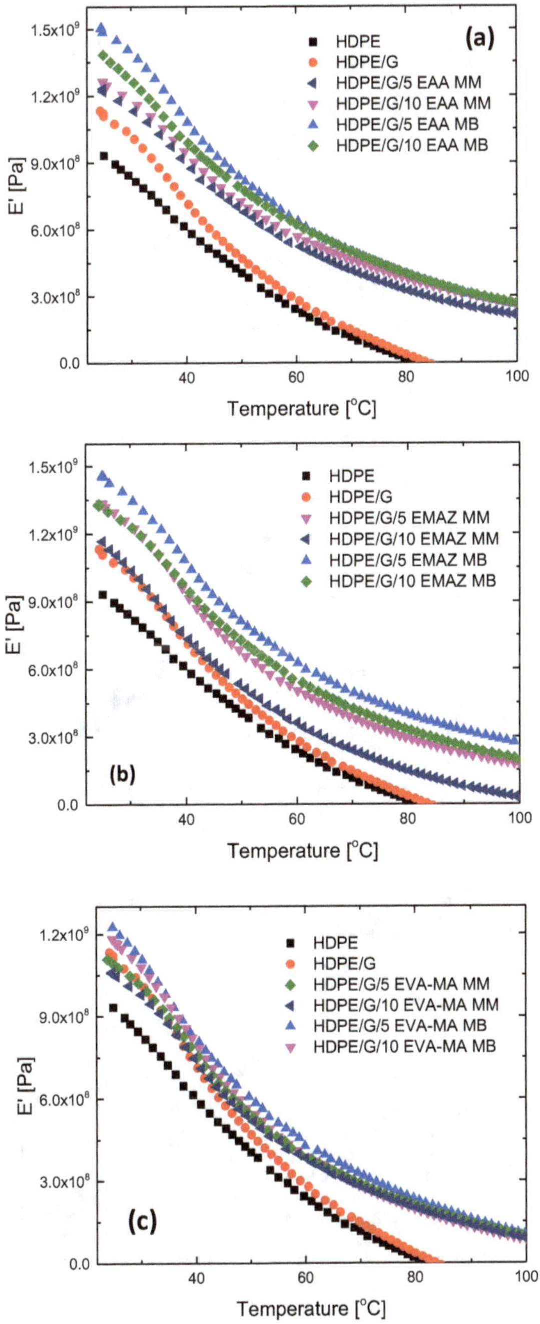

Figure 10.2 Dynamic mechanical profiles of the composites compatibilized with (a) EAA, (b) EMAZ and (c) EVA-MA.

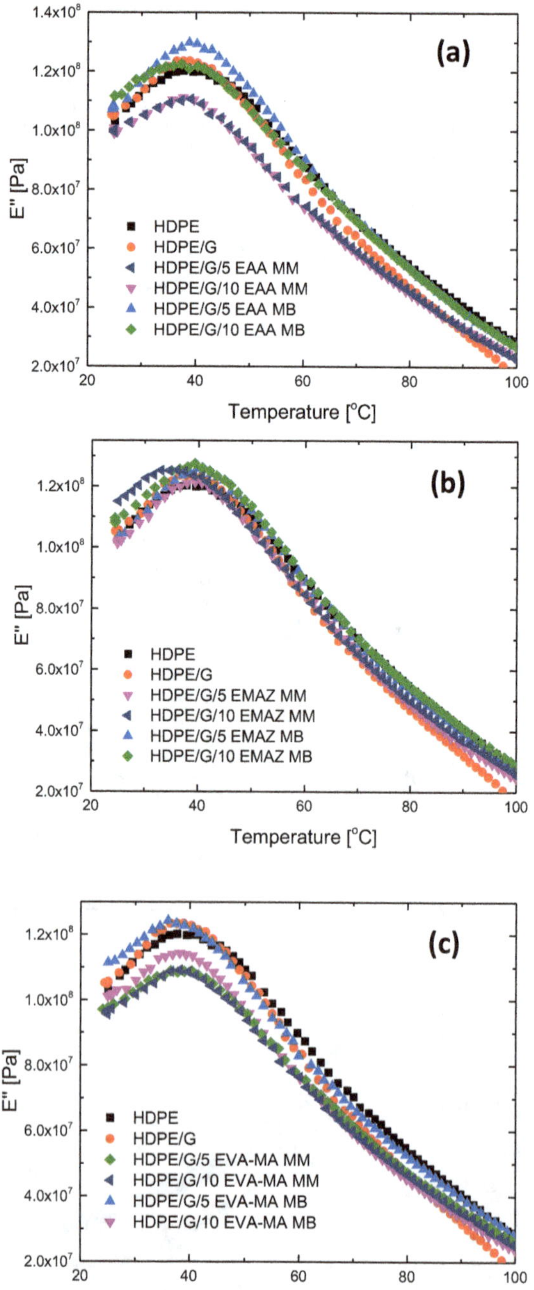

Figure 10.3 Loss modulus profiles of the composites compatibilized with (a) EAA, (b) EMAZ and (c) EVA-MA.

was observed to be higher than the melt mixed composites as well as pure polymer. This indicted that the masterbatch composites had higher energy dissipation capacity.

Figure 10.4 also shows the rheological behavior of the nanocomposites as a function of angular frequency. All the composites exhibited higher shear modulus than the pure polymer at all the angular frequencies. The trend in the increment in the shear modulus was also similar to the tensile tests. Composites generated by masterbatch approach and 5% compatibilizer content were observed to have nearly an order of magnitude higher shear modulus than HDPE at lower frequency values. The increment in the modulus was also retained at higher frequencies for EAA and EMAZ compatibilizer systems, but the modulus curves merged with each other for EVA-MA composites at higher frequency values. Complex viscosity of the nanocomposites with 5% compatibilizer fraction is also plotted in Figure 10.5 as a function of angular frequency. Similar to shear modulus, the viscosity of the composites was much higher than the HDPE at all angular frequencies. HDPE/G/5 EAA/MB exhibited the highest increment in viscosity. Moreover, the composites generated with masterbatch approach had much higher viscosity than the composites prepared with direct melt mixing method, also indicating better filler dispersion leading to higher degree of resistance to flow at the same filler content. It is also to be noted that even the viscosity of the composites was enhanced significantly, the processability of these composites was not affected as similar processing conditions as the pure polymer were still suitable for these nanocomposites.

Figure 10.6 compares the morphology of the HDPE/G, HDPE/G/5 EAA/MM and HDPE/G/5 EAA/MB composites at different magnifications. Both composites had much improved filler dispersion as compared to HDPE/G composite [40]. Comparing the two, the composite generated with direct melt mixing exhibited less optimal filler dispersion than the composite generated with masterbatch approach. Filler tactoids of higher thickness were observed for the HDPE/G/5 EAA/MM composite, whereas the tactoid thickness was relatively reduced for the HDPE/G/5 EAA/MB composite. No single exfoliated graphene platelets were observed for the HDPE/G/5 EAA/MM composite, whereas the HDPE/G/5 EAA/MB composite exhibited presence of single graphene layers. Also, the extent of intercalation was observed to be qualitatively higher in the composite generated with masterbatch approach. In addition, no specific filler

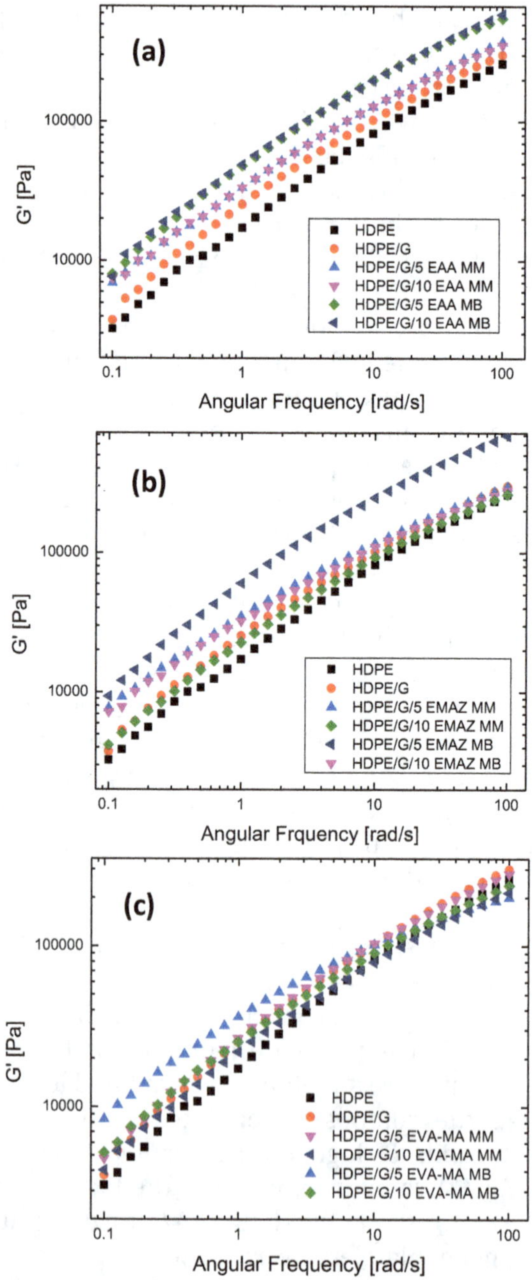

Figure 10.4 Shear modulus of the nanocomposites as a function of angular frequency with (a) EAA, (b) EMAZ and (c) EVA-MA compatibilizers.

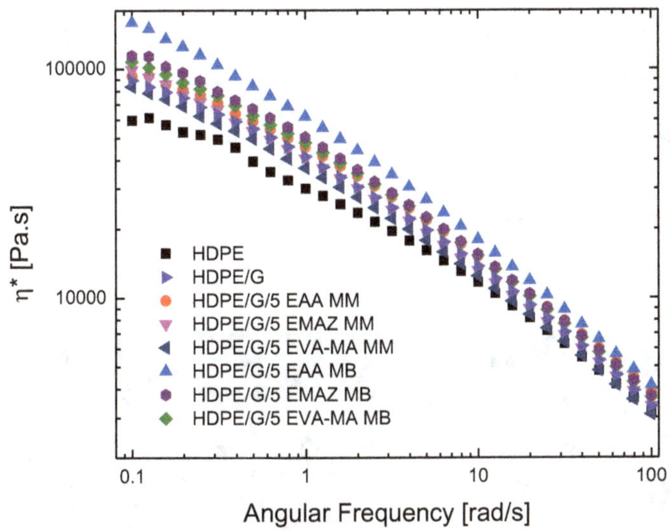

Figure 10.5 Complex viscosity of the nanocomposites as a function of angular frequency.

orientation was observed in the composites at any magnification indicating no flow/stress induced orientation of the filler platelets took place in the molding process. Presence of only 1% filler content is also expected to result in such anisotropic morphology. Other composites also exhibited similar morphology, thus, further confirming the potential of masterbatch method in enhancing the filler dispersion in HDPE matrix. Though no chemical interaction was observed between the compatibilizers and the filler surface, masterbatch method was successful in mixing the filler platelets efficiently with the compatibilizer solution, subsequently enhancing the dispersion of compatibilizer/graphene masterbatch in the HDPE matrix. The findings from tensile, rheological and morphological analysis were further confirmed from the surface resistivity analysis of the samples (Figure 10.7). Pure HDPE exhibited a surface resistivity of 14.01e10 Ωm, which was decreased to 1.04e9 Ωm for HDPE/G composite. HDPE/G/5 EAA/MB, HDPE/G/5 EMAZ/MB and HDPE/G/5 EVA-MA/MB composites exhibited surface resistivity values of 2.08e6, 3.27e7 and 5.38e7 Ωm respectively. On the other hand, HDPE/G/5 EAA/MM, HDPE/G/5 EMAZ/MM and HDPE/G/5 EVA-MA/MM had surface resistivity of 3.75e7, 1.67e8 and 6.39e8 Ωm respectively, indicating better filler dispersion for masterbatch method as well as EAA compatibilizer.

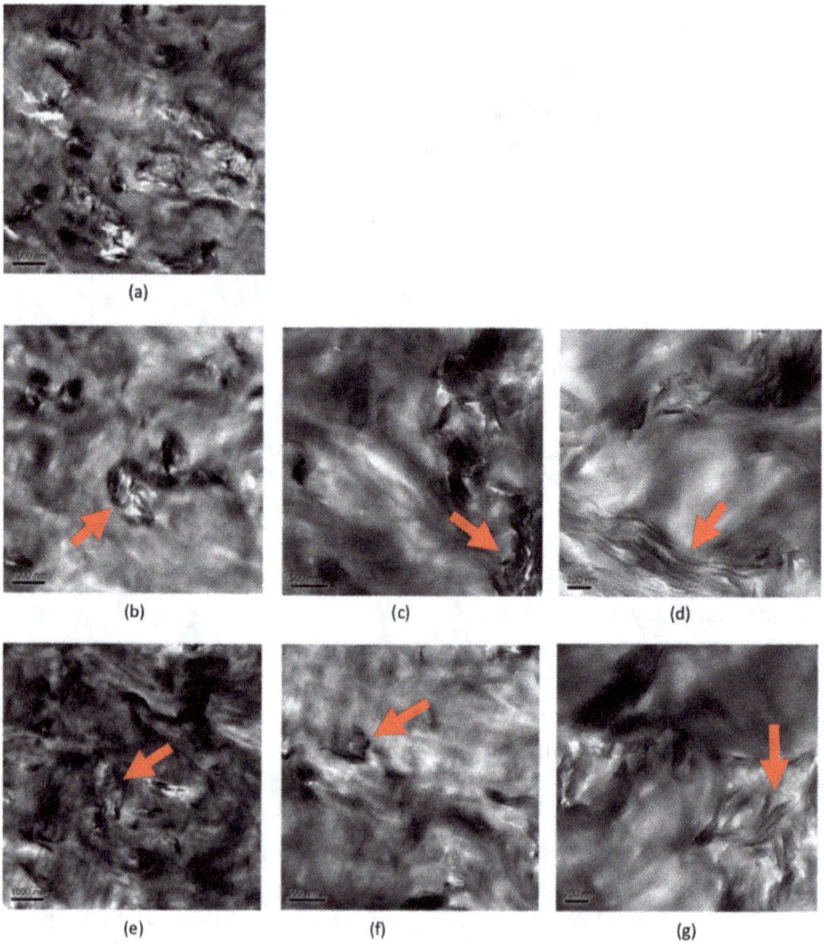

Figure 10.6 TEM micrographs of (a) HDPE/G; (b-d) HDPE/G/5 EAA MM composite at different magnifications and (d-f) HDPE/G/5 EAA MB composite at the same magnifications as (b-d) for comparison.

Table 10.3 describes the calorimetric properties of the nanocomposites. The composites exhibited single broad melting transition, indicating that the compatibilizers were miscible with HDPE. A 3-4 °C decrease in the peak melting temperature was observed in the composites as compared to pure HDPE. In addition, the composites generated with masterbatch approach had 1-2 °C lower peak melting temperatures than the corresponding composites generated by direct melt mixing. Significant changes were observed in the melt enthalpy values (corrected to polymer amount in the composites)

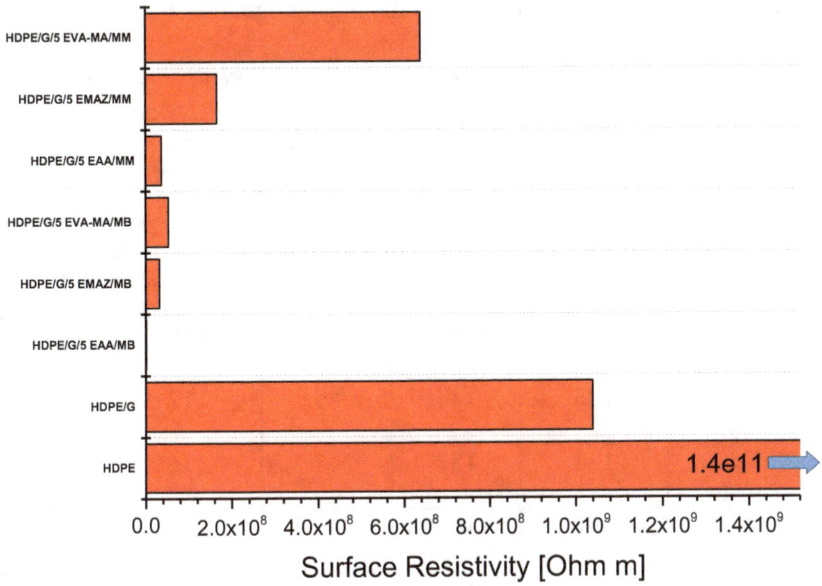

Figure 10.7 Surface resistivity values of the pure polymer and the composites.

depending on the compatibilizer and the composite generation method. HDPE/G/5 EAA/MM and HDPE/G/10 EAA/MM composites exhibited degree of crystallinity of 52% and 41% respectively, as compared to 54% for HDPE. This indicated that the increase in compatibilizer content resulted in hindrance to HDPE crystallization, even though the higher amount of compatibilizer resulted in enhancement in the tensile modulus due to increased extent of filler dispersion. In comparison, HDPE/G/5 EAA/MB and HDPE/G/10 EAA/MB composites had 54% and 45% degree of crystallinity respectively. It again underlined that due to better interaction of the compatibilizer with the graphene platelets in the masterbatch method, overall reduction in the crystallinity at higher compatibilizer fractions was less. Similar observations could also be observed for EMAZ and EVA-MA compatibilizer systems. Peak crystallization temperature of HDPE was also observed to remain constant in the composites, which indicated that the filler platelets did not have any nucleating effect on the polymer. The rheological behavior was also analyzed using criteria reported for compatibility by Han and Chuang [44,45] as shown in Figure 10.8(a,b). When G′ is plotted vs. G″, such analysis generates composition independent correlation for

Table 10.3 Calorimetric properties of the nanocomposites

Composite	Peak meting temperature, °C	Melt enthalpy, J/g	Degree of crystallinity, %	Peak crystallization temperature, °C	Crystallization enthalpy, J/g
HDPE	136	157	54	116	156
HDPE/G	133	149	51	117	151
HDPE/G/5 EAA MM	134	153	52	116	150
HDPE/G/10 EAA MM	134	119	41	116	121
HDPE/G/5 EMAZ MM	135	144	49	117	139
HDPE/G/10 EMAZ MM	134	137	47	116	140
HDPE/G/5 EVA-MA MM	135	128	44	116	128
HDPE/G/10 EVA-MA MM	134	115	39	116	113
HDPE/G/5 EAA MB	133	157	54	117	163
HDPE/G/10 EAA MB	133	132	45	116	138
HDPE/G/5 EMAZ MB	132	155	53	117	159
HDPE/G/10 EMAZ MB	132	139	47	117	137
HDPE/G/5 EVA-MA MB	133	134	46	116	134
HDPE/G/10 EVA-MA MB	133	121	41	116	119

compatible systems, whereas the correlation is composition dependent for incompatible systems. For EAA system, concentration independent correlation was observed except for some deviation at lower frequency values indicating the miscibility of the system. In EVA-MA system, a small degree of deviation was observed for the composites with 10% EVA-MA content. It also indicated that the extent of acrylic acid in EAA was suitable for miscibility with HDPE, however, EVA-MA and EMAZ may have suffered due to the high polar content, which resulted in varying degree of miscibility with HDPE.

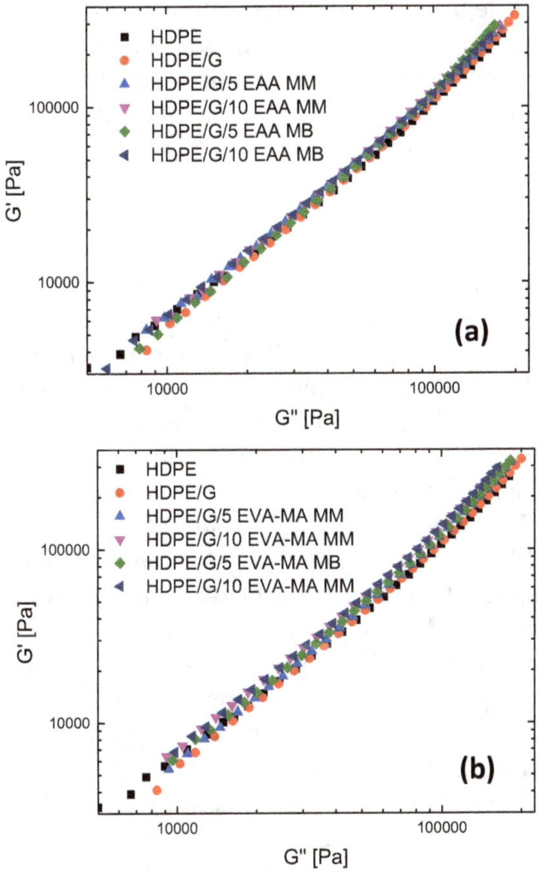

Figure 10.8 Han-Chuang (G' vs. G") plots of (a) EAA and (b) EVA-MA nanocomposites.

In the thermal stability analysis of the nanocomposites, the onset and peak degradation were observed to be significantly delayed in all composites as compared to pure polymer. Due to better filler dispersion, the composites generated with masterbatch method exhibited better thermal performance than the corresponding direct melt mixed composites, thus, resulting in efficient heat transfer to the filler platelets. In addition, increasing compatibilizer content resulted in slight reduction in thermal stability due to low temperature stability of the compatibilizer, however, the degradation still occurred at temperatures higher than the degradation of pure polymer. Synergistic improvement in the thermal stability was observed for HDPE/G/5 EAA/MB and HDPE/G/5 EAA/MM composites as the thermal degradation was even superior to the HDPE/G composite. In other composites, the degradation lied between the HDPE and HDPE/G composites. The dynamic TGA also revealed similar patterns as static TGA, when the time needed to lose 1% of the organic material at fixed temperature was compared.

10.4 Conclusions

In the current study, polyethylene-thermally reduced graphene nanocomposites were generated by direct melt mixing as well as masterbatch (solution mixing followed by melt mixing) methods. The composites were generated using compatibilizers differing in their functionality and amount in the composite. Composites generated with masterbatch approach exhibited much higher mechanical properties than the corresponding composites generated by direct melt mixing. For instance, the composite with 5% EAA had an increment of 41% and 20% respectively for the masterbatch and melt mixed composites. Other compatibilizers showed similar improvements, however, EAA showed the highest extent of property enhancement followed by EMAZ and EVA-MA, thus, confirming the effect of filler-compatibilizer interactions along with the nanocomposite generation method on the resulting properties. Increasing the compatibilizer content to 10% caused matrix plasticization in most of the composites. Composites generated by masterbatch approach and 5% compatibilizer content were observed to have nearly an order of magnitude higher storage modulus than HDPE at lower frequency values. DMA also confirmed that the high temperature dimensional stability of the composites was enhanced significantly on addition of graphene. Morphological studies confirmed the findings from mechanical analysis as extensive delamination was observed in the composites generated by masterbatch approach. Due to better interaction of the compatibilizer with the graphene platelets in the masterbatch method, overall reduction in the crystallinity of HDPE at higher compatibilizer fractions was less. Thermal stability of the nanocomposites also improved on addition of graphene, with 5% EAA composites showing thermal performance even better than the HDPE/G composite.

Summary of the Results

Two routes of nanocomposite generation viz. masterbatch method and direct melt mixing were compared in the current study in relation with the properties of the polyethylene-thermally reduced graphene nanocomposites compatibilized with functional polymers. Filler dispersion was observed to improve in the masterbatch method due to preferential interactions of compatibilizer with the filler, before melt mixing with the matrix polymer. This also translated into improved mechanical properties as more than 2 times increment in tensile modulus was observed in the composite HDPE/G/5 EAA/MB (41%, masterbatch method) as compared to HDPE/G/5 EAA/MM (20%, melt mixing method). In addition, different compatibilizers also influenced the properties to different extent owing to their physical interactions with the filler, however, composites generated with masterbatch were always superior in extent of property enhancements. Increasing the compatibilizer content to 10% also caused varying degree of matrix plasticization. The storage modulus as well as the complex viscosity of the HDPE/G/5 EAA/MB composite was the highest as compared to HDPE/G/5 EMAZ/MB and HDPE/G/5 EVA-MA/MB composites. HDPE/G/5 EAA/MB and HDPE/G/5 EMAZ/MB composites did not show any decrease in degree of crystallinity as compared to pure polymer, whereas the corresponding composites generated with melt mixing exhibited slight decrease in the crystallinity. The composites were observed to have enhanced thermal stability as compared to HDPE. 5% EAA based composites exhibited even higher thermal stability than HDPE/G composites due to improved filler dispersion. Composites generated with masterbatch approach had also higher degree of phase mixing than the corresponding melt mixed composites.

Acknowledgement

The definitive version of this work has been published in Colloid and Polymer Science (2016), 294, 1659-1670. The work has been reproduced here with permission from Springer Nature.

References

1. Alexandre, M., and Dubois, P. (2000) Polymer-layered silicate nanocomposites: Preparation, properties and uses of a new class of materials. *Materials Science and Engineering: R: Reports*, **28**, 1-63.

2. Okada, A., and Usuki, A. (1995) The chemistry of polymer-clay hybrids. *Materials Science and Engineering, C,* **3**(2), 109-115.
3. Osman, M. A., Mittal, V., Morbidelli, M., and Suter, U. W. (2003) Polyurethane adhesive nanocomposites as gas permeation barrier. *Macromolecules,* **36**, 9851-9858.
4. Osman, M. A. Mittal, V. Morbidelli, M. and Suter, U. W. (2004) Epoxy-layered silicate nanocomposites and their gas permeation properties. *Macromolecules.* **37**, 7250-7257.
5. Brechet, Y., Cavaille, J. Y., Chabert, E., Chazeau, L., Dendievel, R., Flandin, L., and Gauthier, C. (2001) Polymer based nanocomposites: Effect of filler-filler and filler-matrix interactions. *Advanced Engineering Materials,* **3**, 571-577.
6. Pavlidoua, S., and Papaspyrides, C. D. (2008) A review on polymer–layered silicate nanocomposites. *Progress in Polymer Science,* **33**, 1119-1198.
7. Mark, J. E. (1996) Ceramic-reinforced polymers and polymer-modified ceramics. *Polymer Engineering & Science,* **36**, 2905-2920.
8. Liu, H., and Brinson, L. C. (2008) Reinforcing efficiency of nanoparticles: A simple comparison for polymer nanocomposites. *Composites Science and Technology,* **68**, 1502-1512.
9. *Advances in Diverse Industrial Applications of Nanocomposites,* Reddy, B. (ed.), InTech, UK (2011).
10. Hui, L., Smith, R. C., Wang, X., Nelson, J. K., and Schadler, L. S. (2008) Quantification of Particulate Mixing in Nanocomposites, *2008 Annual Report Conference on Electrical Insulation and Dielectric Phenomena,* Canada, doi: 10.1109/CEIDP.2008.4772831.
11. Manas-Zloczower, I., and Cheng, H. (1996) Analysis of mixing efficiency in polymer processing equipment. *Macromolecular Symposia,* **112**(1), 77-84.
12. Lee, S. H., Cho, E., Jeon, S. H., and Youn, J. R. (2007) Rheological and electrical properties of polypropylene composites containing functionalized multi-walled carbon nanotubes and compatibilizers. *Carbon,* **45**, 2810-2822.
13. Serageldin, M. A., and Wang, H. (1987) Effect of operating parameters on time to decomposition of high density polyethylene and chlorinated polyethylenes. *Thermochimica Acta,* **117**, 157-166.
14. Lee, C., Wei, X., Kysar, J. W., and Hone, J. (2008) Measurement of the elastic properties and intrinsic strength of monolayer graphene. *Science,* **321**, 385-388.
15. Zhao, H., Min, K., and Aluru, N. R. (2009) Size and chirality dependent elastic properties of graphene nanoribbons under uniaxial tension. *Nano Letters,* **9**, 3012-3015.
16. Lier, G. V., Alsenoy, C. V., Doren, V. V., and Greelings, P. (2000) Ab initio study of the elastic properties of single-walled carbon nanotubes and graphene. *Chemical Physics Letters,* **326**, 181-185.

17. Robertson, D. H., Brenner, D. W., and Mintmire, J. W. (1992) Energetics of nanoscale graphitic tubules. *Physical Review B*, **45**, 12592.
18. Itkis, M. E. Borondics, F. Yu, A. Haddon, R. C. (2007) Thermal conductivity measurements of semitransparent single-walled carbon nanotube films by a bolometric technique. *Nano Letters*, **7**, 900-904.
19. Park, S., and Rouff, S. (2009) Chemical methods for the production of graphenes. *Nature Nanotechnology*, **4**, 217-224.
20. Song, M., and Cai, D. (2012) Graphene functionalization: A review. In Polymer Graphene Nanocomposites, Mittal, V. (ed.)., Royal Society of Chemistry, UK, pp. 1-51.
21. Mukhopadhyay, P., and Gupta, R. K. (2011) Trends and frontiers in graphene- based polymer nanocomposites. *Plastics Engineering*, **67**, 32-42.
22. Geim, A. K., and Novoselov, K. S. (2007) The rise of graphene. *Nature Materials*, **6**, 183-191.
23. Compton, O. C., and Nguyen, S. B. T. (2010) Graphene oxide, highly reduced graphene oxide, and graphene: versatile building blocks for carbon-based materials. *Small*, **6**, 711-723.
24. Mittal, V. (2014) Functional polymer nanocomposites with graphene: A review. *Macromolecular Materials and Engineering*, **299**, 906-931.
25. Mittal, V., Luckachan, G. E., and Matsko, N. B. (2014) PE/chlorinated-PE blends and PE/chlorinated-PE/graphene oxide nanocomposites: Morphology, phase miscibility, and interfacial interactions. *Macromolecular Chemistry and Physics*, **215**, 255-268.
26. Mittal, V., Chaudhry, A. U., and Luckachan, G. E. (2014) Biopolymer–thermally reduced graphene nanocomposites: structural characterization and properties. *Materials Chemistry and Physics*, **147**, 319-332.
27. Supova, M., Martynkova, G. S., and Barabaszova, K. (2011) Effect of nanofillers dispersion in polymer matrices: A review. *Science of Advanced Materials*, **3**, 1-25.
28. Kuilla, T., Bhadra, S., Yao, D., Kim, N. H., Bose, S., and Lee, J. H. (2010) Recent advances in graphene based polymer composites. *Progress in Polymer Science*, **35**, 1350-1375.
29. Fang, M., Wang, K., Lu, H., Yang, Y., and Nutt, S. (2009) Covalent polymer functionalization of graphene nanosheets and mechanical properties of composites. *Journal of Materials Chemistry*, **19**, 7098-7105.
30. Stankovich, S., Piner, R. D., Chen, X., Wu, N., Nguyen, S. T., and Ruoff, R. S. (2006) Stable aqueous dispersions of graphitic nanoplatelets via the reduction of exfoliated graphite oxide in the presence of poly (sodium 4-styrenesulfonate). *Journal of Materials Chemistry*, **16**, 155-158.

31. Stankovich, S., Dikin, D. A., Dommett, G. H. B., Kohlhaas, K. M., and Zimney, E. J. (2006) Graphene-based composite materials. *Nature*, **442**, 282-286.
32. Castelain, M., Martinez, G., Marco, C., Ellis, G., and Salavagione, H. J. (2013) Effect of click-chemistry approaches for graphene modification on the electrical, thermal, and mechanical properties of polyethylene/graphene nanocomposites. *Macromolecules*, **46**, 8980-8987.
33. Vasileiou, A. A., Kontopoulou, M., and Docoslis, A. (2014) A non-covalent compatibilization approach to improve the filler dispersion and properties of polyethylene/graphene composites. *ACS Applied Materials & Interfaces*, **6**, 1916-1925.
34. Yu, B., Wang, X., Qian, X., Xing, W., Yang, H., Ma, L., Lin, Y., Jiang, S., Song, L., Hu, Y., and Lo, S. (2014) Functionalized graphene oxide/phosphoramide oligomer hybrids flame retardant prepared via in situ polymerization for improving the fire safety of polypropylene. *RSC Advances*, **4**, 31782-31794.
35. Seo, H. M., Park, J. H., Dao, T. D., and Jeong, H. M. (2013) Compatibility of functionalized graphene with polyethylene and its copolymers. *Journal of Nanomaterials*, **2013**, Article ID 805201.
36. Wang, J., Xu, C., Hu, H., Wan, L., Chen, R., Zheng, H., Liu, F., Zhang, M., Shang, X., and Wang, X. (2011) Synthesis, mechanical, and barrier properties of LDPE/graphene nanocomposites using vinyl triethoxysilane as a coupling agent. *Journal of Nanoparticle Research*, **13**, 869-878.
37. Kim, S., Do, I., and Drzal, L. T. (2009) Multifunctional xGnP/LLDPE nanocomposites prepared by solution compounding using various screw rotating systems. *Macromolecular Materials and Engineering*, **294**, 196-205.
38. Ansari, S., and Giannelis, E. P. (2009) Functionalized graphene sheet-Poly (vinylidene fluoride) conductive nanocomposites. *Journal of Polymer Science, Part B: Polymer Physics*, **47**, 888-897.
39. Potts, J. R., Dreyer, D. R., Bielawski, C. W., and Ruoff, R. S. (2011) Graphene-based polymer nanocomposites. *Polymer*, **52**, 5-25.
40. Chaudhry, A. U., and Mittal, V. (2013) High-density polyethylene nanocomposites using master batches of chlorinated polyethylene/graphene oxide. *Polymer Engineering and Science*, **53**, 78-88.
41. McAllister, M. J., Li, J. L., Adamson, D. H., Schniepp, H. C., Abdala, A. A., Liu, J., Herrera-Alonso, M., Milius, D. L., Car, R., Prudhomme, R. K., and Aksay, I. A. (2007) Single sheet functionalized graphene by oxidation and thermal expansion of graphite. *Chemistry of Materials*, **19**, 4396-4404.
42. Hummers, W. S., and Offeman, R. E. (1958) Preparation of graphitic oxide. *Journal of the American Chemical Society*, **80**, 1339.
43. Mittal, V., and Chaudhry, A. U. (2015) Polymer-graphene nanocom-

posites: Effect of polymer matrix and filler amount on properties. *Macromolecular Materials and Engineering*, **300**(5), 510-521.

44. Chuang, H. K., and Han, C. D. (1984) Rheological behavior of polymer blends. *Journal of Applied Polymer Science*, **29**, 2205-2229.
45. Han, C. D., and Chuang, H. K. (1985) Criteria for rheological compatibility of polymer blends. *Journal of Applied Polymer Science*, **30**, 4431-4454.

11

Effect of Amine Functionalized Reduced Graphene Oxide on the Thermal, Mechanical, Rheological and Electrical Properties of CPE Compatibilized HDPE

11.1 Introduction

Polymer nanocomposites, in which one dimension of the filler is <100 nm, exhibit significant improvements in mechanical, gas barrier, thermal and fire retardancy properties, among others [1]. As a result, a large number of nanocomposite systems have been developed by reinforcing the polymer matrices with alumino-silicates and carbon nanotubes based fillers. In recent years, graphene reinforced polymer nanocomposites have attracted considerable research attention due to the potential of graphene in imparting extraordinary properties to the composites as compared to the pure polymer matrices [2-4]. Graphene layers are atomically thick two-dimensional (2-D) sheets composed of sp^2 carbon atoms arranged in a hexagonal lattice with large aspect ratio. Graphene possesses Young's modulus of 1 TPa, ultimate strength of 130 GPa, thermal conductivity of 5000 W/(m.K) and electrical conductivity up to 6000 S/cm, thus, underlining the role as a superior nanofiller in polymer matrices as compared to conventional fillers [4,5-8]. However, lack of polymer-graphene interfacial bonding, which originates due to the atomically smooth and non-reactive surface of graphene, inhibits the load transfer from the matrix to the graphene sheets across the polymer-filler interface. In addition, the graphene sheets tend to aggregate in the polymer matrices, thus, resulting in poor dispersion [9-11]. On the other hand, the dispersion of polar graphene oxide in non-polar polymers like polyethylene also suffers due to the absence of positive interactions between them. In order to overcome this problem, compatibilizers have been commonly used to enhance the interfacial adhesion between the filler and host polymer, thus, resulting in improved filler dispersion [12,13]. Moreover, the functionalized graphene sheets provide multiple bonding sites with the polymer matrix or compatibilizer. Functionalization of graphene oxide (GO) sheets is even much

M. R. Vengatesan, H. Y. Al Asafen, A. M. Varghese and V. Mittal
Khalifa University of Science and Technology, Abu Dhabi, UAE

easier as compared to pristine graphene due to hydroxyl and epoxide functional groups on their basal planes, in addition to carbonyl and carboxyl groups located at the sheet edges [14]. Kim *et al.* [15] also reported the importance of functionalized polyethylene as well as blending methods for exfoliation of graphene in the polyethylene matrix. Similar work by Ren *et al.* [16] also reported the development of high density polyethylene (HDPE) nanocomposites using functionalized graphene.

In the present study, the effect of amine functionalized reduced graphene oxide (r-AGO) on the crystallization, thermal stability and mechanical properties of HDPE has been studied, using chlorinated polyethylene (CPE) as compatibilizer. The objective was to develop HDPE/graphene composite materials with enhanced mechanical, thermal, rheological and electrical properties by generating covalent interactions between the compatibilizer and functionalized filler to enhance the degree of filler exfoliation.

11.2 Experimental

11.2.1 Materials

High-density polyethylene BB2581 was kindly supplied by Abu Dhabi Polymers Company Limited (Borouge), UAE. It had a melt flow index of 0.35 g/10 min (190 °C, 2.16 kg). Its melt temperature ranged between 170-200 °C. Chlorinated polyethylene (CPE) Weipren 6025 (25% chlorine content, denoted as CPE25) was obtained from Lianda Corporation, USA. The CPE was in the form of white granules (specific gravity = 1.1-1.3). The melt flow index (190 °C, 2.16 kg) for CPE25 was 1.8 g/10 min and the heat of fusion was 45 J/g [17]. Graphene oxide was synthesized in the lab using Tour's method [18]. 1-(3-Dimethylaminopropyl)-3-ethylcarbodiimide hydrochloride (EDC.HCl), ethylene diamine (EA), dimethylformamide (DMF) and acetone were procured from Sigma Aldrich and were used without further purification.

11.2.2 Preparation of Amine Functionalized Reduced Graphene Oxide

r-AGO was prepared by following the reported literature with slight modification [8]. In brief, 900 mg of GO was dispersed by sonication and ultrasonic probe in 450 mL of DMF for 1 h at room temperature.

10 mL of EA and 200 mg of EDC.HCl were subsequently added to the suspension. The mixture was stirred at 110 °C for 24 h. The product, r-AGO, was diluted with 1000 mL of ethanol and left to settle for 24 h. Later, the superannuate was removed, and the product was filtered and washed with acetone to ensure the removal of any excess diamine. Finally, the products were dried at 50 °C overnight under vacuum.

11.2.3 Nanocomposite Preparation

Melt mixing was used to prepare nanocomposites of r-AGO with CPE25 and HDPE using mini twin conical screw extruder (MiniLab HAAKE Rheomex CTW5, Germany). A mixing temperature of 180 °C for 15 min at 100 rpm with batch size of 5 g was used. The screw length and screw diameter were 109.5 and 5/14 mm conical, respectively. The amount of CPE in the nanocomposites was fixed at 5.0 wt%, whereas r-AGO content was varied as 0.5, 1.0, 1.5, 3.0 and 5.0 wt%). Scheme 11.1 also represents the schematic of nanocomposite preparation. After extrusion, the samples were processed using injection molding (Thermo Scientific) at 190 °C with 400 bar injection pressure for 10 s in order to generate test specimens for mechanical and rheological analysis. In order to confirm the covalent bonding between the r-AGO and CPE, a hybrid of r-AGO/CPE was generated. In brief, 100 mg of r-AGO was dispersed in 100 mL of *o*-xylene using an ultra-sonication bath at 30 °C for 1 h. Subsequently, 100 mg of CPE was added to the dispersion and the mixture was stirred for 2 h at 130 °C. It was followed by the evaporation of the solvent under vacuum at 100 °C, and the resultant product was termed as r-AGO/CPE hybrid. For comparison purposes, HDPE/CPE blend was also generated by melt mixing, using the same processing conditions mentioned above.

11.2.4 Characterization Techniques

Fourier transform infrared (FT-IR) spectra were collected on Nicolet iS10 spectrometer equipped with SmartiTR diamond ATR accessory (angle of incidence of 45°), DTGS KBr detector and KBr beam splitter. It had a diamond ATR crystal (index of refraction = 2.4 at 1000 cm^{-1}) and a depth penetration of 2 μm at 1000 cm^{-1} for sample with refractive index of 1.5. Spectra were recorded by the OMNIC software in the 4000-400 cm^{-1} region with a resolution of 4 cm^{-1} from 32 scans. For

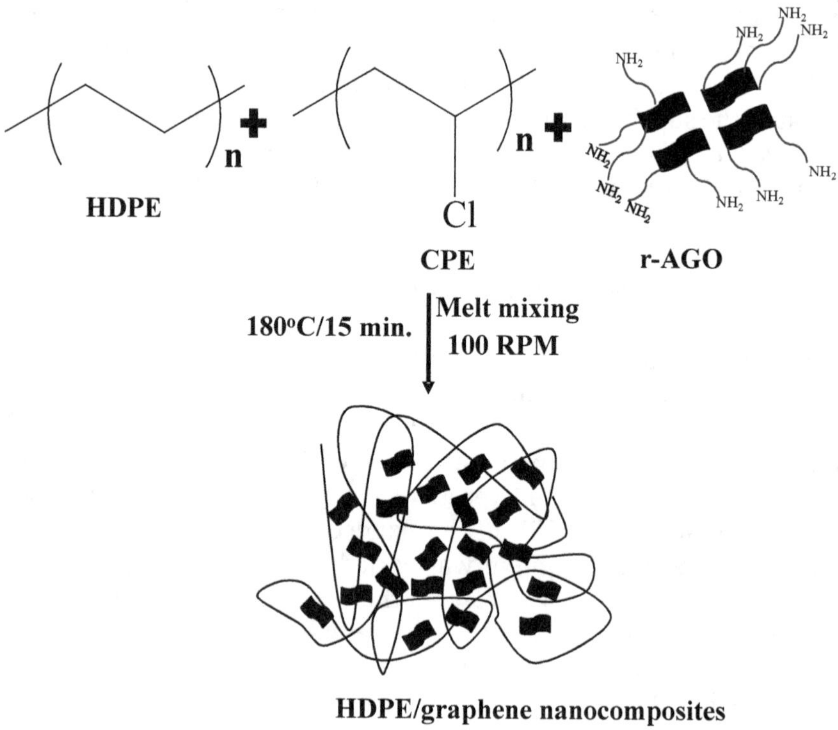

Scheme 11.1 Schematic representation of generation of HDPE/graphene nanocomposites.

the microscopy analysis, ultrathin sections (30-70 nm) of the composite samples were obtained using a PowerTome equipped with a diamond knife at -30 °C. The sections were collected on 400 mesh formvar electron microscopy grids (coated with copper), and were examined in a FEI electron microscope (TECNAI) at 200 kV at room temperature without staining. Wide-angle X-ray diffraction (WAXRD) analysis of the filler and composites was performed using analytical powder diffractometer (X'Pert PRO) using Cu Kα radiation (λ = 1.5406 Å) in reflection mode. A zero-background holder was used to minimize the noise. The samples were step-scanned from 2θ = 5-60° at room temperature using a step size of 2θ = 0.02° and a step time of 10 s.

Discovery TGA from TA Instruments was used for the thermogravimetric analysis (TGA) of the samples. The test samples were heated from 35 to 700 °C under nitrogen atmosphere with a heating rate of 10 °C/min. Differential scanning calorimetric (DSC) analysis of the composites was performed on a TA Discovery DSC under nitrogen atmosphere. The scans were obtained from 35-200 °C using a heating rate of 10 °C/min. The second heating runs were carried out in the same temperature range and were used for calculation of the crystallinity [19]. 4-7 mg of the samples were used for the both DSC and TGA analysis. The mechanical testing of the nanocomposites was performed on universal testing machine (Instron) (ASTM D 638). The dumbbell shaped samples were used, and an average of five samples was recorded. A loading rate of 10 mm/min was employed, and the tests were carried out at room temperature. Bluehill Tree Analysis software was used for the calculation of tensile modulus, ultimate tensile strength (UTS) and tensile strain. The PRS-801 Resistance system was used to measure the surface resistance of the samples. It is capable of getting precision resistance measurements from 0.1 up to 2.0×10^{14} ohms with an overall measurement tolerance of ± 5%. Geometric averages of the resistance measured from 3 to 6 different spots from each side of the sample were recorded.

The rheological properties of the nanocomposites were measured by AR 2000 rheometer from TA Instruments. The disc shape samples of 25 mm diameter and 2 mm thickness was used to measure at 180 °C using a gap opening of 1.2 mm. Strain sweep scans were recorded at $\omega = 1$ rad/s from 0.1 to 100% strain and the samples were observed to be shear stable up to 10% strain. Frequency sweep scans of the samples were recorded at 4% strain from $\omega = 0.1$ to 500 rad/s

11.3 Results and Discussion

The FTIR spectra of CPE and CPE/r-AGO hybrid are presented in Figure 11.1. In the CPE spectrum, the peak at 668 cm^{-1} corresponded to the C-Cl stretching frequency. In addition, the peaks at 2926 and 2853 cm^{-1} corresponded to asymmetric and symmetric characteristic $-CH_2$ stretching vibrations of CPE. For the hybrid, a reduction in the intensity of the peak at 668 cm^{-1} for C-Cl stretching frequency was observed, which indicated the reaction between r-AGO and CPE25. Furthermore, the peak at 1029 cm^{-1}, corresponding to –CNO- stretching frequency, also represented the nucleophilic reaction between the amine groups in r-AGO and Cl group in CPE [18].

The calorimetric values are also depicted in Table 11.1. The heat of fusion of pure crystalline HDPE was taken as 293 J/g and was used to determine the percentage crystallinity in the polymer nanocomposites [8]. The crystallinity of the samples decreased from 59% to 54%

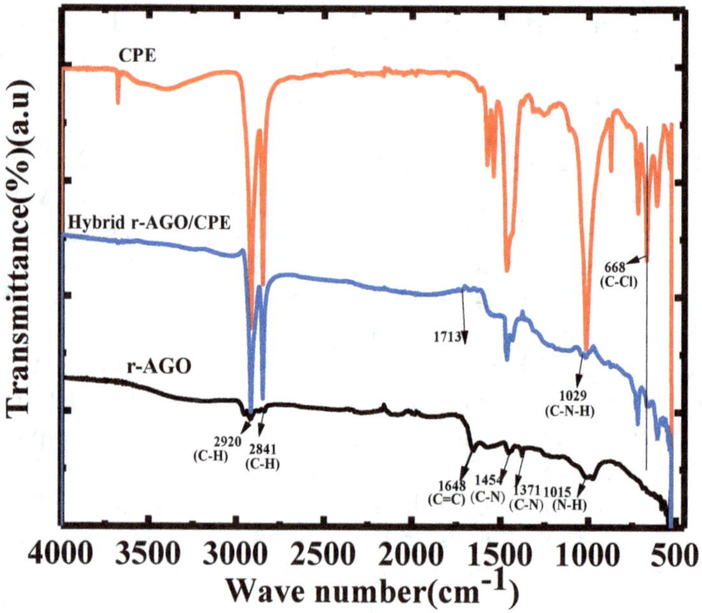

Figure 11.1 FT-IR spectra of r-AGO, r-AGO/CPE and CPE.

Table 11.1 Calorimetric properties of HDPE/r-AGO nanocomposites

Sample	ΔH (J/g)	T_m (°C)	Crystallinity (%)	T_c (°C)
Pure HDPE	174	135	59	119
HDPE/5.0 wt% CPE	158	135	54	122
HDPE/CPE/0.5 wt% r-AGO	174	133	60	120
HDPE/CPE/1.0 wt% r-AGO	175	135	60	120
HDPE/CPE/1.5 wt% r-AGO	175	135	61	119
HDPE/CPE/3.0 wt% r-AGO	174	135	61	119
HDPE/CPE/5.0 wt% r-AGO	161	136	58	119

on the addition of 5 wt% CPE25. This indicated that the CPE25 component did not act as a nucleating agent for the polyethylene chains. The melt enthalpy increased in the nanocomposites with r-AGO content till 3 wt% filler content. For the composite with 5 wt% r-AGO, the melt enthalpy decreased to a value of 161 J/g as compared to 174 J/g for pure HDPE. Thus, the degree of crystallinity increased with increase in r-AGO content up to 3 wt%, which indicated that the r-AGO acted as a nucleating agent for the polyethylene chains. However, a slight reduction in the degree of crystallinity for the 5 wt% r-AGO composite indicated that at high filler concentration, r-AGO tended to aggregate in HDPE and resulted in a decrease in crystallinity. Thermal degradation temperature for 10 and 40% weight loss as well as char yield value at 700 °C are also presented in Table 11.2. From the table, it was observed that the addition of r-AGO resulted in almost similar degradation behavior as HDPE/CPE blend, but was significantly improved as compared to pure HDPE. The char yield of the nanocomposites increased with an increase in the r-AGO content. For example, the char yield of 5 wt% r-AGO filled HDPE was 3.2% as compared to 0% for the HDPE/CPE blend at 700 °C. This indicated that r-AGO acted as thermal barrier and reduced the thermal degradation of the polymer at higher temperature [20].

Table 11.2 Thermal stability and char yield of HDPE/r-AGO nanocomposites

Sample	T_d^{10} (°C)	T_d^{40} (°C)	Char yield at 700 °C (%)
HDPE	428	462	0.00
HDPE/5 wt% CPE	447	473	0.00
HDPE/CPE/0.5 wt% r-AGO	447	473	0.37
HDPE/CPE/1.0 wt% r-AGO	447	476	1.25
HDPE/CPE/1.5 wt% r-AGO	450	476	1.26
HDPE/CPE/3.0 wt% r-AGO	450	476	2.30
HDPE/CPE/5.0 wt% r-AGO	451	478	3.20

T_d^{10} - 10% weight loss temperature
T_d^{40} - 40% weight loss temperature

As presented in Table 11.3, the tensile modulus of pure HDPE was observed to be 1033 MPa. Addition of 5 wt% CPE reduced the modulus of HDPE by about 8% probably due to matrix plasticization by the lower molecular weight CPE chains. An enhancement of the tensile modulus was realized with increase in content of r-AGO up to 3 wt%. Further increasing the content of r-AGO (5 wt%) resulted in a decrease in the modulus value probably due to filler aggregation. Overall, the observed results indicated that the functional groups on the graphene surface effectively enhanced the interfacial adhesion between the graphene and polymer matrix, due to the covalent bonding of the ethylene diamine modified GO (r-AGO) with the HDPE/CPE blend *via* nucleophilic reaction, thus, resulting in improved composite properties. The observation is also in good agreement with the reported literature [21]. Therefore, the strong interfacial adhesion between the HDPE and r-AGO led to improved interfacial stress transfer mechanism and restricted mobility of polymer chains in the nanocomposites. The addition of r-AGO didn't negatively impact the tensile strength of the nanocomposites till 3 wt%, however, the addition of 5 wt% of r-AGO reduced the tensile strength from 29 MPa to 22 MPa. As mentioned earlier, higher loading of r-AGO may form aggregates in the matrix interface and decrease the interfacial adhesion. This is in good agreement with expanded graphite-high density polyethylene nanocomposites reported earlier [23].

Table 11.3 Mechanical properties of HDPE/r-AGO nanocomposites

Sample	Tensile strength (MPa)	Young's modulus (MPa)
Pure HDPE	28.3 (±5)	1003 (±10)
HDPE/5 wt% CPE	29.2 (±5)	949 (±10)
HDPE/CPE/0.5 wt% r-AGO	27.8 (±5)	1009 (±10)
HDPE/CPE/1.0 wt% r-AGO	29.1 (±5)	1026 (±10)
HDPE/CPE/1.5 wt% r-AGO	28.6 (±5)	1051 (±10)
HDPE/CPE/3.0 wt% r-AGO	27.6 (±5)	1122 (±10)
HDPE/CPE/5.0 wt% r-AGO	24.2 (±5)	1005 (±10)

The rheological behavior of the HDPE/r-AGO nanocomposites was studied to understand the network structure formation, dispersion level of r-AGO within the HDPE matrix and interfacial interactions between the r-AGO and HDPE matrix [24,25]. The storage (G') and loss moduli (G") of the HDPE matrix and nanocomposites are presented in Figure 11.2(a-b). Addition of 5 wt% of CPE to HDPE resulted in a decrease in the values of both G' and G", as the addition of CPE enhanced the shear behavior of HDPE during molecular mixing [26]. G' and G" values increased with increase in the content of r-AGO up to 1 wt% and further increase in the r-AGO content resulted in a gradual decrease. Addition of 5 wt% of r-AGO exhibited a significant decrease in the storage and loss modulus values. A well-defined plateau of G' was observed for the HDPE nanocomposites at lower frequency which indicating a transition from the liquid-like to pseudo solid-like viscoelastic behavior, thus, leading to percolation network structure [27,28]. Decrease in G' of 5 wt% r-AGO nanocomposites as compared to the pure HDPE was probably due to the sheet stacking of the r-AGO phase at higher loading. Figure 11.2(c) shows the complex viscosity (η*) as a function of angular frequency. The horizontal curves of the η* values of the HDPE and nanocomposites represented a Newtonian fluid, and the η* values decreased with an increase in angular frequency owing to shear thinning [29]. The η* of the nanocomposites increased up to 1.0 wt% of r-AGO content. Similar to the modulus, further increasing the content above 1.0 wt% resulted in a gradual decrease of η*. This can be attributed to the selective physico-absorption of polymer chains on the r-AGO surface, along with de-entanglement effect [30]. The addition of 5 wt% r-AGO exhibited a significant decrease in η*. As mentioned earlier, the higher r-AGO loading led to sheet stacking due to hydrodynamic effect, and the applied shear had a lesser effect on the re-entanglement due to higher extent of agglomeration of the r-AGO phase [31,32].

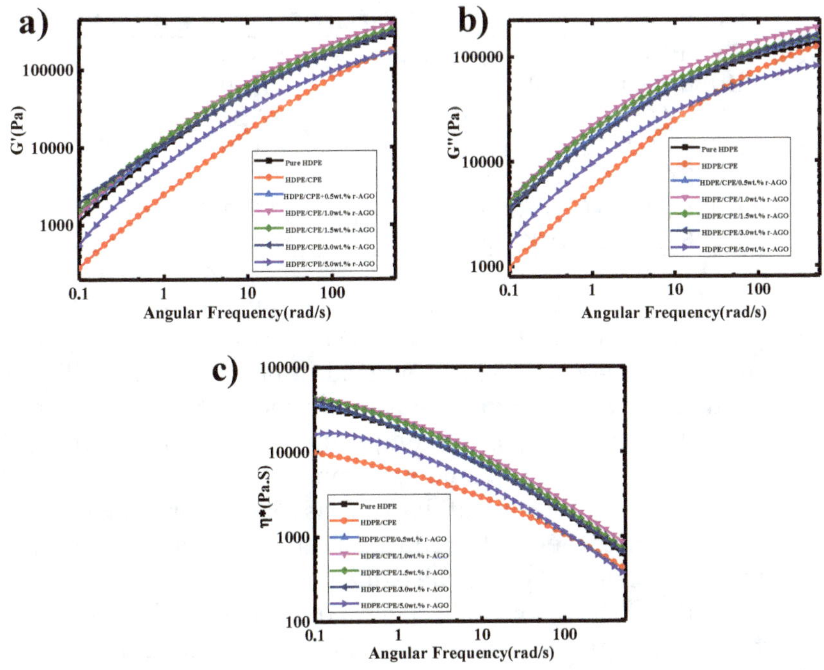

Figure 11.2 (a) Storage modulus (G'), (b) loss modulus (G") and (c) complex viscosity (η*) of HDPE/r-AGO nanocomposites.

The incorporation of the conducting nanofillers in the polymer matrix leads to the generation of conducting pathways for electrical conduction in the polymer nanocomposites *via* percolation of filler [15]. To analyze such an effect, surface resistivity of the samples was measured and the values are presented in Table 11.4. Pure HDPE and HDPE/CPE had high surface resistivity due to the lack of any conducting pathways [33]. Addition of r-AGO decreased the surface resistivity of the nanocomposites up to 3.0 wt%, indicating that the composites had enhanced conductivity. In fact, the resistance decreased by two orders of magnitude on the addition of 3.0 wt% of filler. Further increase in the content of r-AGO did not enhance the conductivity. The rapid transition and/or unchanged conductivity represented the achievement of percolation threshold [34]. This indicated that 3 wt% filler fraction was suitable to form a conducting network in the HDPE matrix.

Table 11.4 Surface resistivity of HDPE/r-AGO nanocomposites

Sample	Surface Resistance (Ohm)
Pure HDPE	$14x10^{10}$
HDPE + 5.0 wt% CPE	$15x10^{10}$
HDPE / CPE/ 0.5 wt% r-AGO	$12x10^{9}$
HDPE / CPE/ 1.0 wt% r-AGO	$9.3x10^{9}$
HDPE/CPE/1.5 wt% r-AGO	$3.4x10^{9}$
HDPE/CPE/3.0 wt% r-AGO	$11.7x10^{8}$
HDPE/CPE/5.0 wt% r-AGO	$10.3x10^{9}$

In addition, Table 11.5 shows the surface resistivity of HDPE/CPE/3 wt% r-AGO nanocomposites with other graphene based polymer nanocomposites reported in literature [4,21,35-37]. The conductivity of the nanocomposites varied with the nature and amount of filler, thickness of the composite film and applied voltage. The conductivity of the graphene nanocomposites mainly depended on the nature of graphene nanofiller such as reduced graphene oxide, expanded graphite, thermally reduced graphene oxide, functionalized graphene oxide and graphene oxide.

Figure 11.3(a) shows the XRD spectra of pure GO and r-AGO. Pure GO exhibited an intense peak at $2\theta = 10.0°$ corresponding to (002) with *d*-spacing of 8.8 Å. After reduction of GO using EDA, the intensity of peak at 10.0° reduced and a broad peak appeared at $2\theta = 26.5°$ was observed. This indicated that the majority of the oxide and water molecules were removed and the EDA intercalated in between the GO layers [35]. Figure 11.3(b) shows the diffraction patterns of HDPE, HDPE/CPE blend and HDPE/r-AGO nanocomposites. The peaks at 2θ

Table 11.5 Comparison of the electrical properties of HDPE/CPE/r-AGO nanocomposites with other polymer/graphene nanocomposites reported in literature

Polymer nanocomposites	Graphene	Filler content (wt%)	Surface resistivity (Ω)/electrical conductivity (S/m or S/cm)	Ref.
Graphene/polyethylene nanocomposites	Thermally reduced graphene oxide	3	4.0×10^{6} Ω	[4]
Hexamethylene diamine functionalized graphene oxide/polypropylene nanocomposites	hexamethylene diamine functionalized graphene oxide	2.0	1×10^{5} Ω	[21]
Polyethylene/reduced graphite oxide nanocomposites	Reduced graphite oxide	3.1	1.1×10^{5} S cm^{-1}	[35]
Graphene based unsaturated polyester resin composites	Graphene nanosheets	0.1	1×10^{15} Ω	[36]
Poly(methyl methacrylate) (PMMA)-reduced graphene oxide composite	Reduced graphene oxide	2.7	1×10^{2} S m^{-1}	[37]
HDPE/CPE/r-AGO	Ethylene diamine functionalized reduced graphene oxide(r-AGO)	3.0	11.7×10^{6} Ω	Present work

= 22.0° (110) and 24.3° (200) were attributed to characteristic diffraction peaks of pure polyethylene. From the diffraction patterns of the nanocomposites, it was observed that the addition r-AGO did not

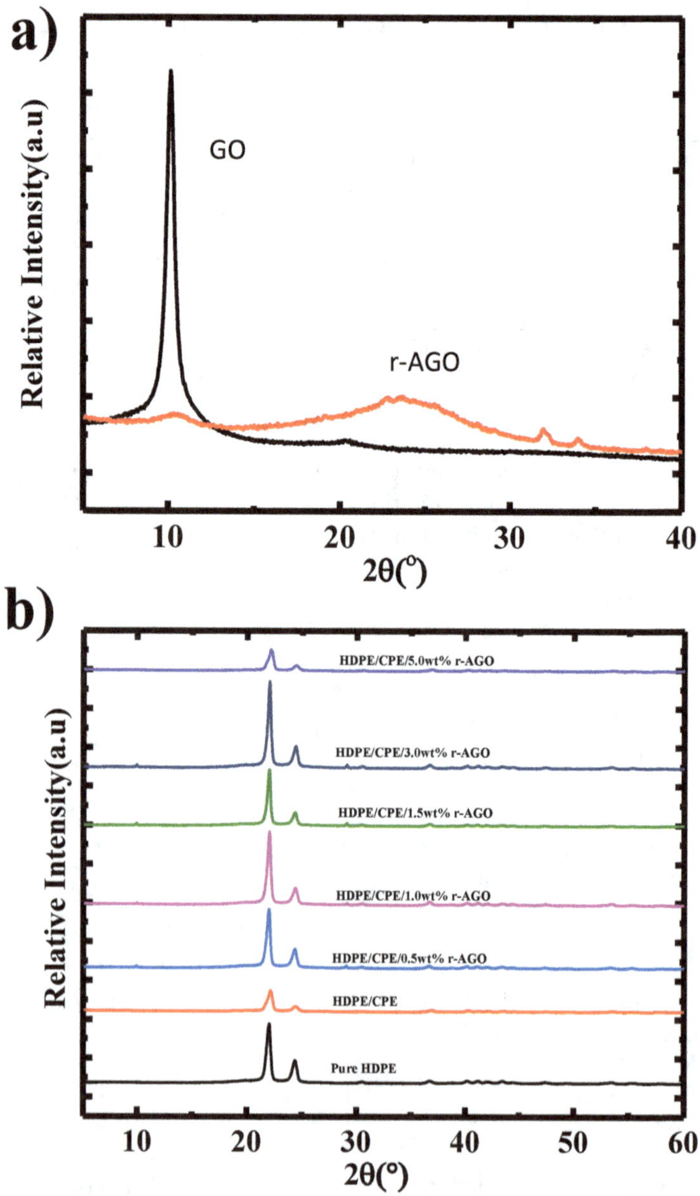

Figure 11.3 XRD spectra of (a) GO and r-AGO; (b) HDPE/r-AGO nanocomposites.

affect the crystal structure of HDPE. In addition, the peak at 26.5° was not observed in the diffraction patterns of HDPE/r-AGO nanocomposites. It indicated that r-AGO was exfoliated in the HDPE matrix, thus, also confirming the earlier findings. In addition, the peak intensity also increased with addition of r-AGO up to 3 wt% qualitatively indicating the increase in percentage of crystallinity. This suggested that the r-AGO acted as a nucleating agent in HDPE nanocomposites, which also confirmed the earlier DSC findings. Figure 11.4 also shows the TEM images of the nanocomposites. It was observed that the r-AGO was well dispersed in the polymer matrix till 3 wt% filler content. However in the case of 5 wt% AGO, aggregates were formed which resulted in poor dispersion in the polymer matrix.

a) b)

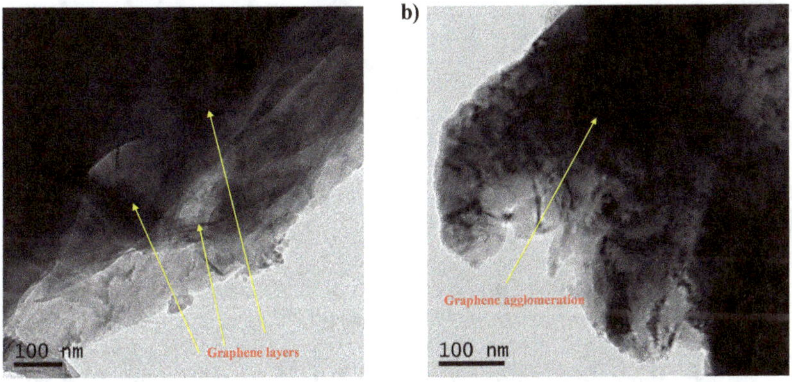

Figure 11.4 TEM images of composites with (a) 1.0 wt% and (b) 5.0 wt% filler content.

11.4 Conclusions

In this study, HDPE-graphene nanocomposites were developed using amine functionalized reduced graphene oxide (r-AGO) as nanofiller and CPE as compatibilizer. r-AGO covalently interacted with CPE *via* nucleophilic reaction, thus, enhancing the filler dispersion in the polymer. Owing to the improved interfacial interactions, enhancement in the thermal, mechanical and electrical properties of the resulting nanocomposites was observed. The Young's modulus of the composites increased marginally up to 3 wt% r-AGO content. The surface resistivity of the nanocomposites decreased from 14×10^{10} Ω to 11.7×10^{8} Ω with an increase in the content of r-AGO up to 3.0 wt%. The storage modulus and complex viscosity increased up to 1 wt% filler content and gradually reduced with further increase in the content of r-AGO, though significant reduction was observed for 5 wt%

composite. Addition of r-AGO did not impact the crystalline structure of the polymer. TEM analysis exhibited uniform dispersion of filler in the composites up to 3 wt% content. The developed nanocomposites have the potential for applications in the field of microelectronics and packaging industries.

Authors' Contributions

M. R. Vengatesan and H. Y. Al Asafen contributed equally to this work.

Acknowledgements

The authors sincerely thank the PI Research Centre, The Petroleum Institute for the financial support.

References

1. Koo, J. H. (2006) *Polymer Nanocomposites: Processing, Characterization, And Applications*, Mcgraw-Hill, Nanoscience and Technology Series, USA.
2. Stankovich, S., Dikin, D. A., Dommett, G. H. B., Kohlhaas, K. M., Zimney, E. J., Stach, E. A., Piner, R. D., Nguyen, S. T., and Ruoff, R. S. (2006) Graphene-based composite materials. *Nature*, **442**, 282-286.
3. Cai, D. Y., and Song, M. (2010) Recent advance in functionalized graphene/polymer nanocomposites. *Journal of Materials Chemistry*, **20**, 7906-7915.
4. Kim, H., Abdala, A. A., and Macosko, C. W. (2010) Graphene/polymer nanocomposites. *Macromolecules*, **43**, 6515-6530.
5. Lee, C., Wei, X., Kysar, J. W., and Hone, J. (2008) Measurement of the elastic properties and intrinsic strength of monolayer graphene. *Science*, **321**, 385-388.
6. Balandin, A. A., Ghosh, S., Bao, W., Calizo, I., Teweldebrhan, D., Miao, F., and Lau, C. (2008) Superior thermal conductivity of single-layer graphene. *Nano Letters*, **8**, 902-907.
7. Tombros, N., Veligura, A., Junesch, J., Guimares, M., Veramarun, I., Jonkman, H., and Wees, B. (2011) Quantized conductance of a suspended graphene nanoconstriction. *Nature Physics*, **7**, 697-700.
8. Bunch, J., Verbridge, S., Alden, J., van der Zande, A., Parpia, J., Craighead, H., and McEuen, P. (2008) Impermeable atomic membranes from graphene sheets. *Nano Letters*, **8**, 2458-2462.
9. He, Q., Sudibya, H. G., Yin, Z., Wu, S., Li, H., Boey, F., Huang, W., Chen, P., and Zhang, H. (2010) Centimeter long and large scale micropat-

terns of reduced graphene oxide films: Fabrication and sensing applications. *ACS Nano*, **4**, 3201-3208.
10. Shen, J., Huang, W., Wu, L., Hu, Y., and Ye, M. (2007) Thermophysical properties of epoxy nanocomposites reinforced with amino functionalized multiwalled carbon nanotubes. *Composites Part A*, **38**, 1331-1336.
11. Wakabayashi, K., Pierre, C., Dikin, D. A., Ruoff, R. S., Ramanathan, T., Brinson, C., and Torkelson, J. M. (2008) Polymer graphite nanocomposites: Effective dispersion and major property enhancement via solid state shear pulverization. *Macromolecules*, **41**, 1905-1908.
12. Mittal, V., and Al Zaabi, K. (2013) Biodegradable polyester nanocomposites: Phase miscibility and properties. *Journal of Applied Polymer Science*, **130**(1), 516-525.
13. Schniepp, H. C., Li, J. L., McAllister, M. J., Sai, H., Herrera-Alonso, M., Adamson, D. H., Prudhomme, R. K., Car, R., Saville, D. A., and Aksay, I. A. (2006) Functionalized single graphene sheets derived from splitting graphite oxide. *Journal of Physics Chemistry B*, **110**, 8535-8539.
14. Chen, Y., Qi, Y., Tai, Z., Yan, X., Zhu, F., and Xue, Q. (2012) Preparation, mechanical properties and biocompatibility of graphene oxide/ ultrahigh molecular weight polyethylene composites. *European Polymer Journal*, **48**, 1026-1033.
15. Kim, H., Kobayashi, S., AbdurRahim, M. A., Zhang, M. J., Khusainova, A., Hillmyer, M. A., Abdala, A. A., and Macosko, C. W. (2011) Graphene/polyethylene nanocomposites: Effect of polyethylene functionalization and blending methods. *Polymer*, **52**, 1837-1846.
16. Ren, P., Wang, H., Huang, H., Yan, D., and Li, Z. (2014) Characterization and performance of dodecyl amine functionalized graphene oxide and dodecyl amine functionalized graphene/high-density polyethylene nanocomposites: A comparative study. *Journal of Applied Polymer Science*, **131**, doi: 10.1002/app.39803.
17. Mittal, V., Luckachan, G. E., and Matsko, N. B. (2013) PE/chlorinated-PE blends and PE/chlorinated-PE/graphene oxide nanocomposites: morphology, phase miscibility, and interfacial interactions. *Macromolecular Chemistry and Physics*, **215**, 255-268.
18. Marcano, D. C., Kosynkin, D. V., Berlin, J. M., Sinitskii, A., Sun, Z., Slesarev, A., Alemany, L. B., Lu, W., and Tour, J. M. (2010) Improved synthesis of graphene oxide. *ACS Nano*, **4**, 4806-4814.
19. Jiang, X., and Drzal, L. T. (2012) Multifunctional high-density polyethylene nanocomposites produced by incorporation of exfoliated graphene nano platelets 2: crystallization, thermal and electrical properties. *Polymer Composites*, **33**, 636-642.
20. Zhao, S., Chen, F., Zhao, C., Huang, Y., Dong, J. Y., and Han, C. C. (2013) Interpenetrating network formation in isotactic polypropylene/graphene composites. *Polymer*, **54**, 3680-3690.

21. Ryu, S. H., and Shanmugharaj, A.M. (2014) Influence of hexamethylene diamine functionalized graphene oxide on the melt crystallization and properties of polypropylene nanocomposites. *Materials Chemistry and Physics*, **146**, 478-486.
22. Achaby, M. E., Arrakhiz, F.-E., Vaudreuil, S., Qaiss, A. K., Bousmina, M., and Fassi-Fehri, O. (2012) Mechanical, thermal, and rheological properties of graphene-based polypropylene nanocomposites prepared by melt mixing. *Polymer Composites*, **33**, 733-744.
23. Lee, S. H., Cho, E. N., Jeon, S. H., and Youn, J. R. (2007) Rheological and electrical properties of polypropylene composites containing functionalized multi-walled carbon nanotubes and compatibilizers. *Carbon*, **45**, 2810-2822.
24. Potschke, P., Fornes, T. D., and Paul, D. R. (2002) Rheological behavior of multiwalled carbon nanotube/polycarbonate composites. *Polymer*, **43**, 3247-3255.
25. Chaudhry, A. U., and Mittal, V. (2013) High-density polyethylene nanocomposites using masterbatches of chlorinated polyethylene/graphene oxide. *Polymer Engineering and Science*, **53**, 78-88.
26. Achaby, M. E., and Qaiss, A. (2013) Processing and properties of polyethylene reinforced by graphene nanosheets and carbon nanotubes. *Materials and Design*, **44**, 81-89.
27. Horst, M. F., Quinzani, L. M., and Failla, M. D. (2014) Rheological and barrier properties of nanocomposites of HDPE and exfoliated montmorillonite. *Journal of Thermoplastic Composite Materials*, **27**, 106-125.
28. Li, Y., Zhu, J., Wei, S., Ryu, J., Sun, L., and Guo, Z. (2011) Poly (propylene)/graphene nanoplatelet nanocomposites: melt rheological behavior and thermal, electrical, and electronic properties. *Macromolecular Chemistry and Physics*, **212**, 1951-1959.
29. She, Y., Chen, G., and Wu, D. (2007) Fabrication of polyethylene/graphite nanocomposite from modified expanded graphite. *Polymer International*, **56**, 679-685.
30. Tang, Y., Gao, P., Ye, L., and Zhao, C. (2010) Organoclay-modified thermotropic liquid crystalline polymers as viscosity reduction agents for high molecular mass polyethylene. *Journal of Materials Science*, **45**, 5353-5363.
31. Tuteja, A., Mackay, M. E., Hawker, C. J., and Van Horn, B. (2005) Effect of ideal, organic nanoparticles on the flow properties of linear polymers: Non-Einstein like behavior. *Macromolecules*, **38**, 8000-8011.
32. Thomas, S. P., De, S. K., and Hussein, I. A. (2013) Impact of aspect ratio of carbon nanotubes on shear and extensional rheology of polyethylene nanocomposites. *Applied Rheology*, **23**, 23635.
33. Kaushik, A., Singh, P., and Bhagat, S. (2009) Preparation and characterization of graphite flakes-filled polyester composites. *Polymer-*

Plastics Technology and Engineering, **48**, 802-807.
34. Perez, L. D., Zuluaga, M. A., Kyu, T., Mark, J. E., and Lopez, B. L. (2009) Preparation, characterization, and physical properties of multiwall carbon nanotube/elastomer composites. *Polymer Engineering and Science*, **49**, 866-874.
35. Pavoski, G., Maraschin, T., Milani, M. A., Azambuja, D. S., Quijada, R., Moura, C. S., Basso, N. S., and Galland, G. B. (2015) Polyethylene/reduced graphite oxide nanocomposites with improved morphology and conductivity. *Polymer*, **81**, 79-86.
36. Swain, S. (2013) Synthesis and characterization of graphene based unsaturated polyester resin composites. *Transactions on Electrical and Electronic Materials*, **14**, 53-58.
37. Pham, V. H., Dang, T. T., Hur, S. H., Kim, E. J., and Chung, J. S. (2012) Highly conductive poly(methyl methacrylate) (PMMA)-reduced graphene oxide composite prepared by self-assembly of pmma latex and graphene oxide through electrostatic interaction. *ACS Applied Materials & Interfaces*, **4**, 2630-2636.
38. Navaee, A., and Salimi, A. (2015) Efficient amine functionalization of graphene oxide through the Bucherer reaction: an extraordinary metal-free electrocatalyst for the oxygen reduction reaction. *RSC Advances*, **5**, 59874-59880.

12

Polyethylene-Copolymer Blends: Morphology and Properties

12.1 Introduction

Polymer blends are generated by the mixing of two or more polymers together in order to combine the characteristics of individual polymers. The compatibilization of the phases in the blends is also achieved by a number of physical or chemical means. Polymer blends are widely used materials in the modern polymer industry owing to their wider range of properties such as performance combinations, aid in processing, low capital costs as compared with development of new monomers and polymers of required properties, etc. As a result, polymer blends find application in numerous fields such as adhesion, colloidal stability and design of composite materials [1]. The dispersed phase in the many blend systems can also acquire a specific morphology which is beneficial for applications such as barrier performance achieved by the dispersed phase acquiring the platy morphology. Both solution as well as melt mixing techniques to generate the polymer blends have found their application, though melt mixing is more environmentally friendly method due to the absence of any additional solvent and suitability with the existing processing technology. On the other hand, the melt mixing method requires the use of high temperature and shear for the generation of well-mixed polymer blends, which may degrade the heat sensitive polymers, thus, requiring optimization of the mixing conditions so as to achieve an optimal mixing, however, also avoiding an extensive thermal degradation. The studies on the polymer blends have significantly increased in recent years leading to the generation of functional blend systems with superior mechanical and thermal properties [2-10]. Both miscible and immiscible blends have been studied in recent years, along with new pathways to achieve compatibilization of the phases in the immiscible blends.

Vikas Mittal[a], M. Suria[a,] and Nadejda Matsko[b]*
[a]Khalifa University of Science and Technology, Abu Dhabi, UAE; [b]Graz Centre of Electron Microscopy, Graz, Austria
**Current address: University of Waterloo, Canada*

Polyolefins represent the largest group of industrial thermoplastics in terms of production and have superior rank among commodity plastics owing to their use in a large number of applications. The number of individual polyolefin members is limited, and the number of applications based on such materials is constantly increasing, along with the increasing stringent needs the polymers are expected to fulfill. Thus, various polyolefin-based blends and composites are required to meet the increasing demand of materials with required properties and processing suitable for specific applications [11-13]. For instance, Wang *et al.* [7] reported polyolefin based blends with styrene polymers and engineering plastics and developed effective strategy to compatibilize the blends by using a polyolefin based multiphase compatibilizer. Chlorinated polyethylene (CPE) blends with different polymers have also been commonly reported for many purposes such as improvements in the toughening properties of host polymers [14], as a compatibilization agent through enhancement of interfacial interactions between two polymers [15,16], improved processing properties [17], ameliorated adhesion properties [18], enhanced ignition resistance properties [19], and modified gas transport properties [20]. In a recent study, Mittal and Matsko [21] studied the morphology of polyethylene-CPE blends for the purpose of generating materials for reduced low temperature brittleness. CPE materials of two different chlorination levels (25% and 35%) were employed at 5 wt% content. Figure 12.1 shows the AFM height images of the PE and PE-CPE block faces prepared by microtoming at -140 °C. Surface topography analysis (surface height or roughness variation) at 3 different places on the sample surface was also performed. It is noteworthy that all parameters related to hardness of conventional PE, such as the elastic constants, tensile strength, and crushing strength increased with decreasing temperature. Also, the elastic region of the stress-strain curve was extended to higher stress values with decreasing temperature. These results were consistent with the experience that cryo-sectioning of polycrystalline PE becomes more difficult at lower temperature. Incorporation of the amorphous CPE (CPE35) to PE, however, resulted in completely different behavior. Higher height variation in its micrograph indicated higher degree of compression which represents compressibility and flexibility of polymer materials during sectioning. The extent of plasticization was also higher in the case of amorphous CPE35 polymer where the height variation was significantly large as compared with PE blend with CPE25.

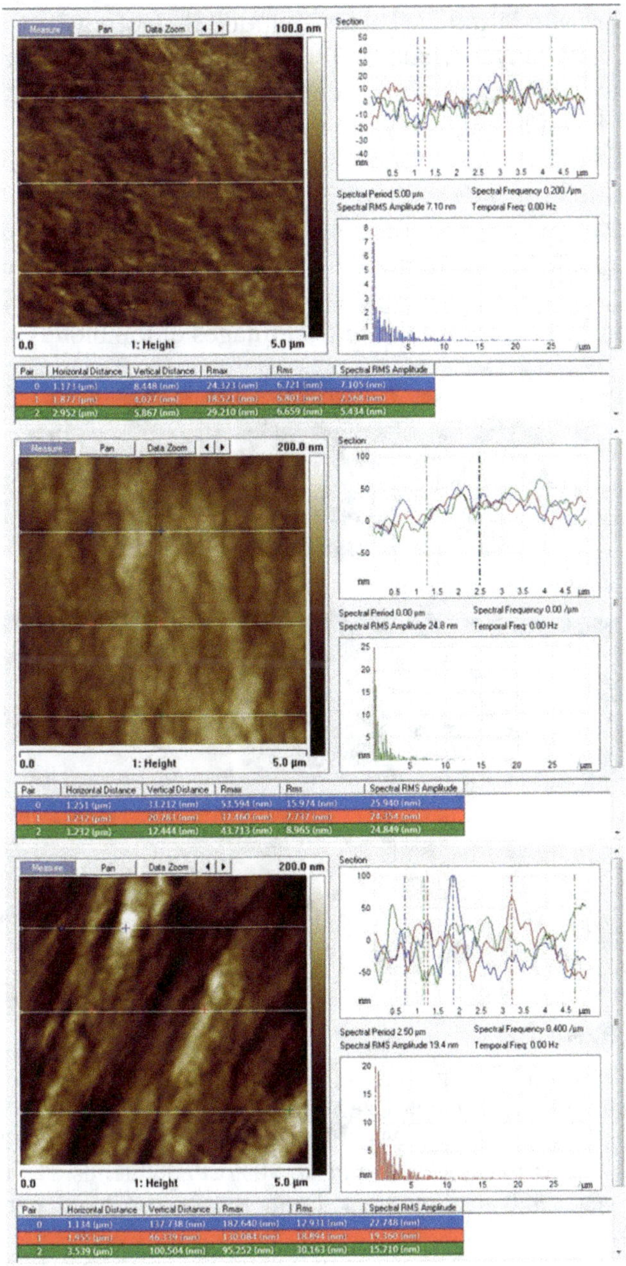

Figure 12.1 AFM height images and corresponding height corrugation analysis for PE (top image), PE+5%CPE25 (middle row) and PE+5%CPE35 (bottom row) when sectioned at -140 °C. Reproduced from Reference 21 with permission from Springer.

Chaudhry and Mittal [22] also expanded their study on the blends of high density polyethylene (HDPE) with CPE in the weight fraction range from 1% to 30%. The rheological analysis of the data using Cole-Cole, Han-Chuang and van Gurp plots confirmed the miscibility of CPE25 blends (except for 30% CPE25 blend at lower frequency) whereas CPE35 blends with 10-30% CPE content were immiscible. Highest increase in the rheological properties (complex moduli) was observed at 2% CPE content. The optical microscopy also confirmed the better mixing of CPE25 polymers in HDPE than CPE35. For instance, Figure 12.2 shows the optical images of the blends with both

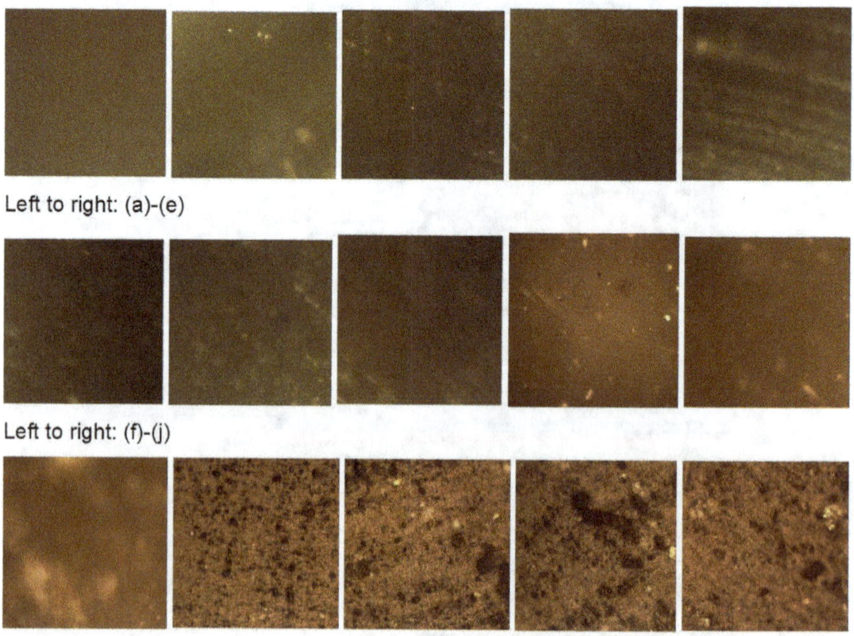

Figure 12.2 Optical micrographs of (a) pure HDPE, (b) 1% CPE25, (c) 2% CPE25, (d) 5% CPE25, (e) 10% CPE25, (f) 15% CPE25, (g) 20% CPE25, (h) 30% CPE25, (i) 1% CPE35, (j) 2% CPE35, (k) 5% CPE35, (l) 10% CPE35, (m) 15% CPE35, (n) 20% CPE35, and (o) 30% CPE35 blends. The width of the images reads 500 μm. Reproduced from Reference 22 with permission from Wiley.

CPE25 and CPE35 as a function of blend composition. The CPE25 blends with 1%, 2%, and 5% CPE concentration (Figure 12.2b-d) were similar to pure HDPE (Figure 12.2a), thus, indicating molecular mixing of the two phases. The blends with 10%, 15%, 20%, and 30%

CPE25 (Figure 12.2e-h, respectively) had small whitish regions indicating CPE25 phase, but these were observed to be dispersed homogenously. In the case of CPE35 blends, the lower concentrations of CPE did lead to good mixing (Figure 12.2i-k for 1%, 2%, and 5%, respectively). However, beyond 5%, the immiscibility of the two phases was extensive and large domains of CPE35 (as black areas in the bright field images of Figure 12.2l-o for 10%, 15%, 20%, and 30% CPE35 content) were clearly visible. The mechanical properties of the CPE25 blends were also superior than the corresponding CPE35 blends especially at higher CPE concentration where effects of immiscibility as well as matrix plasticization played a role.

In order to further explore the development of polyethylene based functional materials, especially for low temperature flexibility, blends of polyethylene with a variety of copolymers have been generated in the current study. The copolymers employed for this purpose were polyethylene-*g*-maleic anhydride copolymer (PE-MA), ethylene methacrylic acid copolymer with zinc ion (EMAZ) and ethylene vinyl acetate-*g*-maleic anhydride (EVA-MA). The amount of the copolymers was also varied in polyethylene in order to study the effect on miscibility and resulting blend properties. In addition, the effect of copolymer molecular weight was also analyzed in the blends by employing PE-MA copolymers of two different molecular weights. Melt mixing process was employed for the generation of blends due to its industrial acceptability. The compounding parameters were also optimized to ensure effective mixing without thermal degradation of the polymers.

12.2 Experimental

12.2.1 Materials

HDPE (BB2581) of specific gravity 0.958 was supplied by Abu Dhabi Polymers Company Limited (Borouge), UAE. Various copolymers of different specifications and functionalities were procured from suppliers, as mentioned in Table 12.1. The polymer materials were used as obtained.

12.2.2 Generation of Blends

Polymer blends were prepared by melt mixing of the copolymers with HDPE using mini twin conical screw extruder (MiniLab HAAKE

Table 12.1 Specifications of HDPE and copolymers

Polymer/ Compatibilizer	Supplier and trade name	MFI, g/10 min (2.16 kg, 190 °C)	Density, g/cm³	Peak melting temperature, °C
PE	Borouge BB2581	0.35	0.958	147
EVA-MA	Dupont Fusabond C190	16	0.950	71
EMAZ	Dupont Surlyn 9020	1	0.96	85
PE-MA1	Aldrich PE-*g*-MA	500 cP viscosity (140 °C)	--	--
PE-MA2	Aldrich PE-*g*-MA	1700-4500 cP viscosity (140 °C)	--	--

Rheomex CTW5, Germany). A mixing temperature of 185 °C for 7 min at 80 rpm with batch size of 5 g was used. The screw length and screw diameter were 109.5 mm and 5/14 mm conical, respectively. Blends of copolymers in weight percentages 5%, 10%, and 15%, were generated. The disc and dumbbell-shaped test samples were prepared by mini injection molding machine (HAAKE MiniJet, Germany) at a processing temperature of 185 °C. The injection pressure was 700 bar for 6 s, whereas holding pressure was 400 bar for 3 s. The temperature of the mold was kept at 55 °C.

12.2.3 Characterization Techniques

Thermal properties of the blends were recorded using Netzsch thermogravimetric analyzer (TGA). Nitrogen was used as a carrier gas and the scans were obtained from 50 to 700 °C at a heating rate of 10 °C/min. Calorimetric properties of blends were recorded on a Netzsch DSC under nitrogen atmosphere. The scans were obtained from 30-180-30 °C using heating and cooling rates of 10 °C/min and 5 °C/min, respectively. The heat enthalpies, useful to estimate the extent of crystallinity, were recorded in a narrow error range (±0.1%), which were also confirmed by repeated runs.

AR 2000 rheometer from TA Instruments was used to characterize the rheological properties of the blends such as storage modulus (G′), loss modulus (G″), and complex viscosity (η*). Disc shaped samples of 25 mm diameter and 2 mm thickness were measured at 185 °C using a gap opening of 1.2 mm. Strain sweep scans were recorded at ω = 1 rad/s from 0.1 to 100% strain and the samples were observed to be shear stable up to 10% strain. For the comparison, frequency sweep scans (dynamic testing) of all the samples were recorded at 2% strain from ω = 0.1 to 100 rad/s.

The mechanical testing of the blends was performed on universal testing machine (Testometric, UK). The dumbbell-shaped samples with 53 mm length, 4 mm width, and 2 mm thickness were used. A loading rate of 5 mm/min was employed and the tests were carried out at room temperature. WinTest Analysis software was used for the calculation of tensile modulus and yield stress properties of the blends. An average of five values was calculated.

For the microscopy analysis, the samples were mounted in special holders that at the same time fit in the microtome and were also suitable for the examination of the block face by atomic force microscopy (AFM). The block faces of specimens after cryo-ultramicrotomy were investigated at ambient conditions using Dimension 3100 AFM/SPM (Veeco, USA) atomic force microscope. AFM images were collected in tapping mode using silicon nitride cantilevers with natural frequencies in the 300 kHz range (force constant = 20 $N.m^{-1}$, tip radius = 10 nm (NT-MDT, Russia)). AFM image processing was performed using Nanoscope v720 software (Veeco, USA). The block faces of specimens after cryo-ultramicrotomy were also investigated at ambient conditions using ZEISS Axioplan light microscope (Zeiss, Germany) equipped with Zeiss Axio Cam ICc 1 CCD camera in reflected polarized light.

Wide-angle X-ray diffraction (WAXRD) analysis of the blends was performed using a Panalytical powder diffractometer (X'Pert PRO) using Cu Kα radiation (λ = 1.5406 Å) in reflection mode. A zero-background holder was used to minimize the noise. The samples were step-scanned from 2Θ = 10-70° at room temperature using a step size of 2Θ = 0.02° and a step time of 10 s. Time studies were also carried out by measuring the same samples after 90 days so as to ascertain any change in the crystal structure.

12.3 Results and Discussion

The choice of the copolymers in the study was made on the basis of functional groups. The nonpolar component of the compatibilizers was based on polyethylene in order to attain compatibility with the matrix polymer. The miscibility of these copolymers with PE can also subsequently lead to the compatibilization of PE with reinforcing fillers through the polar functional groups. Many studies have reported the blending of these copolymers with PE, however, no systematic study was performed where the effect of these copolymers was studied on the morphology as well as properties as a function of their molecular architecture and content in the blend. For instance, the effect of maleic anhydride on the PE microstructure and properties was studied through EVA-MA as well as PE-MA copolymers. In addition, the ionomer EMAZ was used to study the effect of ionic species on the polymer matrix. The melt flow indices of the compatibilizers varied between 1 and 20, thus, also indicating a wide variation in their molecular weight characteristics. As mentioned earlier, PE-MA of two different molecular weight characteristics were employed to study the molecular weight effect on the blend morphology and properties. Comparison with the MFI of PE of 0.35 also revealed that the copolymers had much lower molecular weight than PE. In addition, the copolymers were semi-crystalline in nature with significantly lower crystallinity and peak melting points than pure PE.

Figure 12.3 demonstrates the TGA thermograms of PE blends with PE-MA1, PE-MA2, EMAZ and EVA-MA as a function of blend composition. Owing to their low molecular weight, no enhancement in the thermal stability of PE was expected on the addition of copolymers, however, it was necessary to ensure that the thermal stability of the PE is not significantly reduced on the addition of these copolymers, which would subsequently affect the high temperature applications. At 5% and 10% content of PE-MA1 in the blend, slight reduction in the onset and peak degradation temperature of PE was observed. Further enhancing the content of PE-MA1 in the blend led to significant reduction in the onset of degradation temperature of PE, as observed in Figure 12.3a. However, all the blends were still stable till 350 °C. In addition, the degradation pattern of the polymer was not impacted on the addition of PE-MA1 copolymer as confirmed by the single degradation step for PE and blends. Addition of higher molecular weight PE-MA2 to PE exhibited better thermal performance as compared to PE-MA1. No change in the onset and peak degradation temperature of PE was observed even after the incorporation of 15% of PE-MA2, thus, confirming superior thermal performance of PE/PE-MA2 blends. Similar to PE-MA1 blends, the PE-MA2 blends also exhibited single degradation pattern. In the case of EMAZ based blends, the thermal degradation of PE was retained till 10% EMAZ content in the blend. However, a reduction in the onset and peak degradation temperature was observed on further enhancing the EMAZ content to 15%. Similar to PE-MA1, it indicated that the optimal amount of copolymer for addition in PE was 10%, beyond which the thermal performance reduction was significant. Similarly, addition of 5% EVA-MA copolymer to PE also exhibited no effect on the thermal degradation of PE. However, addition of 10% and 15% copolymer to PE correspondingly reduced the onset and peak degradation temperature of PE. Thus, it can be concluded that even though all the blends were still stable up to 350 °C, 10% amount in PE was optimal as it largely retained the thermal performance.

Table 12.2 and Figures 12.4-12.5 represent the calorimetric behavior of PE blends with PE-MA1, PE-MA2, EMAZ and EVA-MA. In the

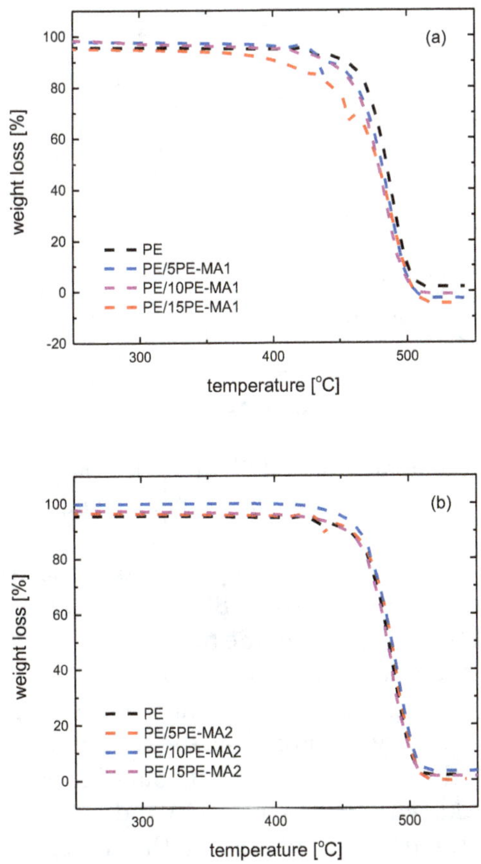

Figure 12.3 (continued on next page)

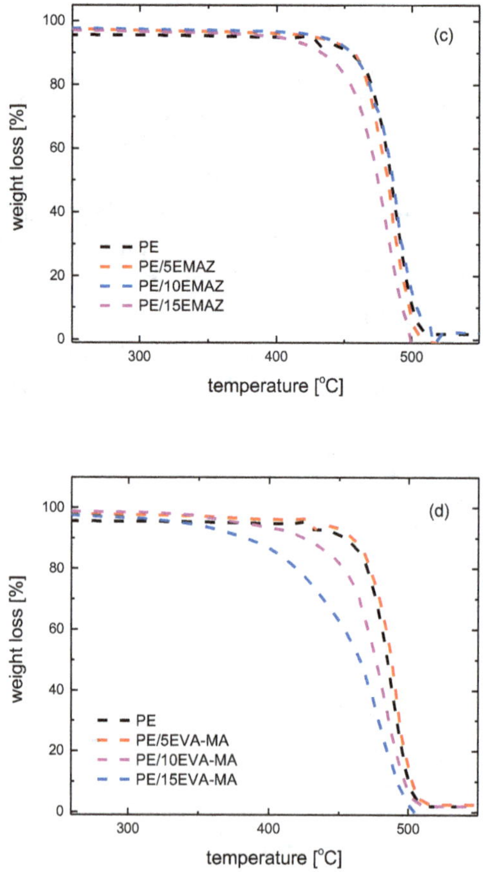

Figure 12.3 TGA thermograms of (a) PE-MA1, (b) PE-MA2, (c) EMAZ and (d) EVA-MA blends (increasing copolymer content on moving downwards).

blends with EMAZ and EVA-MA, a slight hump at the corresponding copolymer melting temperatures of 85 °C and 71 °C was also observed in the melt thermograms of the blends, as seen in Figure 12.4c and d. The magnitude of the copolymer peak was also observed to increase with increasing copolymer amount in the blend, thus, indicating phase immiscibility on enhancing the copolymer content. In an earlier study [23], the melt transition attributed to PE was reported to exhibit dual peaks with PE-MA, EVA-MA and EMAZ copolymers at 5 wt% content, in the presence of filler. In the current study, this behavior was not observed, though melting transitions corresponding to EMAZ and EVA-MA were observed in the thermograms. PE-MA based blends were observed to have a single melt transition peak, thus, indicating better phase miscibility. It has also to be noted that the observed changes in the current study would have resulted due to the use of different processing conditions in two studies. Though better phase intermixing was achieved in the current study with PE-MA copolymers, however, it can be expected that the copolymers were not molecularly or thermodynamically mixed with PE owing to the absence of any interacting force. The blends were observed to retain their microstructure as no change in the calorimetric behavior

Table 12.2 Calorimetric properties of PE blends

Material	Peak melting temperature, °C	Melt enthalpy, J/g	Peak crystallization temperature, °C
PE	138.4	154	117.0
PE/5PE-MA1	137.8	148	116.1
PE/10PE-MA1	136.6	136	115.7
PE/15PE-MA1	135.0	130	115.4
PE/5PE-MA2	136.7	148	116.6
PE/10PE-MA2	136.1	139	116.2
PE/15PE-MA2	135.1	120	116.2
PE/5EMAZ	138.2	132	115.8
PE/10EMAZ	137.2	120	115.2
PE/15EMAZ	136.6	116	115.1
PE/5EVA-MA	138.3	129	115.5
PE/10EVA-MA	136.9	121	115.1
PE/15EVA-MA	135.1	114	114.4

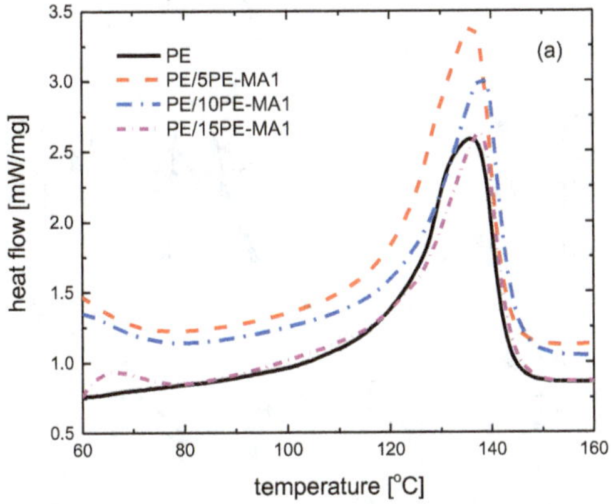

Figure 12.4 (continued on next page)

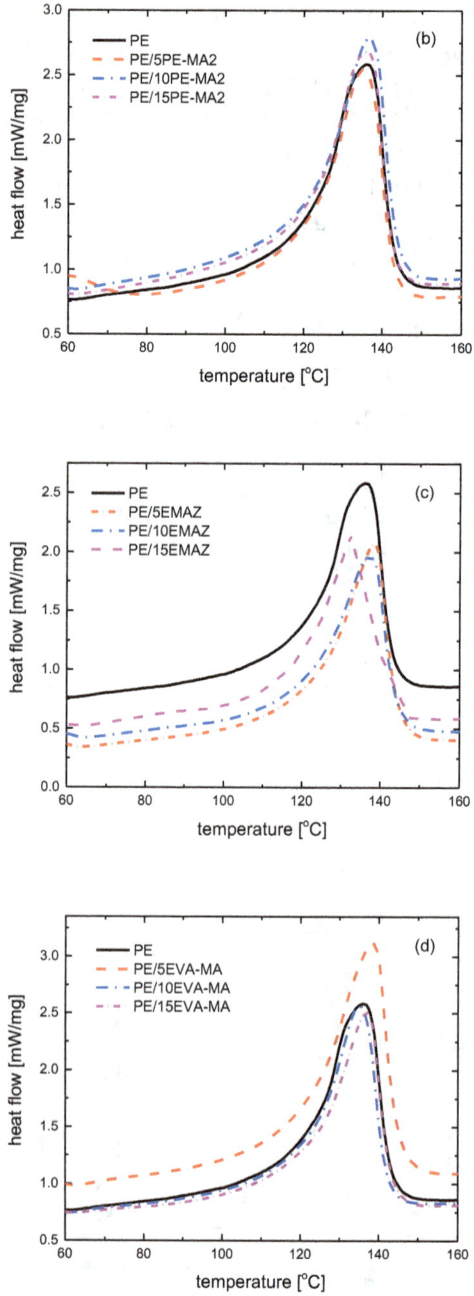

Figure 12.4 Melting thermograms of (a) PE/PE-MA1, (b) PE/PE-MA2, (c) PE/EMAZ and (d) PE/EVA-MA blends.

was observed even after few processing cycles.

All copolymers led to a slight reduction in the peak melting temperature of PE. In addition, the extent of reduction was a function of the copolymer content in the blend. Similar to the peak melting point, the melt enthalpy of PE was also observed to decrease with the addition of copolymers, thus, indicating the hindrance to polymer crystallization in the presence of copolymer chains. Maximum reduction in the melt enthalpy was observed in the case of EVA-MA and EMAZ copolymers. For instance, the blends with 15% EMAZ and EVA-MA exhibited the melt enthalpy values of 116 J/g and 114 J/g, in comparison with 154 J/g for pure PE. On the other hand, PE-MA1 and PE-MA2 based blends with 15% copolymer content exhibited the melt enthalpy values of 130 J/g and 120 J/g, respectively. As both EMAZ and

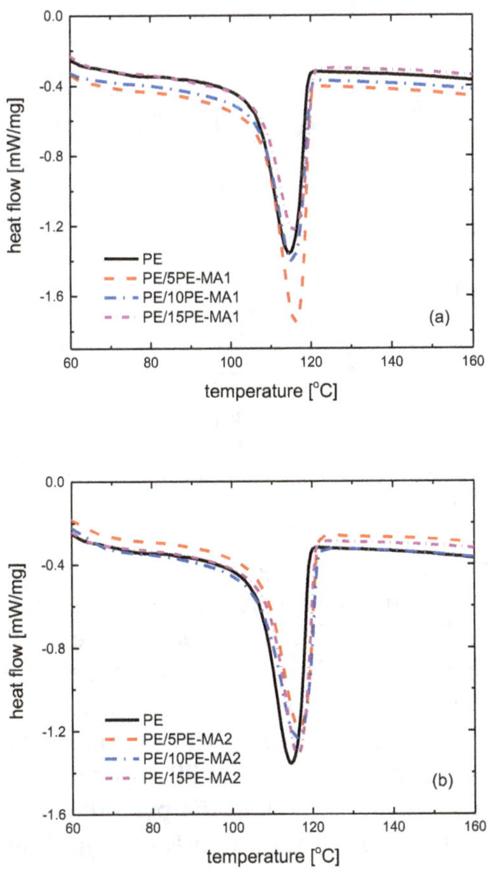

Figure 12.5 (continued on next page)

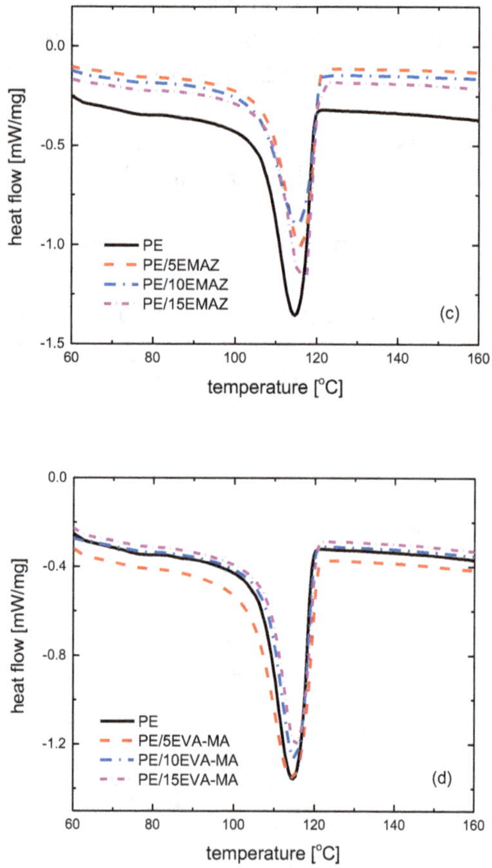

Figure 12.5 Crystallization thermograms of (a) PE/PE-MA1, (b) PE/PE-MA2, (c) PE/EMAZ and (d) PE/EVA-MA blends.

EVA-MA copolymers were also observed to have lower degree of miscibility with PE, it would have resulted in higher extent of hindrance to polymer crystallization, thereby decreasing the degree of crystallinity of PE. The peak crystallization temperature of PE though decreased slightly from 117 °C to 114-116 °C in the blends, but the amount of copolymer did not affect it, thus, indicating no nucleation effect. The DSC studies carried out after 90 days of blend preparation also revealed no change in the melt enthalpy and peak melting/crystallization temperatures. As both polymers in the blend are above their respective glass transition temperatures (T_g) at room temperature, thus, no contraction/expansion of the polymer pairs would be expected. In the case of polymer blends with one component above and other below T_g at room temperature, the contraction/expansion phenomenon has been reported to cause significant changes in the microstructure and crystallinity of the phases in the blends, also called physical ageing [24].

The microstructure of the blends was also studied by X-ray diffraction, as shown in Figure 12.6. The incorporation of copolymers did not affect the major diffraction peaks attributed to PE crystals, as also observed earlier from DSC analysis. However, small changes in the 17-22° 2Θ angle range were observed in the case of EMAZ and EVA-MA containing blends, which may also confirm the findings from DSC about more significant changes in the crystallinity and morphology due to EVA-MA and EMAZ as compared to both PE-MA1 and PE-MA2

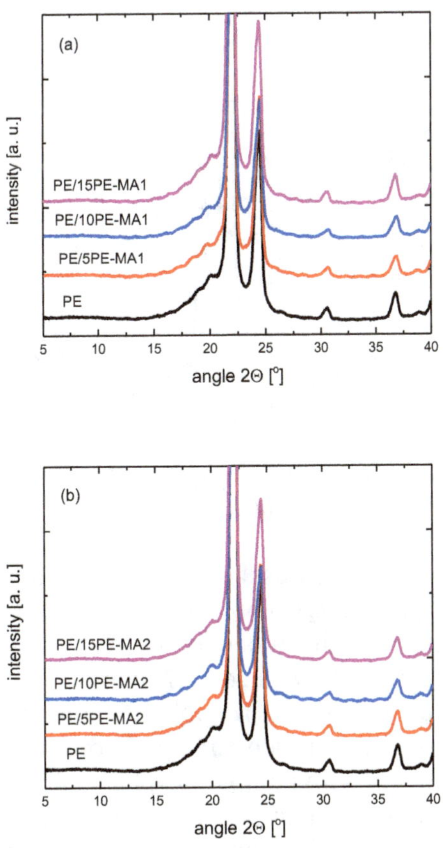

Figure 12.6 (continued on next page)

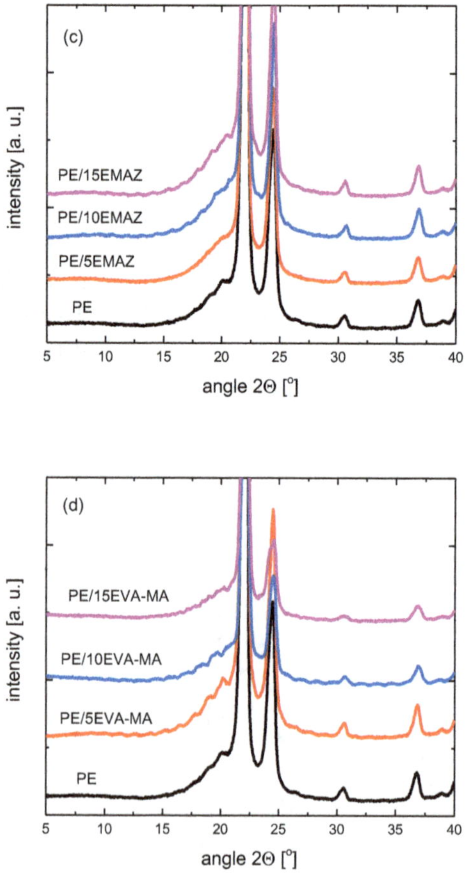

Figure 12.6 X-ray diffractograms of (a) PE/PE-MA1, (b) PE/PE-MA2, (c) PE/EMAZ and (d) PE/EVA-MA blends.

copolymers, however, these observations were required to be further confirmed through microscopy analysis.

Further insights into the morphology of the blends were also gained through microscopy. Figure 12.7 shows the light microscopy images of the blends with 15 wt% copolymer content at two different magnifications. At the microscale, the morphology of the blends was observed to be uniform with no obvious phase distribution of the PE and copolymer phases. Similar morphology was also observed for other blend compositions. Further morphological features about the blends were observed in the atomic force microscopy analysis of the blends, as demonstrated in Figures 12.8 to 12.11. Specifically, Figure

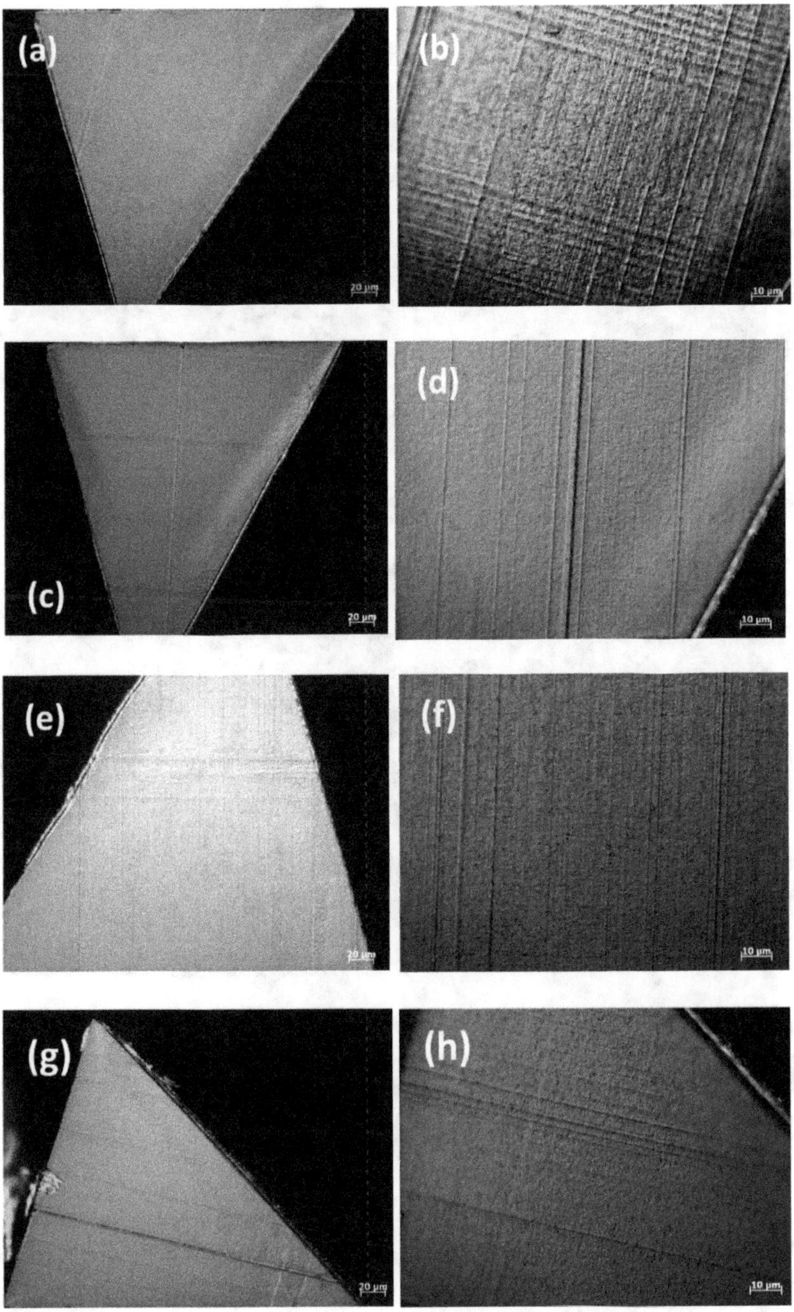

Figure 12.7 Light microscopy images of (a,b) PE/15PE-MA1, (c,d) PE/15PE-MA2, (e,f) PE/15EMAZ and (g,h) PE/15EVA-MA blends.

12.8 represents the height (Figure 12.8a) and phase (Figure 12.8b-d) images of the PE/15PE-MA1 blend at different magnifications. As

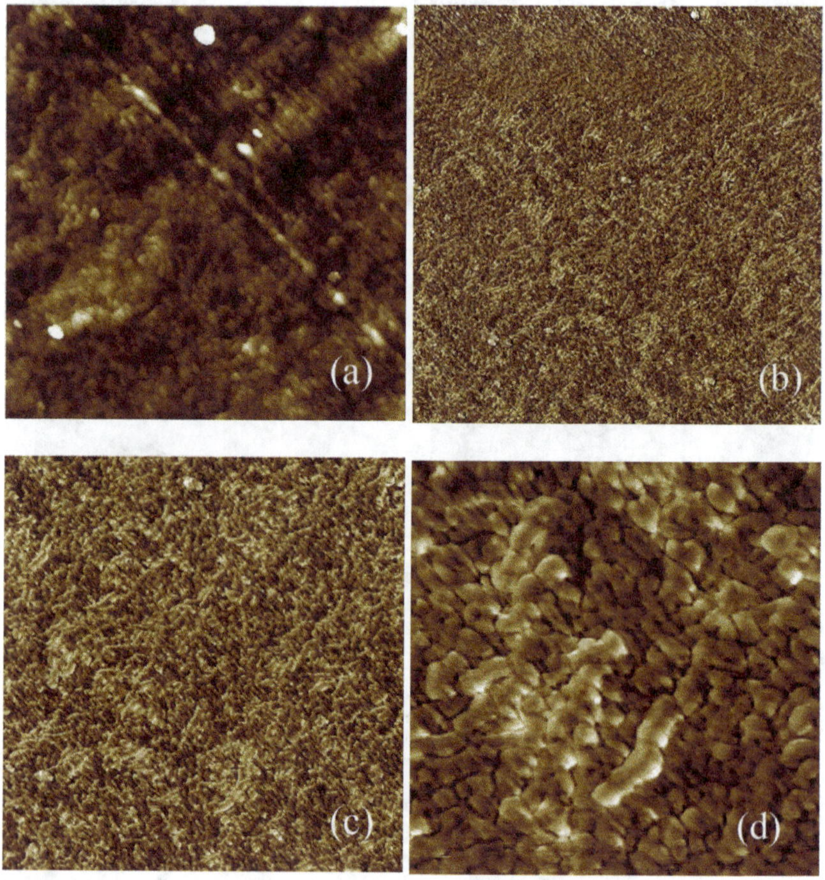

Figure 12.8 AFM images of the PE/15PE-MA1 blend: (a) height image (the width of the image is 10 μm); (b-d) phase images with image width as 20, 10 and 3 μm respectively. Fine dispersion of mixture of hard (bright phase representing crystalline region) and soft (dark phase representing amorphous region) regions is clearly visible.

Observed in the phase images, the PE-MA1 blend exhibited uniform morphology with fine mixture of hard (bright phase representing crystalline region) and soft (dark phase representing amorphous region) regions. It also confirmed the earlier findings about the good degree of phase intermixing of PE-MA1 with PE. Similarly, Figures 12.9-12.11 present the height and phase images of PE/15PE-MA2,

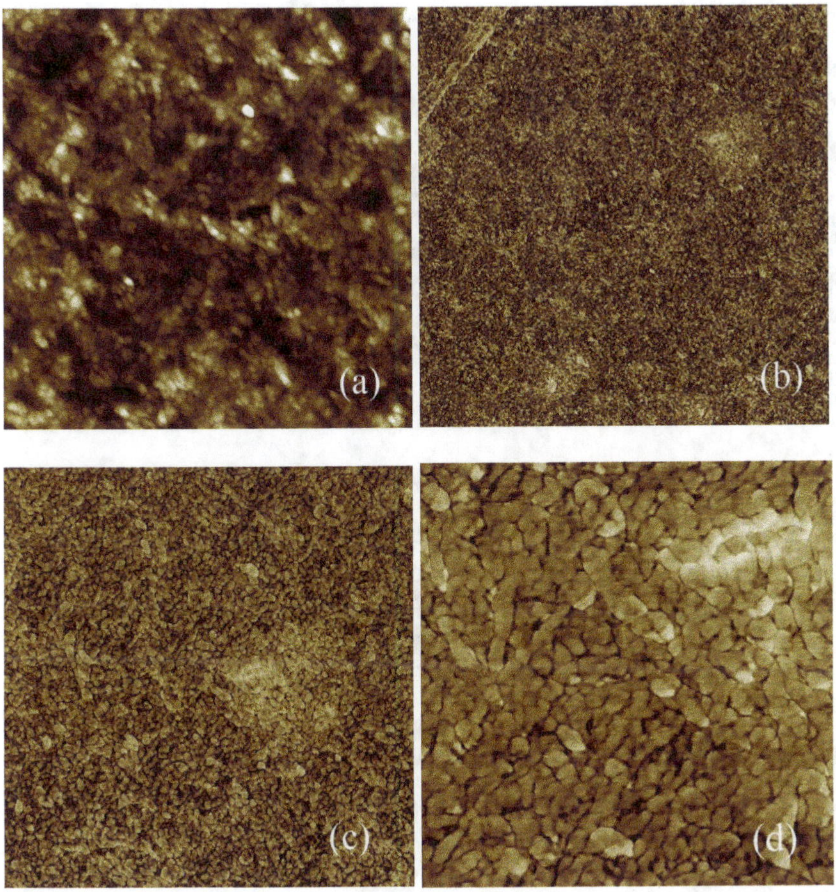

Figure 12.9 AFM images of the PE/15PE-MA2 blend: (a) height image (the width of the image is 10 μm); (b-d) phase images with image width as 20, 10 and 3 μm respectively. Fine dispersion of mixture of hard (bright phase representing crystalline region) and soft (dark phase representing amorphous region) regions is clearly visible.

PE/15EMAZ and and PE/15EVA-MA blends. The PE-MA2 also exhibited similar uniform phase morphology, thus, confirming that both PE-MA1 and PE-MA2 resulted in homogenous samples, even at 15 wt% copolymer content. As observed earlier in DSC and X-ray diffraction, the phase morphology of PE was observed to be significantly affected in the EMAZ and EVA-MA based blends. EVA-MA was observed to exhibit maximum distortion of the phase morphology, which also corroborated with the maximum decrease in melt enthalpy for this material. The hard and soft phases in the EMAZ and EVA-MA based

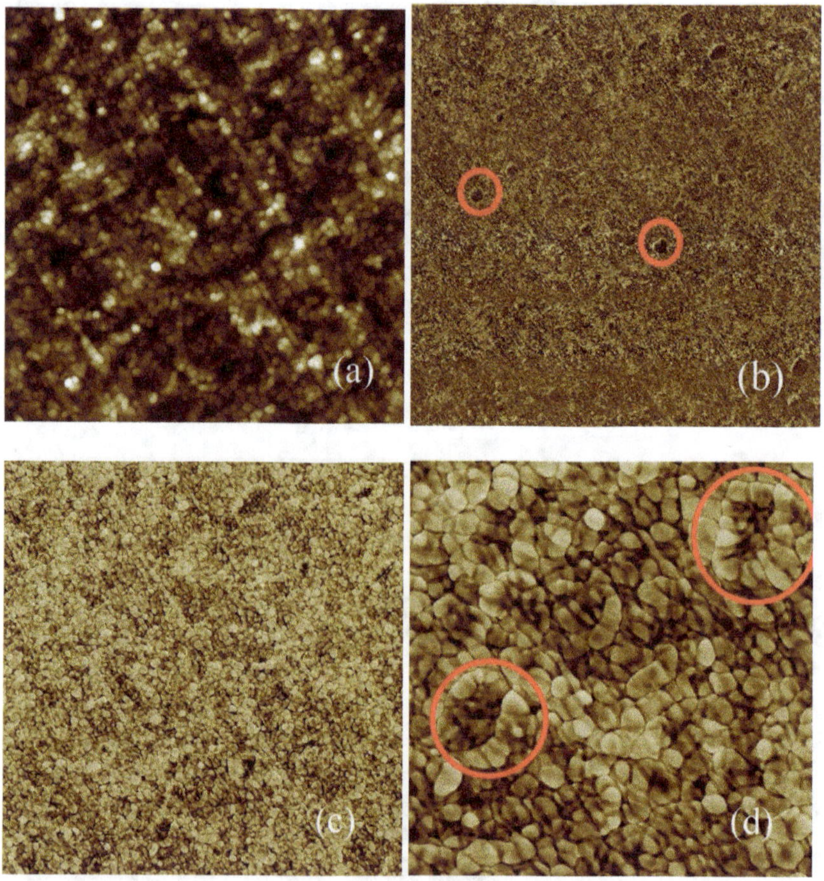

Figure 12.10 AFM images of the PE/15EMAZ blend: (a) height image (the width of the image is 10 μm); (b-d) phase images with image width as 20, 10 and 3 μm respectively. The fine dispersion of mixture of hard (bright phase representing crystalline region) and soft (dark phase representing amorphous region) regions was disturbed representing immiscibility. In addition, the presence of "eye-like" structures with soft central part and dense borders was observed.

blends did not show compatibility with each other. In addition, the presence of "eye-like" structures with soft central part and dense borders were due to the immiscible copolymer phase present as large dispersed domains and surrounded by semi-crystalline PE chains.

Network structure of the polymer blends was evaluated with shear rheology, and storage and loss moduli of the samples as a function of angular frequency are presented in Figures 12.12 and 12.13.

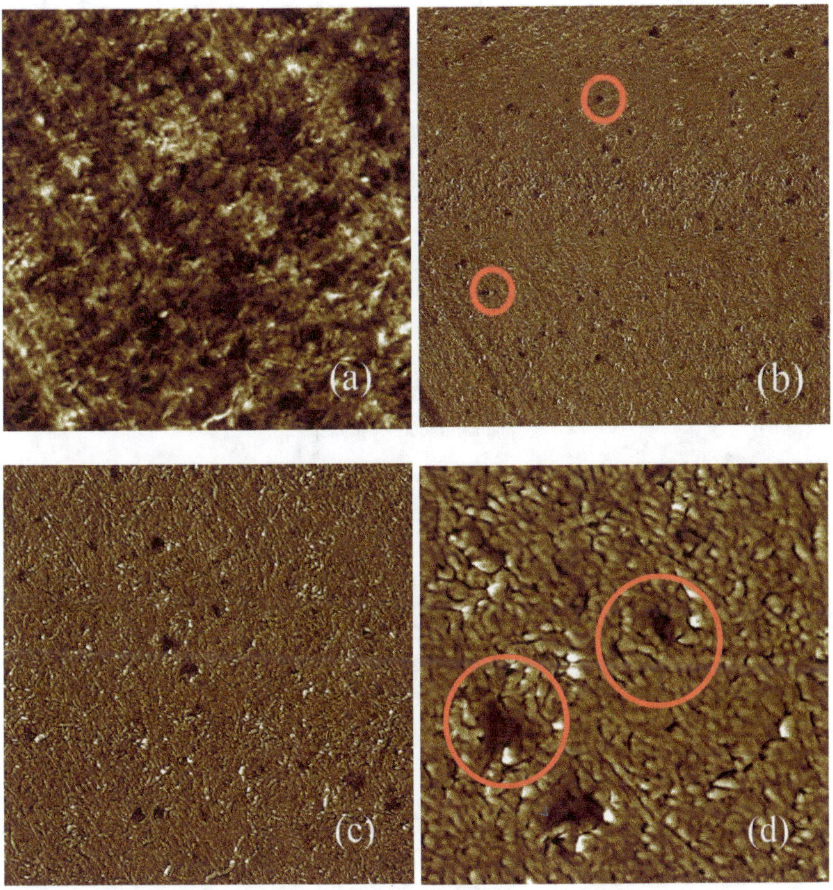

Figure 12.11 AFM images of the PE/15EVA-MA blend: (a) height image (the width of the image is 10 μm); (b-d) phase images with image width as 20, 10 and 3 μm respectively. The fine dispersion of mixture of hard (bright phase representing crystalline region) and soft (dark phase representing amorphous region) regions was disturbed representing immiscibility. In addition, the presence of "eye-like" structures with soft central part and dense borders was observed.

Strain sweep was conducted and samples were found to be safe up to 10% strain. Frequency sweep of the samples was performed with controlled shear strain of 2% using frequency range of 0.1-100 rad/s. All the samples exhibited gradual decline in the extent of modulus enhancement as a function of angular frequency due to the shear thinning effect. In the case of PE-MA1 based blends, the storage modulus of the blend with 5% copolymer content exhibited higher value than

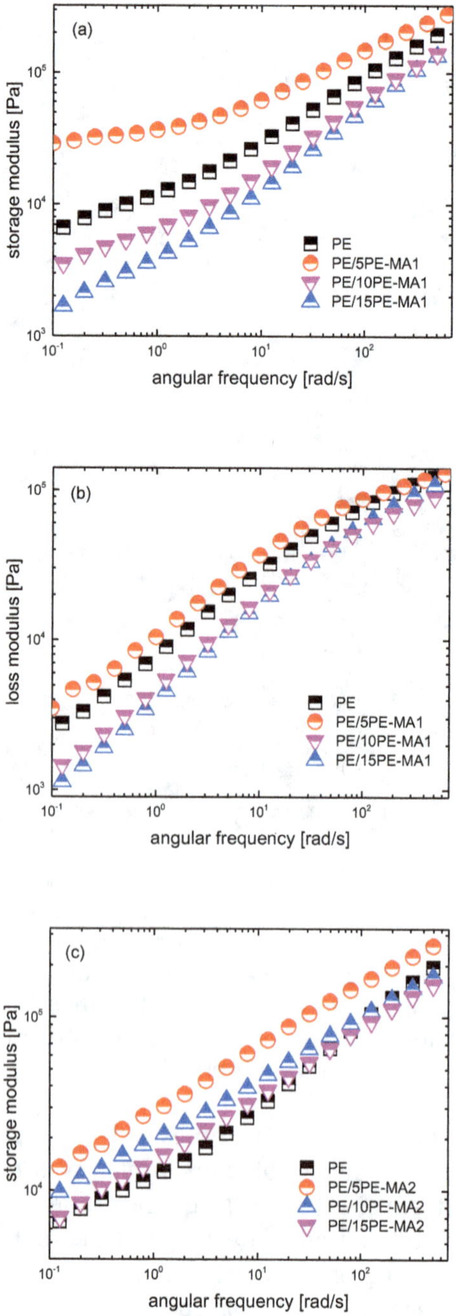

Figure 12.12 (continued on next page)

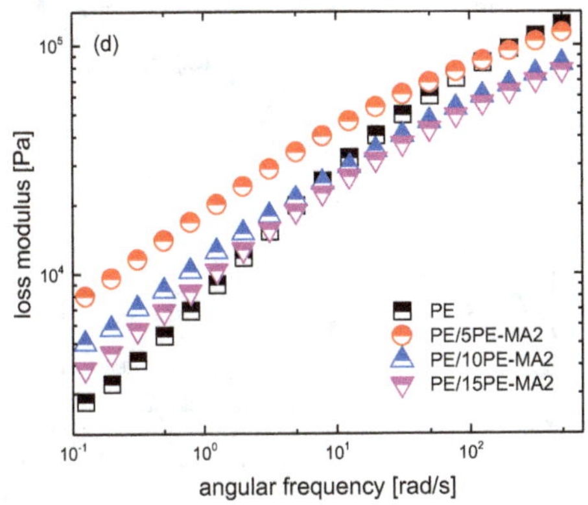

Figure 12.12 (a) Storage (G') and (b) loss (G") modulus of the PE/15PE-MA1 blend; and (c) storage (G') and (d) loss (G") modulus of the PE/15PE-MA2 blend, as a function of angular frequency as well as copolymer concentration in the blends.

the pure polymer over the complete angular frequency range, though the difference was observed to decrease at higher angular frequency. It indicated that the good degree of intermixing of the phases may have enhanced the shear behavior of the polymer. On further increasing the PE-MA1 content, the modulus was observed to decrease due to matrix plasticization in the presence of copolymer. Even though the blends with 10% and 15% PE-MA1 had lower storage modulus than PE, however, the difference in the magnitude was not large. The loss modulus of the PE-MA1 based blends exhibited the similar trend as storage modulus with the modulus curves almost converging into

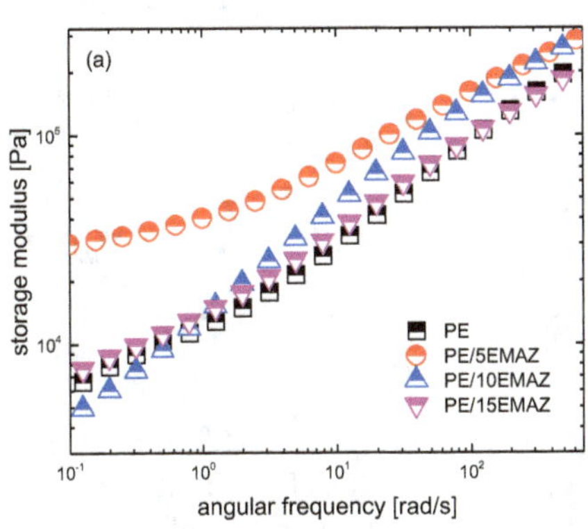

Figure 12.13 (continued on next page)

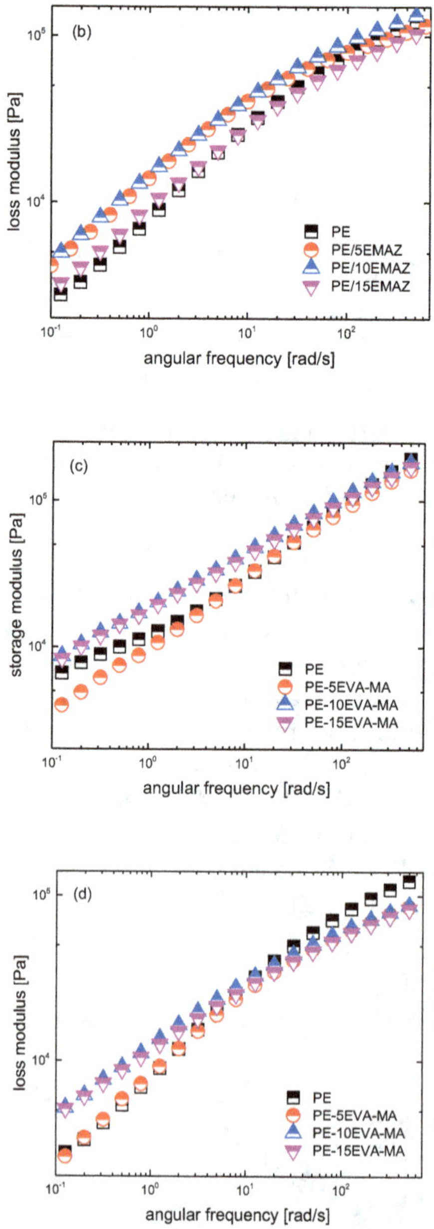

Figure 12.13 (a) Storage (G') and (b) loss (G") modulus of the PE/15EMAZ blend; and (c) storage (G') and (d) loss (G") modulus of the PE/15EVA-MA blend, as a function of angular frequency as well as copolymer concentration in the blends.

a single curve at higher angular frequency. In addition, the loss modulus of the blends was lower than the storage modulus, especially for 5% copolymer blend, thus, indicating that the elastic behavior was dominant in the blends at the measurement conditions. In the case of PE-MA2, the modulus was also observed to improve with the addition of the copolymer, though the extent of increase was not significant. Similar to the PE-MA1 case, the blend with 5% PE-MA2 had the highest modulus. The modulus of the blends with 10% and 15% copolymer was also higher than pure PE in this case, especially at lower angular frequency. Similar trends were also observed for the EMAZ and EVA-MA based blends, where the storage modulus of the blends was slightly improved as compared to the pure polymer. Overall, the PE-MA1 blends were observed to exhibit the highest degree of matrix plasticization due to the lower molecular weight of PE-MA1. Also, there was no significant difference in the bulk rheological performance of the blends due to the differences in the phase morphology at the nano-scale, probably due to the measurements in the melt state.

Figure 12.14 also demonstrates the complex viscosity of the blends. Contrary to the modulus, the viscosity of the blends decreased on increasing angular frequency. Due to the increase in frequency, polymer structure shows temporary network of entanglements which leads to more flexibility (lower viscosity). In many blends, this allowed more deformation energy to be stored resulting in elastic dominance which leads to increased modulus. Thus, it simultaneously reduced the contribution of lost deformation energy and, hence, viscous behavior. Similar to storage modulus, some increments were

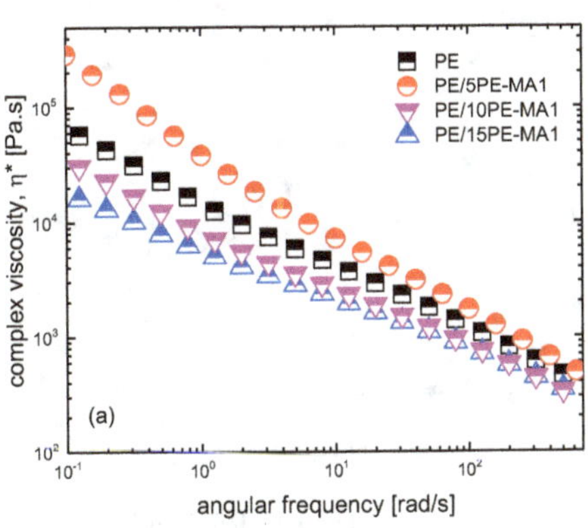

Figure 12.14 (continued on next page)

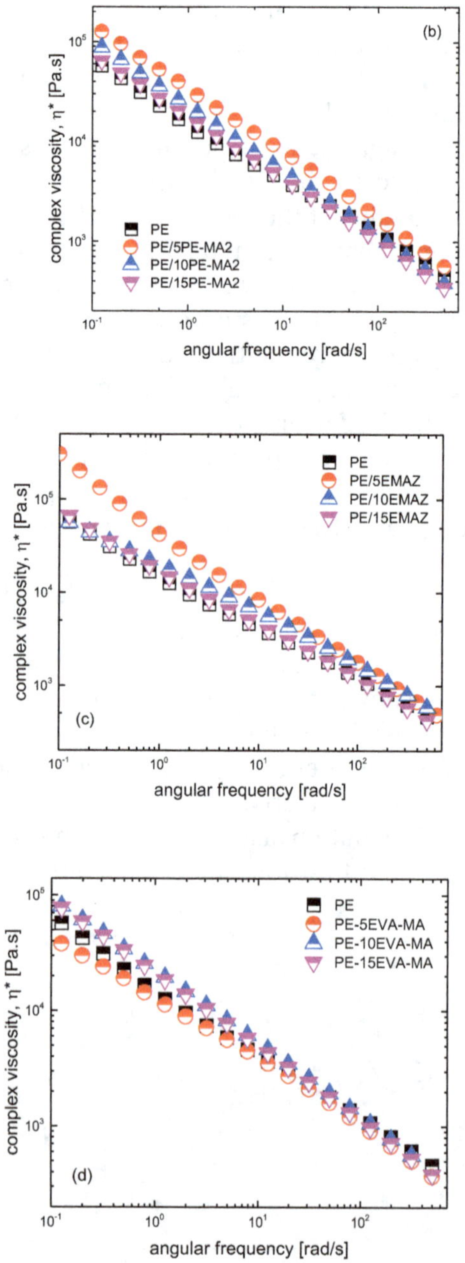

Figure 12.14 Complex viscosity of (a) PE/PE-MA1, (b) PE/PE-MA2, (c) PE/EMAZ, and (d) PE/EVA-MA blends, as a function of angular frequency as well as copolymer concentration in the blends.

observed in the viscosity of the blends with the addition of copolymers. Even though the viscosity of the blends increased in many instances as compared to the pure polymer, however, the extent of increase was not significant and blends could still be processed using the same conditions as pure PE.

The rheological behavior of the blends was also analyzed using the criteria reported for compatibility of polymer blends by Han and Chuang [25,26] as shown in Figure 12.15. As G' is plotted vs. G", such analysis generates composition independent correlation in the case of compatible blends (indicated by the merging of the curves into a common curve), whereas the correlation is composition dependent for incompatible blends. For the PE-MA1 blends, the curves almost converged into single curve at higher modulus (or higher frequency)

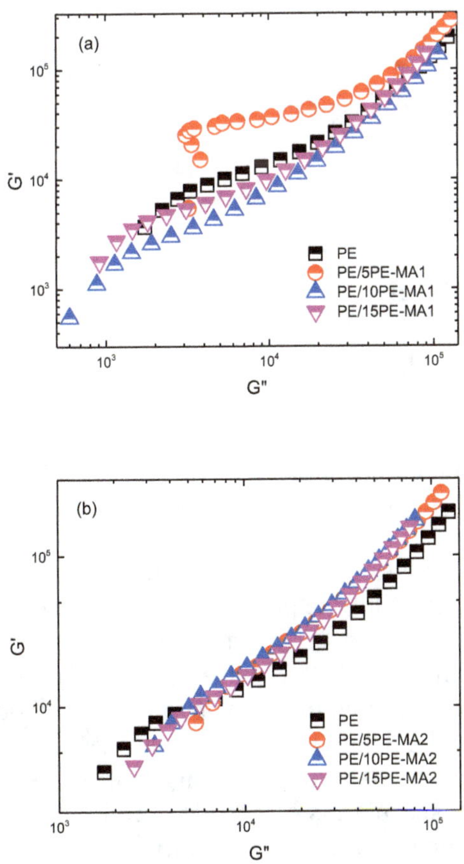

Figure 12.15 (continued on next page)

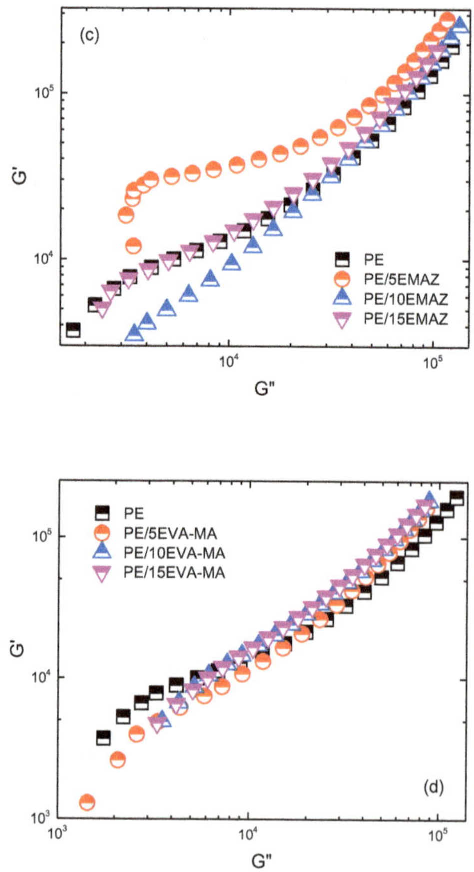

Figure 12.15 Han-Chuang (G' vs. G") plots of (a) PE/PE-MA1, (b) PE/PE-MA2, (c) PE/EMAZ, and (d) PE/EVA-MA blends.

values, however, deviations were observed at lower modulus values. Thus, it also indicated that the phases were not molecularly miscible, even though good degree of intermixing was observed for these blends earlier. Similarly, PE-MA2 based blends also exhibited somewhat concentration independent behavior due to good phase intermixing, as observed earlier. The blends with EMAZ and EVA-MA copolymers exhibited higher extent of concertation dependent behavior than PE-MA based blends at both low and high modulus values due to larger degree of phase immiscibility, which also corroborated with the earlier observations from calorimetric and morphological analysis. Similar insights about the phase miscibility were also gained through Cole-Cole viscosity plots as well as van Gurp plots (by plotting η′ vs. η″ as well as complex modulus, G* with phase angle δ).

Figure 12.16 demonstrates the tensile modulus of the blends as a function of copolymer content in the blends. The modulus of pure polymer was measured to be 1172 MPa. In all the blends, the modulus was observed to decrease as a function of copolymer content owing to reduced crystallinity and matrix plasticization. Largest decrease in the modulus was exhibited by the EVA-MA blends. For instance, the modulus of the blend with 15% EVA-MA was measured as 846 MPa, in comparison with 915 MPa, 936 MPa and 915 MPa for the blends with 15% content of EMAZ, PE-MA2 and PE-MA1. It indicated that the observed immiscible phase morphology as well as largest extent of crystallinity reduction in the PE/EVA-MA blends resulted in corresponding reduction in tensile modulus. Though a lower extent of phase immiscibility was also observed for EMAZ containing blends, however, these blends exhibited almost similar performance as the PE-MA based blends. Overall, though the modulus of PE expectedly decreased with the incorporation of a variety of copolymers, however, the blends still exhibited appreciable modulus. Figure 12.17 also demonstrates the effect of copolymer content on yield and break stress as well as corresponding yield and break strain. Similar to

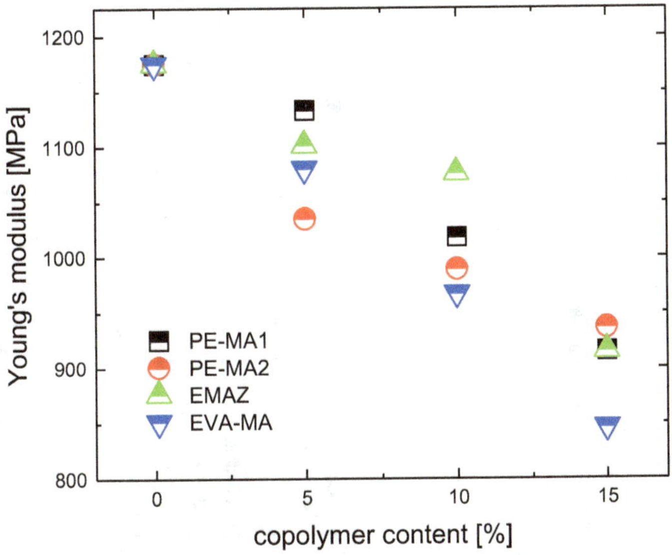

Figure 12.16 Tensile modulus of the blends as a function of copolymer content. The modulus of the blends varied in the range of ±2%.

tensile modulus, the yield stress of the blends decreased with increasing the copolymer content. Among the different copolymers, EVA-MA based blends exhibited largest extent of reduction in the yield stress, thus, re-confirming the effect of reduced crystallinity, plasticization as well as phase immiscibility. The blend with 15% EVA-MA content exhibited a reduction of >25% in the yield stress. Due to enhanced flexibility in the blends, the yield strain was also correspondingly increased with copolymer content. From a value of 6% for pure PE, the yield strain in the blends with 15% copolymer content was observed in the range of 11-12%, thus, exhibiting extensive plasticization of the matrix. Similarly, the stress at break and strain at break in the blends were also affected by the copolymer content, though a large probable error existed in the values of these quantities. As mentioned earlier,

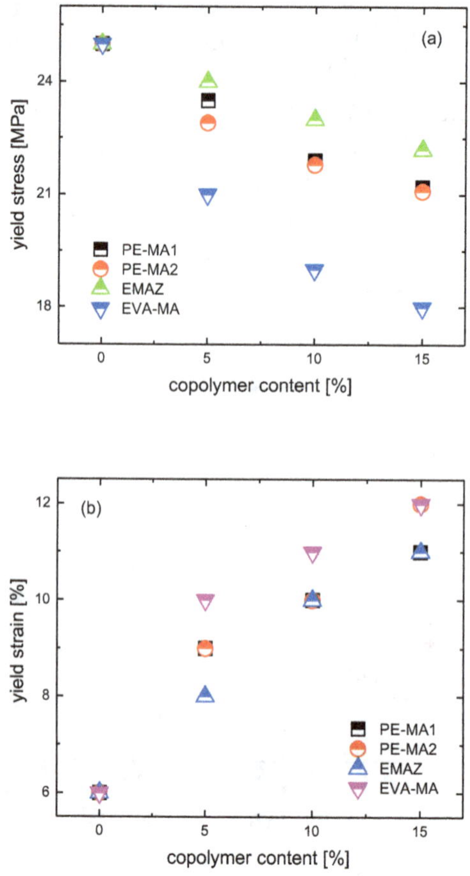

Figure 12.17 (continued on next page)

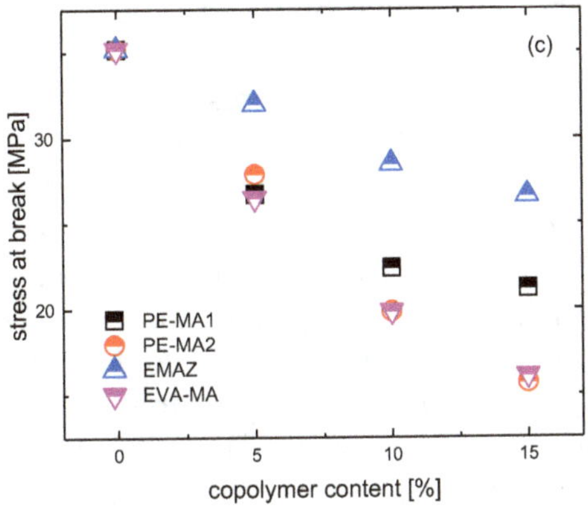

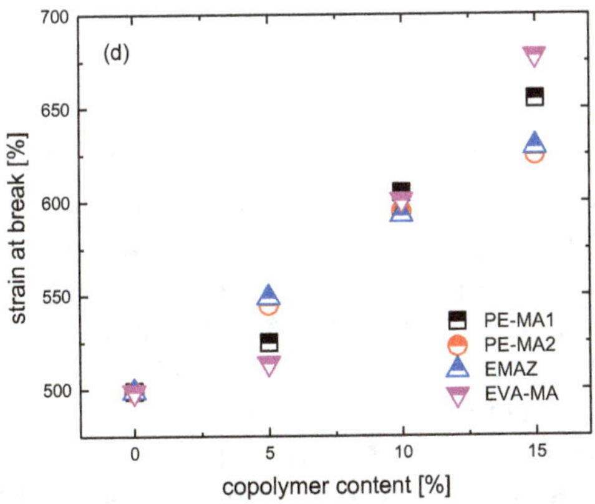

Figure 12.17 (a) Yield stress, (b) yield strain (c) stress at break and (d) strain at break of the blends as a function of copolymer content. The yield stress and yield strain varied in the range of ±2%, whereas the stress at break and strain at break had the relative probable error of ±5%.

the tensile properties of the blends in an earlier study with CPE25 and CPE35 were also significantly affected by the type as well as amount of the CPE used [22]. For instance, the tensile modulus as well as the peak strength of the CPE25 blends had lower decrease in magnitude as compared with CPE35 blends. The decrease was observed only after 5% CPE content and the difference between the two systems enhanced on enhancing the CPE content. These findings were also similar to the tensile properties observed in the current study. Thus, the factors affecting the mechanical performance of the blends are the nature of copolymer, matrix plasticization, effect on crystallinity as well as extent of compatibility of the copolymer with the matrix polymer.

The generated blends provide systematic insights into the effect of a variety of copolymers on the thermal, mechanical, rheological and morphological properties of PE. In addition, the effect could also be quantified as a function of copolymer content. These findings enable the user to choose the appropriate blend system and composition to target the required set of properties. In addition, the insights into the phase miscibility of the copolymers in PE would also benefit the design of advanced composites where the compatibilization of the filler phase with the polymer is necessary. Though these findings are applicable to PE based systems, however, blends and composites based on other polyolefins can also benefit from these, owing to their similar nature.

12.4 Conclusions

In the current study, development of polyethylene based functional blends with a variety of copolymers has been demonstrated. The blends of polyethylene-*g*-maleic anhydride copolymer (PE-MA), ethylene methacrylic acid copolymer with zinc ion (EMAZ) and ethylene vinyl acetate-*g*-maleic anhydride (EVA-MA) copolymer in the weight fraction of 5%, 10% and 15% were generated. The thermal degradation of the PE was affected by the addition of the copolymers (except PE-MA2), as the onset and peak degradation temperatures were reduced, though the blends were still stable up to 350 °C. The incorporation of copolymers also reduced the peak melting point of PE in the blends, which was also a function of copolymer content. The melt enthalpy of PE was also observed to decrease owing to the hindrance to PE chains, with largest reduction in enthalpy exhibited by EVA-MA. The EVA-MA and EMAZ based blends exhibited phase immiscibility due to the presence of separate melt transition of these copolymers in the DSC thermograms of the blend. The X-ray diffraction analysis also indicated some changes in the diffraction peaks in the EVA-MA and EMAZ copolymer based blends, though major diffraction peaks of PE were not affected in intensity as well as peak position. The morphological analysis of the blends confirmed the earlier finding of phase immiscibility in EVA-MA based blends. The EMAZ based blends also exhibited some phase inhomogeneity, whereas the PE-MA based blends had uniform morphology. The rheological performance of the blends exhibited generally higher modulus in the 5% copolymer blends and no obvious effect of phase morphology was observed on the properties in melt. The tensile modulus of the blends was a function of reduced crystallinity, matrix plasticization as well as phase morphology. As a result, the EVA-MA based blends exhibited lower tensile values than other copolymers. Overall, even though the mechanical properties of the blends deteriorated after the inclusion of copolymers, these were still appreciable. The generated blends lead to valuable insights about the properties and morphology of PE based blends with different copolymers, which are subsequently useful for the design of advanced composites where compatibilization with the filler is additionally required.

References

1. Lipatov, Y. S. (2002) Polymer blends and interpenetrating polymer networks at the interface with solids. *Progress in Polymer Science*, **27**, 1721-1801.
2. Fekete, E., Foldes, E., and Pukanszky, B. (2005) Effect of molecular interactions on the miscibility and structure of polymer blends. *European Polymer Journal*, **41**, 727-736.
3. Litmanovich, A. D., Plate, N. A., and Kudryavtsev, Y. V. (2002) Reactions in polymer blends: Interchain effects and theoretical problems. *Progress in Polymer Science*, **27**, 915-970.
4. Sundararaj, U., and Macosko, C. W. (1995) Drop breakup and coalescence in polymer blends: The effects of concentration and compatibilization. *Macromolecules*, **28**, 2647-2657.
5. Macosko, C. W., Guegan, P., Khandpur, A. K., Nakayama, A., Marechal, P., and Inoue, T. (1996) Compatibilizers for melt blending: Premade block copolymers, *Macromolecules*, **29**, 5590-5598.
6. Rameshwaram, J. K., Yang, Y.-S., and Jeon, H. S. (2005) Structure-property relationships of nanocomposite-like polymer blends with ultrahigh viscosity ratios. *Polymer*, **46**, 5569-5579.
7. Wang, D., Li, Y., Xie, X.-M., and Guo, B.-H. (2011) Compatibilization and morphology development of immiscible ternary polymer blends. *Polymer*, **52**, 191-200.
8. Higgins, J. S., Tambasco, M., and Lipson, J. E. G. (2005) Polymer blends; stretching what we can learn through the combination of experiment and theory. *Progress in Polymer Science*, **30**, 832-843.
9. Menyhard, A., Varga, J., Liber, A., and Belina, G. (2005) Polymer blends based on the β-modification of polypropylene. *European Polymer Journal*, **41**, 669-677.

10. Gopakumar, T. G., Ponrathnam, S., Lele, A., Rajan, C. R., and Fradet, A. (1999) In situ compatibilisation of poly(phenylene sulphide)/wholly aromatic thermotropic liquid crystalline polymer blends by reactive extrusion: Morphology, thermal and mechanical properties. *Polymer*, **40**, 357-364.
11. Jancar, J. (1999) *Mineral Fillers in Thermoplastics I: Raw Materials and Processing*, Springer-Verlag, Germany.
12. Mittal, V. (2012) *Functional Polymer Blends: Synthesis, Properties and Performance*, CRC Press, USA.
13. Hippi, U. (2008) *Polyolefin Composites*, John Wiley & Sons, USA.
14. Zhou, L., Wang, X., Lin, Y., Yang, J., and Wu, Q. (2003) Comparison of the toughening mechanisms of poly(vinyl chloride)/chlorinated polyethylene and poly(vinyl chloride)/acrylonitrile-butadiene-styrene copolymer blends. *Journal of Applied Polymer Science*, **90**, 916-924.
15. Eastwood, E. A., and Dadmun, M. D. (2002) Compatibilization of poly(vinyl chloride) and polyolefin elastomer blends with multiblock/blocky chlorinated polyethylenes. *Polymer*, **43**, 6707-6717.
16. He, P., Huang, H., Xiao, W., Huang, S., and Cheng, S. (1997) A study on PVC/LLDPE blends with solid-state-chlorinated polyethylene as compatibilizer. *Journal of Applied Polymer Science*, **64**, 2535-2541.
17. Stoeva, S. (2006) Study of poly(vinyl chloride) blends with solid-state chlorinated polyethylene. *Journal of Applied Polymer Science*, **101**, 2602-2613.
18. Waddington, S., and Briggs, D. (1991) Adhesion Mechanisms between polymer coatings and polypropylene studied by XPS and SIMS. *Polymer Communications*, **32**, 506-508.
19. Ogoe, S. A., Liou, D. W., Ainsworth, O. C., Bethea, J. R., Flores, D. R., Ellebracht, S. R., Muskopf, J. W., Werling, C. L., Wilson, E. J., Cocup, J. W., and Allison, J. L. (1996) Blends of Polycarbonate and Chlorinated Polyethylene, US patent 5,525,651.
20. Barrie, J. A., and Webb, W. D. (1989) Gas transport in miscible blends of ethylene-vinyl acetate copolymer and chlorinated polyethylene. *Polymer*, **30**, 327-332.
21. Mittal, V., and Matsko, N. (2013) PE-CPE blends and their graphene oxide nanocomposites with reduced low temperature brittleness. *Colloid and Polymer Science*, **291**(8), 1949-1961.
22. Chaudhry, A. U., and Mittal, V. (2014) Blends of high-density polyethylene with chlorinated polyethylene: Morphology, thermal, rheological, and mechanical properties. *Polymer Engineering and Science*, **54**(1), 85-95.
23. Mittal, V., and Chaudhry, A. U. (2015) Effect of amphiphilic compatibilizers on the filler dispersion and properties of polyethylene-thermally reduced graphene nanocomposites. *Journal of Applied Polymer Science*, **132**(35), doi:10.1002/app.42484.

24. Matsko, N. B., and Mittal, V. (2014) Thermotropic dynamic processes in multiphase polymer systems by (cryo-)AFM. *Soft Matter*, **10**, 5478-5488.
25. Chuang, H. K., and Han, C. D. (1984) Rheological behavior of polymer blends. *Journal of Applied Polymer Science*, **29**, 2205-2229.
26. Han, C. D., and Chuang, H. K. (1985) Criteria for rheological compatibility of polymer blends. *Journal of Applied Polymer Science*, **30**(11), 4431-4454.

Index

A

B

C

D

E

F

G

H

N

O

P

R

S

T

U

V

W

X

Y

Z

www.ingramcontent.com/pod-product-compliance
Lightning Source LLC
LaVergne TN
LVHW011517200926
848484LV00001B/55